Modern Techniques of Gel Electrophoresis

Modern Techniques of Gel Electrophoresis

Edited by **Jill Clark**

New York

Published by Callisto Reference,
106 Park Avenue, Suite 200,
New York, NY 10016, USA
www.callistoreference.com

Modern Techniques of Gel Electrophoresis
Edited by Jill Clark

© 2015 Callisto Reference

International Standard Book Number: 978-1-63239-466-8 (Hardback)

Contents

Preface

I am honored to present to you this unique book which encompasses the most up-to-date data in the field. I was extremely pleased to get this opportunity of editing the work of experts from across the globe. I have also written papers in this field and researched the various aspects revolving around the progress of the discipline. I have tried to unify my knowledge along with that of stalwarts from every corner of the world, to produce a text which not only benefits the readers but also facilitates the growth of the field.

As a fundamental concept, gel electrophoresis is a biotechnology method in which macromolecules such as DNA, RNA or protein are fragmented according to their physical characteristics such as molecular mass or charge. These molecules are enforced throughout a porous gel matrix in the influence of electric field enabling countless functions and utilizations. This book is not all-inclusive but still covers a majority of applications of this technique in the varied fields of medical and life sciences. This book is divided into four sections: Electrophoresis Application in Enzymology, Temporal Temperature Gel Electrophoresis, Two-Dimensional Gel Electrophoresis (2-DE) and Other Applications of Gel Electrophoresis Technique. We have made an attempt to keep the data of the book detailed and wide-ranging, and we hope that it will benefit the readers.

Finally, I would like to thank all the contributing authors for their valuable time and contributions. This book would not have been possible without their efforts. I would also like to thank my friends and family for their constant support.

Editor

Part 1

Electrophoresis Application in Ecological and Biotechnological Aspects

Proteomics in Seaweeds: Ecological Interpretations

Loretto Contreras-Porcia and Camilo López-Cristoffanini
Universidad Andrés Bello, Faculty of Ecology and Natural Resources
Department of Ecology and Biodiversity, Santiago
Chile

1. Introduction

Macro and micro-algae are fundamental components of coastal benthic ecosystems and are responsible for a large part of the coastal primary production (Lobban & Harrison, 1994). Adverse effects on these groups caused by natural or anthropogenic phenomena, can affect directly or indirectly organisms of higher trophic levels and the integrity of entire ecosystems. In this context, both the ecological and economic importance of many algal species justifies the need to expand our knowledge on the molecular biology of these organisms.

The distribution and abundance of algal species occurring in the marine zone results from the interplay of biotic (i.e. competition and herbivore pressure) and abiotic (i.e. tolerance to extreme and fluctuating environments) factors (Abe et al., 2001; Burritt et al., 2002; Davison & Pearson, 1996; Pinto et al., 2003; van Tamelen, 1996). For example, the distribution of macroalgal species at the upper limit of the rocky intertidal zone is principally determined by abiotic factors such as UV radiation, light, salinity, temperature changes, nutrient availability and desiccation (e.g. Aguilera et al., 2002; Burritt et al., 2002; Cabello-Pasini et al., 2000; Contreras-Porcia et al., 2011a; Véliz et al., 2006). On the other hand, the microalgae diversity is maintained by a combination of variable forces - environmental oscillations (e.g. habitat instability), more severe disturbances and recovery from catastrophic forcing - backed by the powerful dispersive mobility of this group (Reynolds, 2006). The richness, relative abundance and occasional dominances of the phytoplankton in successive years, depends on water movements, thermal stress and carbon fluxes, but mainly on nutrient enrichment of the sea (Hodgkiss & Lu, 2004; Holm-Hansen et al., 2004; Reynolds, 2006; Wang et al., 2006; Zurek & Bucka, 2004).

Superimposed on the natural abiotic oscillations, algae are also exposed to various other sources of stress, particularly those resulting from human industrial, urban and agricultural activities. Among these is copper mining, whose wastes have reportedly caused severe and negative effects on the coasts of England (Bryan & Langston, 1992), Canada (Grout & Levings, 2001; Marsden & DeWreede, 2000), Australia (Stauber et al., 2001) and Chile (Correa et al., 1999). Although copper is a micronutrient for plants and animals, occurring naturally in coastal seawater at levels at or below 1 µg L⁻¹ (Apte & Day, 1998; Batley, 1995; Sunda, 1989), at higher concentrations it becomes highly toxic. The phenomenon of toxicity in algae is strongly influenced by the speciation of this metal (Gledhill et al., 1997), and within the cell it likely operates through the Haber-Weiss reaction, characterized by a heavy metal-catalyzed production of hydroxyl radicals from hydrogen peroxide (Baker & Orlandi, 1995). For example, in northern Chile, mine wastes originated at a copper mine pit are disposed of directly into the sea. The rocky intertidal zone along the impacted coasts shows a severe reduction in species richness, and the macroalgal assemblage is reduced to the opportunistic algae *Ulva compressa* (Plantae, Chlorophyta) and *Scytosiphon lomentaria* (Chromista, Ochrophyta) (Medina et al., 2005). This negative effect on the biota has been widely recognized as the result of the persistent high levels of copper in the water, by far the most important metal brought into the system by mine wastes (Medina et al., 2005). Many macroalgae species are absent, such as *Lessonia nigrescens* complex (Chromista, Ochrophyta), which are key components in structuring the intertidal zone (Ojeda & Santelices, 1984). As for microalgae, an example is a mine effluent that contained high levels of copper, which was disposed in a reservoir named Venda Nova in northern Portugal. There, a phytoplankton survey was carried out between the years 1981-1982. A shift in the dominant species was demonstrated when compared with an uncontaminated area, Alto Rabagão. More than 50% of the algal species developed lower populations. Also, at the most polluted zone, phytoplankton density, biomass and richness were strongly reduced (Oliveira, 1985).

In macro and micro-algae it is possible to determine that under natural abiotic factors, a common cellular response could involve the over-production of reactive oxygen species (ROS) (Andrade et al., 2006; Contreras et al., 2005, 2007b, 2009; Contreras-Porcia et al., 2011a; Kumar et al., 2010; Lee & Shin, 2003; Liu et al., 2007; Rijstenbil, 2001). ROS are ubiquitous by-products of oxidative metabolism that are also involved in intracellular signalling processes (e g. Blokhina & Fagerstedt, 2010; Rhee, 2006). ROS are produced directly by the excitation of O_2 and the subsequent formation of singlet oxygen, or by the transfer of one, two or three electrons to O_2. This results in the formation of superoxide radicals, hydrogen peroxide or hydroxyl radicals, respectively (Baker & Orlandi, 1995). Oxidative damage to cellular constituents such as DNA/RNA, proteins and lipids may occur (e g. Contreras et al., 2009; Vranová et al., 2002) when ROS levels increase above the physiological tolerance range. However, a coordinated attenuation system can be activated in order to eliminate this ROS over-production, and therefore, the oxidative stress condition (e. g. Burritt et al., 2002; Ratkevicius et al., 2003; Rijstenbil, 2001). For example, in the coastal zones of northern Chile it has been demonstrated that the high copper levels in the seawater generate in sensitive species a high oxidative stress condition, which appears as the starting point for a series of molecular defense responses. In first place, the condition of oxidative stress has been demonstrated by the direct production of ROS and oxidized lipid in individuals living at an impacted site as well as in those transplanted from control sites to the impacted site (Contreras et al., 2005; Ratckevicius et al., 2003). Compared with high tolerant species such as *Ulva* and *Scytosiphon*, in low tolerant species such as *L. nigrescens* the ROS production by copper, specifically superoxide anions, is poorly attenuated, which is reflected in i) higher levels of oxidized lipids, ii) the generation of cellular alterations and iii) negative effects on early developmental stages of the life cycle (Andrade et al., 2006; Contreras et al., 2007a; 2009). Thus, ecophysiological differences are evident between diverse algal species. This is also true for microalgal species since there are species-specific responses to oxidative stress caused by high levels of copper. For example, it was demonstrated that 4 species of phytoplankton under high concentrations of copper only grew up to 80-95% of that observed in the control condition (Bilgrami & Kumar, 1997). Furthermore, a study including two microalgae species exposed to copper stress showed significant differences between them. In the high tolerant species, *Scenedesmus vacuolatus,* in comparison to the low tolerant species, *Chlorella kessleri,* the chlorophyll a/chlorophyll b ratio was partially reduced. Likewise, both the antioxidant enzyme activity and protein content were progressively increased (Sabatini et al., 2009).

Another environmental factor that affects the abundance and distribution in macroalgae is desiccation. It is an important stress factor faced by living organisms because, as cells lose water, essential macromolecules are induced to form non-functional aggregates and organelles collapse (Alpert, 2006). Some animals (Clegg, 2005) and plants are well adapted to significant water losses, displaying full physiological recovery during rehydration (Alpert, 2006; Farrant, 2000). Compared to vascular plants or animals, in macroalgae the effects of desiccation on the physiology and the molecular mechanisms involved in its tolerance are poorly understood. For example, in one of the few reports available, the activation of different antioxidant enzymes, such as ascorbate peroxidase (AP) and glutathione reductase (GR) was recorded in the upper intertidal macroalga *Stictosiphonia arbuscula* (Plantae, Rhodophyta) (Burritt et al., 2002) as a response to desiccation-mediated oxidative stress. The remaining studies have focused on assessing the capacity to tolerate desiccation displayed by measuring the photosynthetic apparatus activity in *Porphyra, Gracilaria, Chondrus,* and *Ulva* species among others (Abe et al., 2001; Ji & Tanaka, 2002; Smith et al., 1986; Zou & Gao, 2002). Presently, the only study using molecular approaches to unravel the desiccation tolerance responses, found that genes encoding for photosynthetic and ribosomal proteins are up-regulated in *Fucus vesiculosus* (Chromista, Ochrophyta) (Pearson et al., 2001, 2010). Additionally, independent studies have shown that diverse physiological parameters are altered by desiccation including the lipid and protein levels (Abe et al., 2001), photosynthetic alterations (*Fv/Fm*) as well as cellular morphology and ontogenetic changes (e.g. Contreras-Porcia et al., 2011b; Varela et al., 2006). Moreover, in microalgae it has been shown that salt (i.e. changes in water osmolarity) and temperature stress can be highly stressful and may finally trigger a programmed cell death (PCD) (Kobayashi et al., 1997; Lesser, 1997; Takagi et al., 2006; Zuppini et al., 2010). In these species the effects of both types of stress have been widely studied, and have been reported to provoke photosynthetic alterations, ROS production and ultimately cell death (Liu et al. 2007; Lesser, 1996; Mishra & Jha, 2011; Vega et al., 2006).

Recently, the red species *Porphyra columbina* Montagne (Plantae, Rhodophyta) was recognised among the macroalgae that are highly tolerant to natural desiccation stress. *P. columbina* is highly seasonal and grows abundantly along the upper intertidal zone (Hoffmann & Santelices, 1997; Santelices, 1989). This alga is well adapted to the extreme fluctuating regimes of water/air exposure, as demonstrated by the formation of sporophytic thalli from monoecious fronds (*n*) during long daily periods of desiccation stress due to its position in the intertidal zone (Contreras-Porcia et al., 2012). Additionally, desiccation in *P. columbina* induces morphological and cellular alterations accompanied by a loss of ca. 96 % of the water content (Contreras-Porcia et al., 2011b). Specifically, under natural desiccation stress, the production of ROS (i.e. H_2O_2 and O_2) in *P. columbina* is significantly induced (Contreras-Porcia et al., 2011b). However, during the high tide, ROS quickly returned to basal levels because *P. columbina* displays an efficient antioxidant system. In addition, at biomolecular level, only a low production of oxidized proteins is recorded during desiccation, due to the efficient antioxidant system of this alga.

The results mentioned above, indicate that desiccation in *P. columbina* causes an over-production of ROS, which is efficiently attenuated. Morphological and photosynthetic changes could be operating as tolerance mechanisms, due to the fact that these responses principally prevent biomolecular alterations, protein aggregation and cellular collapse. For example, it has been proposed that cell wall folding is a cellular strategy used to prevent tearing the plasmalemma from the cell wall during desiccation, ensuring cell integrity (Contreras-Porcia et al., 2011b). The activation of antioxidant enzymes and the photoinhibition of the photosynthetic apparatus help to explain the attenuation of ROS. Thus, ROS excess is buffered by the activation of several physiological and biochemical responses, which suggest a mechanism allowing this plant to tolerate desiccation (Contreras-Porcia et al., 2011b). The ecophysiological responses in this species help, in part, to account for its position and dominance at the highest level in the intertidal zone, and thereby, suggesting desiccation stress tolerance as a determinant trait for explaining that situation. In fact, our recent results demonstrate that the magnitude of the effects generated by desiccation in algae is related to the position of the species in the intertidal zone. Additionally, this work demonstrated the exceptional metabolism of *P. columbina* used to buffer this stress condition. Thus, the determinations of novel metabolic pathways are necessaries in order to fully understand the high desiccation tolerance in this species, for example at the proteomic level. In fact, in this time our forces are concentrated in resolving the proteomic profile of this species under natural hydration and desiccation stress.

Finally, the need to unravel the mechanisms associated with tolerance to different environmental factors by algal species opens the electrophoretic and proteomic approximations as important tools in comprehending and explaining the observed tolerances. However, little information regarding electrophoretic and proteomic analysis is available in algal species. Compared with other group of organisms (e.g. vascular plant or animals) protein extraction in macroalgae has been extraordinary difficult, due principally to the limited knowledge at biochemical and molecular levels. In this context, the present chapter aims to understand the different proteomic approaches utilized in this group of organisms in order to comprehend their ecophysiological behaviour.

2. Proteomic methodology in micro and macroalgae

Sample preparation, in particular the quality of protein extraction, is critical to the successful resolution of 2-DE patterns. In fact, when protein extraction protocols from higher plants are applied to algae, the 2-DE resolution is reduced (Contreras et al., 2008; Hippler et al., 2011). Due to the large variation in cellular biochemical composition among diverse organisms, which affects solubility and recovery of a complex mixture from the sample, there are no 2-DE sample preparation protocols accurate for all organisms. In macro and microalgae the protein extraction protocol must be optimized, due to the high concentration of photosynthetic pigments that are known to interfere with the resolution of the 2-DE gels (e.g. Contreras et al., 2008; Wang et al., 2003; Wong et al., 2006). Particularly, in macroalgae protein extraction is difficult due to a low concentration and the co-extraction of contaminants such as anionic polysaccharides, polyphenols and salts, which are highly concentrated in the tissue (Chinnasamy & Rampitsch, 2006; Cremer & Van de Walle, 1985;

Flengsrub & Kobro, 1989; Mechin et al., 2003). These contaminants pose a significant difficulty for 2-DE, as they cause horizontal and vertical streaking, smearing and a reduction in the number of distinctly resolved protein spots. Thus, the selection of the most appropriate protein extraction method is necessary in order to obtain high quality extracts, and therefore, a high quality 2-DE pattern. For a better understanding and explanation of the current techniques and methodology in algae proteomic, this chapter has been divided in two sections: microalgae and macroalgae methodology.

2.1 Microalgae methodology

It is important to highlight that due to the small size of microalgae, all of the protein extraction protocols for these organisms begin with a centrifugation step in order to pellet cells. This helps to concentrate cells, and consequently allows a correct extraction of the desired proteins.

2.1.1 Early proteomic studies

One of the first proteomics studies on microalgae dates from the year 1972, in which Mets & Bogorad showed alterations in the chloroplast ribosomes proteins of erythromycin-resistant mutants of *Chlamydomonas reinhardtii* (Plantae, Chlorophyta) compared to the wild-type. The ribosomal protein extraction performed on this work was the LiCl-urea method described by Leboy et al. (1971) that was developed for *Escherichia coli* (as cited in Mets & Bogorad, 1972). Thus, the Mets & Bogorad work was a precursor to microalgae proteomic studies. Here, ribosomes are disrupted and freed of RNA by adding LiCl. Then, the samples are centrifuged to precipitate total RNA and the supernatant, which contains the proteins, is retained.

Several studies in the same decade also focused their attention on characterizing ribosomal proteins (e.g. Götz & Arnold, 1980; Hanson et al., 1974). The Hanson et al. (1974) work based their protocols on the Mets & Bogorad (1972) research paper and also used *C. reinhardtii* as model species. Instead, in 1980 Götz & Arnold used a different ribosomal protein extraction after testing several protocols. The procedure chosen was the acetic-acid method in presence of $MgCl_2$ according to Kaltschmidt & Wittmann (1972), method that was also first developed for *E. coli*. In this method, $MgCl_2$ and glacial acetic acid are added to the ribosome suspension, and then the mixture is centrifuged to pellet RNA. For better mixture cleaning, the pellet can be extracted a second time in the same way.

Not all studies from this decade focused their attention on ribosomal proteins as was the case of the work of Piperno et al. (1977), in which the protein mixture came from *Chlamydomona* flagella and axonemes. It is important to highlight this research since the extraction method used was very rustic. After the flagella and axoneme separation, the proteins were dissolved only in SDS and kept for 2-DE analysis.

2.1.2 Current proteomic studies

Recent studies evaluate more complex protein mixtures, so the method chosen must be more accurate in extracting proteins with minimum contaminants and interferents. In fact, a work in *C. reinhardtii* that performed an analysis of all the thylakoid membranes proteins used a more complex protocol (Hippler et al., 2001) than the ones previously discussed in this chapter. This method uses methanol in order to precipitate cell debris and retains proteins in the supernatant. Then, chloroform is added and the sample is vortexed and centrifuged. The upper phase containing DNA is discarded. Afterwards, methanol is added to the sample in order to pellet proteins and leave the RNA in the aqueous phase. Finally, the pellet is washed with methanol in order to remove contaminants.

In 2003, a study tested different protein extraction protocols in the microalga *Haematococcus pluvialis* (Plantae, Chlorophyta) in order to determine which ones yielded better results (Wang et al., 2003). After cell disruption, the samples were dialysed to remove any salt left in the samples, which are known to interfere in the IEF step. After the dyalisis, each sample was treated in three different ways: i) proteins were left to precipitate in a non-denaturing preparation, ii) a mixing of dialysate with acetone was kept at -20 °C o/n to allow complete precipitation and iii) a mixing of dialysate with TCA in acetone containing β-mercaptoethanol also kept at -20°C o/n. Methods ii) and iii) were denaturing procedures but it was procedure iii) the one that yielded 2-DE gels with higher resolution (detailed protocol in Appendix A).

The work by Kim et al. (2005) is interesting since the protein extraction protocol used is relatively simple when compared to others (e.g. Wang et al., 2003; Contreras et al., 2008) (detailed protocol in Appendix A). Proteins of *Nannochloropsis oculata* (Chromista, Ochrophyta) are obtained in very short time compared to the other microalgae protocols, however, not with the same quality as the more complex protocols. In another *C. reinhardtii* work, but this time conducting a whole cell proteomic study (Förster et al., 2006), a protocol described by Mathesius et al. (2001) that is suited for root proteins was used (detailed protocol in Appendix A). This procedure is denaturing and relatively simple, but includes washing steps that help to improve the quality of the protein extracts compared to the one used on *N. oculata* (Kim et al., 2005). A work from 2009 in the microalga *Haematococcus lacustris* also had a denaturing protocol in which pelleted cells were grounded to a fine powder in liquid nitrogen (Tran et al., 2009). Then, they are disrupted with a lysis buffer containing urea, thiourea, DTT, CHAPS, Tris-base and a plant protease inhibitor cocktail tablet. Samples are centrifuged to separate cell debris, and then the pellet is resuspended in acetone to precipitate proteins and remove contaminants. Finally, the samples are again centrifuged, acetone is removed by air-drying and pellet is clean and ready for 2-DE gels.

Chlamydomonas reinhardtii is one of the most studied microalgae worldwide and as noted in this chapter, proteomics studies are no exception. Another protocol for this algae dates from 2011, in this one the cells are disrupted with a lysis buffer containing urea, CHAPS and thiourea (Mahong et al., 2012). The sample is centrifuged and the supernatant retained. To eliminate possible photosynthetic pigments and other hydrophobic compounds, the samples are washed with ice-cold acetone. Then, each sample is centrifuged, and the pellet is ready for electrophoretic processes.

2.1.3 Gel loading: From proteins to gels

Another key step in obtaining 2-DE gels is gel loading and gel running. After protein extraction, the pellet must be resuspended in a rehydration buffer, which is generally the same in all works. Then, proteins are loaded in order to perform the IEF step for their correct horizontal migration, however, the protocols varied according both to the biological model and the protein type extracted (i.e. soluble or membrane proteins). Finally, proteins separated in the IEF step are loaded in to the second dimension (SDS-PAGE). Thus, in this section rehydration buffers, IEF steps and second dimension gels will be analyzed.

2.1.3.1 Early proteomic studies

In the work of Mets & Bogorad (1972), ribosomal proteins were only run in the IEF step at 1.5 mA for 4 h but it was enough to separate them due to the low quantity of proteins that were obtained in this extraction. The second dimension was run at 25 mA, enough time to allow the protein migration, since the 2-DE gel patterns are very clear and well resolved. Also, no vertical or horizontal streaking is present, thereby, permitting clear protein detection. It is not astonishing to observe similar 2-DE patterns in quality terms in the work by Hanson et al. (1974), since both the ribosomal protein extraction and the two-dimensional gel electrophoresis were performed essentially as described by Mets & Bogorad (1972). Therefore, no vertical or horizontal streaking was found, resulting in gels with high resolution. Both protein extraction and gel electrophoresis proved to be very efficient and adequate for protein separation. However, it should be emphasized that the patterns from both works are easier to obtain, since the protein mixture is very simple since it only came from ribosome structures.

Unlike the ribosomal protein mixture, others do not generate 2-DE patterns with the same resolution. One case may be flagella and axonemes of *C. reinhardtii* in which a larger number of proteins are founded. Piperno et al. (1977) compared proteins of this structure from both wild-type and paralyzed mutants strains of this species. The IEF step was performed at 300 V for 18-19 h and followed by 400 V for 1.5 h. The second dimension was first run at 25 mA (initial voltage: 60 V) for 1 h and then it was raised to 50 mA. The run continued until the dye in the molecular weight standard had reached the bottom according to Ames and Nikaido (1976) (as cited in Piperno et al., 1977). The 2-DE gels had minimum vertical streaking, but lot of horizontal streaking and big stains regardless of the sample. The horizontal streaking could be due to a more complex protein mixture; however, the protein extraction protocol of this work is very deficient since it only uses SDS. Regardless of this, some spots were easily detected in the gels allowing for comparison between wild-types and mutant strains. Finally, in the work of Götz & Arnold (1980) ribosomal proteins from eight species were evaluated with two gels showing clear and well-resolved 2-DE patterns. The protein extraction was well suited for all species. Therefore, the $MgCl_2$-acetic acid method proved efficient in a large number of species, but again it was used to extract only ribosomal protein, so minimum contaminants are present.

2.1.3.2 Current proteomic studies

In more recent papers, such as those described in the previous section, the rehydration buffer used to resuspend the proteins prior to gel loading is key for the proper migration of proteins. The most commonly used buffer contains the reagents thiourea, urea, CHAPS, DTT, ampholytes and bromophenol blue. However, the concentrations of the reagents vary among the different works, so choosing the most accurate one is no easy task. As an example, we chose the protocol described by Wang et al. (2003) in which several reagents were tested to determine which one that yielded the best 2-DE pattern (i.e. no streaking and more defined spots) (see Appendix A). The majority of researchers state in their works that after resuspending the proteins, the mixture must be left at room temperature for at least 1 h (e.g. Hippler et al., 2001; Kim et al., 2005; Tran et al., 2009). Likewise, the amount of proteins normally loaded is 500 μg, concentration enough to yield well resolved gels (e.g. Förster et al., 2006; Mahong et al., 2012; Wang et al., 2003).

The IEF profile contains several steps, which vary between the different works, so making comparisons is complicated and not very productive. Nowadays, researchers worldwide use IPG gel strips for a better protein migration, which leads to a better 2-DE pattern (e.g. Mahong et al., 2012; Wang et al., 2003). Having said that, all IPG gel strips must be first rehydrated for at least 10 h before setting the IEF profile. As an example we chose the IEF profile of Wang et al. (2003) which was initiated at 250 V for 15 min, and gradually ramped to 10,000 V over 5 h, and remained at 10,000 V for an additional 6 h.

After the IEF steps and prior to the second dimension, IPG gel strips must be incubated twice in an equilibration buffer containing Tris-HCl, urea, glycerol and SDS. The first time DTT is added to the equilibration buffer in order to denaturate proteins, whereas the second time iodoacetamide is added to alkylate the reduced cysteines and inhibit protein refolding. After equilibration, IPG gel strips are ready to be loaded on to the second dimensional SDS-PAGE for the vertical protein separation (i.e. according to their molecular weight). Gel thickness will vary in each experiment in order to allow the desired protein separation. Regardless of this, gels are run until the bromophenol blue reaches the bottom of the gel since it migrates faster than the proteins. The last step for obtaining the 2-DE gel is gel staining in which two principal stains are used: blue Coomassie and silver nitrate. Regardless of this, generally prior to staining, the gels are washed with deionised water. After staining, the excess of dye is removed with deionised water to obtain well-defined gels with minimum background noise.

Now with the gels stained, we are able to determine which protocol(s) yielded the best 2-DE gel(s) in terms of patterns quality (i.e. minimum or none streaking, spots with defined circles, a maximum spot number and high spot intensity). In the work by Kim et al. (2005) 2-DE gel images show smearing, some vertical streaking and high horizontal streaking specifically in the acidic side of the gel. Also, spots are not well-defined circles and are overlapped among them. Similar were the image gels by Tran et al. (2009), because smearing as well as vertical and horizontal streaking are present in the acidic part of the 2-DE gel. Also several spots were overlapped among them; nevertheless a few of them were well defined. These were the two protocols that yielded the worst results (e.g. poor gel resolution quality) and this must be to the simplicity of the protein extraction protocols used. The two protocols that follow in terms of 2-DE gel quality are those of Hippler et al. (2001) and Förster et al. (2006). In both works 2-DE gels are of high quality, which obviously obey more complex protein extraction protocols. In the oldest work, there are several traits that give this images high quality: i) minimum horizontal streaking, ii) well defined spots (i.e circle shaped), iii) highly stained spots and iv) high number of spots (since only thylakoid membrane proteins were extracted) (Hippler et al., 2001). The high quality of 2-DE gels is probably due to that only a portion of the cell proteins was extracted having less contaminants interfering in both IEF and second dimension. Förster et al. (2006) 2-DE gel images show a high number of spots and most of them are well define with almost no smearing. However, a lot of vertical streaking is observed in the gels, thus the problems must be found in the second dimension since minimum horizontal streaking is present.

Finally, the protocols that yielded the 2-DE images of higher quality were those developed by Wang et al. (2003) and Mahong et al. (2012). In both works, total proteins were extracted from two different microalgae, *H. pluvialis* and *C. reinhardtii* respectively. Highlight that both protocols are the most complex ones among all six analyzed. Gels from both works succeeded in having reduced streaking as well as defined, highly stained and high number of spots. Nevertheless, if one must choose between both, it is Mahong et al. (2012) protocol the one with the best results since gels in this work have minimum background allowing an easier spot detection.

2.2 Macroalgae methodology

The difficulty in obtaining high quality 2-DE gels from macroalgae was first highlighted by Wong et al. (2006), who obtained algal proteins from *Gracilaria changii* (Plantae, Rhodophyta) using four different extraction methods: 1) direct precipitation by trichloroacetic acid/acetone, 2) direct lysis using urea buffer, 3) tris buffer and 4) phenol/chloroform. However, only methods 3) and 4) were compared for their suitability to generate *G. changii* proteins for two-dimensional gel electrophoresis. It was stated in this work that the phenol/chloroform method (detailed protocol in Appendix B) was Ideal for obtaining well resolved 2-DE patterns. Nevertheless, the quality of the 2-DE profiles was poor due to the presence of high amounts of interfering substances accompanied by low protein yield and horizontal and vertical streaking along gels regardless the pH gradient. Thus, this method is not fully accurate for this algal species.

As part of an on-going work focused on unravelling the metabolic processes occurring in physiologically stressed brown macroalgae, a new method for protein extraction that minimizes the co-extraction of non-protein compounds using two structurally distinct brown algal species *Scytosiphon gracilis* (Chromista, Ochrophyta) (Contreras et al., 2007b) and *Ectocarpus siliculosus* (Chromista, Ochrophyta) (Contreras et al., 2008) was developed. In order to do this, several protein extraction methods available in the literature were tested. However, neither of the previous protocols was ideal for obtaining a good quality algal protein extraction, due to high background noise, band distortion, and more importantly, very low protein dissolution. The protocol developed in this work allowed the use of a highly resolving 2-DE protein analyses, providing the opportunity to unravel potentially novel physiological processes unique to this group of marine organisms (see Table 1 and Results section). Specifically, the protocol uses an initial desalting step with Milli Q water – phosphate buffer in order to remove the salt from the algal tissues. Afterwards, the tissue is pulverized using liquid nitrogen and homogenized with sucrose, EDTA and CHAPS. The proteins are extracted using phenol and washed with ammonium acetate. Finally, the quality of the extracted proteins is improved by using the 2-D clean-Up Kit (GE Healthcare).

In another important proteomic work with macroalgae developed by Kim et al. (2008) and published contemporarily with the Contreras et al. (2008) work, using as models the red algae *Bostrychia radicans* and *B. moritziana* (Plantae, Rhodophyta), used a lysis buffer comprised principally by urea and thiourea (detailed protocol in Appendix B). Although these species belong to the same group of red algae like *G. changii*, the simplicity of this method utilized in comparison with the phenol one (Wong et al., 2006) is due to the morphological characteristics of this species (see image in Appendix B).

The work described by Yotsukura et al. (2010) presents a similar protocol to the one described by Contreras et al. (2008). Here, proteins are extracted from the brown alga *Saccharina japonica* (Chromista, Ochrophyta), important kelp described principally on the coastal areas of northern Japan. In this protocol, the protein extraction was improved by using phenol as the principal component in the lysis buffer (detailed protocol in Appendix B). This protocol was also used in *Ecklonia cava* (Chromista, Ochrophyta), other important kelp found on the coast along the Sea of Japan, and also good quality 2-D patterns were obtained (Yotsukura et al., 2012). The use of phenol in the protein extraction described by Contreras et al. (2008) has been recently used in the red alga *Porphyra columbina* in order to identify the proteins that are over-induced during desiccation stress tolerance responses. A highly resolved 2-DE protein was obtained using this method (Fig. 1), with minor modifications (detailed protocol in Appendix B), such as an important rinse of the protein pellet due principally to the over-production of phycocyanin and phycoerythrin. Thus, the phenol protocol developed by Contreras et al. (2008) could be used in macroalgae species from different taxonomic groups.

The first dimension of the 2-DE in the works mentioned above used approximately 200-500 μg of extracted proteins. However, for the isoelectric focusing (IEF) the protocols varied depending on the algal species used. For example, in *Bostrychia radicans* and *B. moritziana*, the voltage was linearly increased from 150-3,500 V during 3 h, followed by a constant 3,500 V, with focusing complete after 96 V. In *Scytosiphon gracilis* and *Ectocarpus siliculosus*, on the other hand, the strips are actively rehydrated for 15 h in IEF buffer containing the proteins and focused at 20°C with the following successive steps: a liner increase from 0 to 250 V for 15 min, a gradient phase from 250 V to 10,000 V for 4 h, and the a hold phase at 10,000 V for a total of 60 kVh. Using this protocol, the IEF for *Porphyra columbina* has some modifications, principally in a total operational voltage of 70 kVh.

3. Results and discussion: From gel to molecular/ecological interpretation

Proteomic analyses have proved to be an important molecular approximation that enables comparisons between species and/or cell variants, and understanding of cell function and stress tolerance (e.g. metals, high salinity, high temperatures, among others) (e.g. Contreras et al., 2010; Kim et al., 2005; Ritter et al., 2010). Due to the particularity of the cellular components (e.g. high content of polysaccharides) of this group of organisms, protein extraction has been the principal problem. However, as stated in the previous sections, some protocols have proved capable of producing high quality protein extracts for 2-DE electrophoresis (microalgae: Mahong et al., 2012; Wang et al., 2003 and macroalgae: Contreras et al., 2008). A high quality protein extract will yield high-resolved 2-DE patterns. Therefore, with a suitable protocol the use of a proteomic approximation appears to be of high importance for understanding various physiological responses in this group of organisms. However, it is imperative to highlight that proteomic works in micro and principally in macroalgae, are considerably lower in comparison with vascular plants and animals. Then, our effort in this chapter was concentrated in describing those important works utilizing as model the algal assemblage.

One of the first proteomic studies in microalgae characterized the chloroplastic ribosomal proteins of wild-type and erythromycin-resistant mutants of *Chlamydomonas reinhardtii* (Mets & Bogorad, 1972). In the mutant *ery-M2d* a protein of the 52S subunit was missing when compared to the wild-type. Nevertheless, low intensity proteins spots with the same

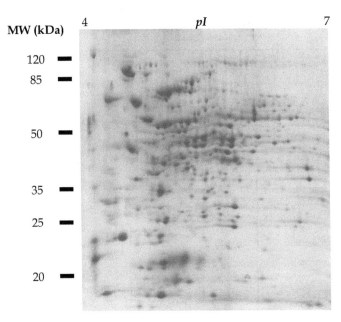

Fig. 1. 2-D proteome of *Porphyra columbina* under natural desiccation. First dimension was performed on a linear gradient IPG strip of pH 4-7 using 600 μg of total proteins. The 12.5% SDS-PAGE gel was stained with colloidal Coomassie blue.

pI but different molecular weight were found, indicating that the *ery-M2* gene is involved in determining the properties of that protein. Hanson et al. (1974) performed 2-DE gels in order to characterize cytoplasmic and chloroplastic ribosomal proteins. Their results showed that the number of proteins in both small and large subunits was higher in cytoplasmic than in chloroplastic ribosomes, indicating that cytoplasmic ribosomes are more complex. Another study did a comparison between the ribosomal proteins of 8 species including *C. reinhardtii* (Götz & Arnold, 1980). The results showed that the number of proteins in both subunits was similar among all species, and that the *Polytoma papillatun* (Plantae, Chlorophyta) proteins were the most similar to those of *C. reinhardtii* in terms of protein homology.

Piperno et al. (1977) analysed the flagella proteins of wild-type and paralyzed mutants of *C. reinhardtii* in order to identify the mutated protein that incapacitates the mobility in mutant strains. In the flagella of *pf* 14, which completely lack radial spokes and associated spokeheads, 12 polypeptides were missing. Also in *pf* 1 flagella, where spokes are clearly present but spokeheads appear to be absent, 6 polypeptides were missing. Then, protein electrophoretic studies confirmed the phenotypical characteristics displayed by both paralyzed mutants, where the missing proteins may be involved in spokes and spokeheads correct morphology. Another work in *C. reinhardtii*, used a proteomic approach to analyse photosynthetic thylakoid membrane proteins isolated from wild-type and mutant strains (Hippler et al., 2001). The two mutant strains were Δ*ycf4* (PSI-deficient) and *crd1* (which is conditionally reduced in PSI and LHCI under copper-deficiency). In this work more than 30 different LHCP spots were identified using a tandem quadrupole mass spectrometer,

Protein	Expression level[a]	n° of peptides analized[b]	Species, n° access[c]	pI, Mw (KDa)
Transferase	over	13	Dechloromonas aromatica (Q47F82)	5.6, 65
tRNA synthetase	over	24	Helicobacter pylori (P56126)	5.9, 60
Phosphomannomutase	over	24	Schizosaccharomyces pombe (Q9UTJ2)	5.3, 53
Proteosome, subunit α	over	24	Oryza sativa (Q10KF0)	5.6, 53
ATP synthase, subunit α	over	13	Syntrophus aciditrophicus (Q2LQZ7)	6.0, 67
Ribulose biphosphate carboxylase large chain	over	22	Porphyra yezoensis (Q760T5)	9.6, 65
Glyceraldehyde 3-phosphate dehydrogenase 1	over	22	Gracilaria verrucosa (P30724)	6.2, 43
Peptidase/Protease	over	19	Methanothermobacter (O27355)	6.3, 42
tRNA binding protein	over	23	Anaplasma (Q2GJX4)	6.2, 38
ATP binding protein	over	24	Methanocaldococcus jannaschi (Q58049)	6.4, 38
Transcriptional regulator	over	24	Mesorhizobium loti (CAD31581.1)	8.7, 29
Carbohydrate kinase	over	35	Salmonella enterica (YP_152740.1)	3, 18.2
RNA binding protein	over	28	Bacillus phage (P06953)	6.4, 25
ABC transporter subunit	over	40	Theileria parva (XP_764551.1)	8, 20.6
RNA polymerase, subunit α	over	19	Francisella tularensis (Q5NHU3)	9.2, 17.5
Peroxiredoxin	over	20	Porphyra purpurea (P51272)	9.5, 18
Chaperonine	over	10	Caulobacter crescentus (P48222)	5.6, 8
ABC transporter subunit	over	45	Desulfitobacterium hafniense (ZP_01371968.1)	6.2, 8.2
ABC transporter subunit	over	35	Janibacter sp. (ZP_00996449.1)	8.6, 7.3

Table 1. Proteins differentially expressed in *S. gracilis* exposed to copper excess. The analysis by MSMS allowed to obtain various protein peptides which were identified by BLASTP (NCBI). (a) Changes in expression level compared with controls. (b) Number of peptides analyzed by LC/MS/MS. (c) NCBI access number of the species with the highest identity obtained by BLASP.

thereby, permitting proteins with transmembrane domains to be separated with high resolution. Here, the results showed that LHCI spots were present on Δycf4 and absent on crd1 mutants.

Proteomics approaches have been helpful in understanding tolerance to naturally or anthropologically occurring environmental factors (e.g. high light, thermal stress and heavy metals respectively) in different species. Due to anthropological activities (e.g. industry and mining), heavy metals such as cadmium (Cd) and copper (Cu) are accumulating in the environment (Vermeer & Castilla, 1991; Medina et al., 2005). At high concentrations these metals are a source of abiotic stress, and can be highly toxic to organisms. In this matter, proteomic approaches are of high utility because they may provide new information regarding mechanisms to cope with the stress induced by the high concentration of metals. For example, in the work developed by Contreras et al. 2010, the copper-tolerance capacity of the brown algae species *Scyotsiphon gracilis* was evaluated by means of the 2-DE approximation. In this work, using the protocol previously described by Contreras et al. 2008 in the Appendix B, 19 over-expressed proteins were identified, including a chloroplast peroxiredoxin, a cytosolic phosphomannomutase, a cytosolic glyceraldehyde-3-phosphate dehydrogenase, 3 ABC transporters, a chaperonine, a subunit of the proteasome and a tRNA synthetase, among others (Table 1). The possible involvement of these over-expressed proteins in buffering oxidative stress and avoiding metal uptake in *S. gracilis* exposed to copper excess is discussed considering this proteomic information. For example, the peroxiredoxine (PRX) is an enzyme involved in the detoxification of hydrogen peroxide and fatty acid hydroperoxides (Dietz et al., 2006). In plants, *prx* transcripts increase in response to different abiotic stresses such as salinity, drought and metals (Dietz, 2003; Wood et al., 2003). Furthermore, PRX in the microalga *C. reinhardtii* and the red macroalga *Porphyra purpurea* (Plantae, Rhodophyta) have shown high similarity with plant PRXs (Baier and Dietz, 1997; Goyer et al., 2002). The expression of PRX in *C. reinhardtii* seems regulated by light, oxygen and redox state (Goyer et al., 2002). Thus, the PRX identified in *S. gracilis* may play an important role in oxidative stress buffering and in lipoperoxides detoxification. In fact, we have recently demonstrated the active participation of this enzyme in copper tolerant species in comparison with sensitive ones, where the over-expression of this enzyme is localized in the cortical cells (Lovazzano et al., personal communication). The proteomic information obtained by Contreras et al. 2010 in *S. gracilis* opens the opportunity of understanding many biological/physiological processes in algae. Using this information and those obtained using a biochemistry approximation, it is possible to strongly suggest a cross-talk between different pathways to re-establish the cellular homeostasis distorted by copper-associated oxidative stress in this species as well as in other tolerant ones (Fig. 2). Thus, the differential ability of each species to deal with oxidative stress resulting from the high copper levels, explains the persistence of tolerant species and the absence of sensitive ones at copper contaminated zones.

Using the method described by Contreras et al. 2008, it was also possible to evaluate differential tolerance in *Ectocarpus siliculosus* strains, originated from habitats with contrasting histories of copper levels (Ritter et al., 2010). Here, the authors showed a differential stress tolerance between 50 and 250 µg L^{-1} of copper. This difference was also observed at the level of the 2-DE proteome profile. For example, in the tolerant strains from a copper contaminated site (i.e. Chañaral, Chile) a specific expression of PSII Mn-stabilizing protein, fucoxanthine chlorophyll a-c binding protein and vanadium-dependent bromoperoxidase proteins, among others, was observed. Thus, the occurrence of the differential proteome profile among the strains could be strongly suggested by the persistence copper driving force in the evolution of *Ectocarpus siliculosus* from the copper contaminated sites (Ritter et al., 2010). In other brown macroalgae such as *Ecklonia cava*, it was possible to observe the effects of temperature on the proteomic profile (Yotsukura et al., 2012). Here, the authors define that the differential protein expression induced by temperature could be considered as an important biomarker of the health individuals in the culture conditions.

In *Saccharina japonica* it was possible to observe differences at the level of the proteome under seasonal variation and pH conditions (Yotsukura et al., 2010; 2012). Under seasonal variation, the specific expression of different proteins was identified, among them the vanadium-dependent bromoperoxidase (Yotsukura et al., 2010). Comparatively, under different pH culture conditions the over-expression of several proteins was described such as: glyceraldehyde-3-phosphate dehydrogenase, actin, phosphoglycerate kinase, elongation factor Tu and ATP synthase subunit β, among others. Thus, different metabolic pathways could be induced in brown macroalgae according to the type of stress factor. In this context, the utilization of the 2-DE approximation has been extraordinarily important in unravelling the tolerance mechanisms associated with environmental variables from natural and anthropogenic sources. In fact, the identification of important enzymes, never before described in algae (i.e. Peroxiredoxine (Contreras et al., 2010) and vanadium-dependent bromoperoxidase (Ritter et al., 2010)), opens the opportunity to further understanding the biology of this group of organisms.

In microalgae, several works have also been reported. For example, Wang et al. (2004) studied the proteome changes of *Haematococcus pluvialis* under oxidative stress induced by the addition of acetate and Fe^{2+} and exposure to excess of high light intensity. About 70 proteins were identified in which 19 were up-regulated (e.g. antioxidant enzymes and sugar synthesis proteins) and 13 were down-regulated (e.g. metabolism and cell growth proteins). Also, transient regulated proteins were identified in which 31 were up-regulated (e.g. antioxidant enzymes) and only 8 were down-regulated (e.g. chloroplastic proteins). In 2006, Förster et al. performed a proteome comparison among wild-type and two very high light-resistant mutants (*VHL^R-S4* and *VHL^R-S9*) under different high light stress. About 1500 proteins were detected in the gel and 83 proteins from various metabolic pathways were identified by peptide mass fingerprinting. The results revealed complex alterations in response to the stress, where total proteins varied drastically in the wild-type compared to the mutants. Nevertheless, the mutant *VHL^R-S4* proved to have better adaptation to high light stress since a more controlled protein regulation was conducted (e.g. up-regulation of several chaperonins and down-regulation of energy metabolism proteins). Another work conducted in *H. pluvialis* analyzed the proteome under high irradiance, but combined with nitrogen starvation (Tran et al., 2009). In the gels, about 900 protein spots were detected of which 13 were down-regulated and 36 up-regulated. Among the up-regulated proteins, a glutathione peroxidase and a translocase from the outer mitochondrial membrane were matched to *C. reinhardtii*; therefore, these stress responses may be common among these microalgae. A study assessing a proteomic analysis on *C. reinhardtii* under a short-term exposure to irradiance revealed significant down regulation of several heat-shock proteins (HSPs) (Mahong et al., 2012) under differential times of exposition to this stress (0 h, 1.5 h, 3 h and 6 h of high light). Spot densities allowed the determination that early rearrangement of the light-harvesting antenna proteins occurs, where this was manifested by the up- and down-regulation of several protein spots identified as LHC-II polypeptides. Moreover, increased expression of proteins involved in carbohydrate metabolism was found, which could help accelerate the utilization of electrons generated, in order to minimize the risk of superoxide formation. Surprisingly, after 6 hours of high light several molecular chaperones were down-regulated and this could result in drastic effects on cell structure and function. Nevertheless, *C. reinhardtii* is normally light-sensitive which could be explained by the down-regulation of molecular chaperones.

In microalgae, the response of species to heavy metal contamination has also been evaluated. A proteomic analysis conducted on *N. oculata* showed differences between protein expression

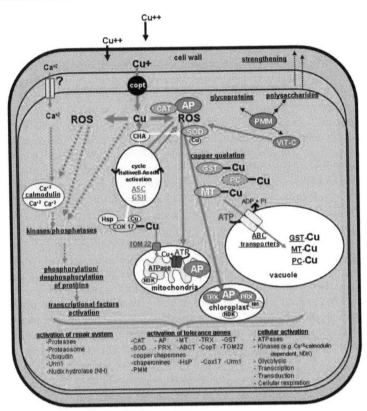

Fig. 2. Cellular events involved in the mechanisms of copper stress tolerance in algae. Dotted arrows indicate routes not directly evidenced in brown algae. The alteration of the state redox, cell damage, and the metal may trigger the antioxidant machine [i.e. compounds and antioxidant enzymes (activation of cycle Halliwell-Asada (MDHAR, DHAR and GP), CAT, SOD, AP, TRX, PRX)] as the activations of protein/genes that form part of various metabolic pathways. Proteins such as HSP or CHA may be involved in the protein protection as in the transport of the metal to proteins that use it as a cofactor, respectively. The sequestration of the metal by different proteins (i.e. MT, GST and PC) is an important homeostatic pathway of tolerance to the metal. The strengthening of the cell wall can increase the resistance to the entry of the metal to the cell. Copts, copper transporter; ROS, reactive oxygen species; MDHAR, monodehydroascorbate reductase; DHAR, dehydroascorbate reducatase; GP, gluthathione peroxidase; AP, ascorbate peroxidase; SOD, superoxide dismutase; CAT, catalase; TRX, tioredoxine; PRX, peroxiredoxine; PMM, phosphomannomutase; VIT-C, vitamin C or ascorbic acid; CHA, copper chaperone; GST, glutathione-s-transferase; PC, phytochelatin; MT, metallothionein; ASC, ascorbate; GSH, glutathione; HSP, heat shock protein; TOM 22; cox 17 transporter; Urm1 (modifier protein type ubiquitin); NDK, nucleoside diphosphate kinase.

of treated (10 µM Cd for 4 days) and untreated (control) cells (Kim et al., 2005). The protocol used in this work, as was discussed before, yielded deficient 2-DE gels, resulting in few proteins detected with only 11 of them with significant changes. Also, the Cd concentration was far from toxic levels suggesting that changes in the protein expression were not needed. This is a non-sequenced species, and therefore, cross-species protein identification was conducted in order to identify those expressed in *N. oculata*. The results showed that malate dehydrogenase and NADH-dehydrogenase were newly induced, whereas glyceraldehyde 3-phosphate dehydrogenase was suppressed. The induction of malate dehydrogenase could be a defense mechanism against Cd toxicity, since at least in *C. reinhardtii* this enzyme controls the malate valve system, which exports reducing power from the chloroplast. Another work assessing Cd toxicity evaluated the proteomic profiles of treated (150 μM Cd) and untreated (control) mutants lacking cell walls of *C. reinhardtii* (Gillet et al., 2006). These mutants are more sensitive to heavy metals due to the lack of a cell wall (Macfie et al., 1994 as cited in Gillet et al., 2006). It was observed that cadmium slowed down the growth rate, and furthermore, induced a 30-50% of growth inhibition. In this work, an elevated number of protein spots were detected and subsequently identified. In fact, 20 proteins were down-regulated in response to Cd stress. Among the down-regulated proteins were those that are involved in amino acid and nitrogen metabolism, chloroplast function and molecule biosynthesis to minimize ROS production. The most variable protein was the RubisCo large subunit, where the protein spot in the control treatment was 15.3 times more intense than in the Cd treatment. It was observed that enzymes with antioxidant properties, chaperonins, and enzymes involved in ATP and carbohydrate metabolism were up-regulated. In addition, in both works chloroplast proteins were found to be down-regulated and proteins involved in antioxidant response to be up-regulated. Therefore, the Cd tolerance mechanism may be similar among different species of microalgae.

4. Conclusions

Micro and macroalgae contain high levels of compounds that interfere with protein extraction. These compounds lead to precipitation of insoluble polymers where the proteome obtainment is almost impossible. However, many efforts have been made in the last years to minimize the coprecipitation of those compounds, and thus now important proteomic protocols are available. For example in macroalgae, it is highlighted the use of phenol during the protein extraction, resulting in consistent electrophoresis runs in several species, conciliating suitable quality and reliability for 2-DE gels and its downstream analysis. The advantage of using phenol as an extracting agent resides in its capacity to disrupt membranes, leaving most of the water-soluble molecules totality in the aqueous phase.

Compared to animals and vascular plants, there is limited information about the use of 2-DE in either micro or macroalgae, both at technical and proteomic level. In fact, low number of published information in the proteomic context can be registered in micro and macroalgae. For example, in this group of organisms only about 42 works can be founded in the www.ncbi.nlm.nih.gov data base, using as search the concepts proteomic or proteome. On the other hand, in vascular plants it is possible to find about 3,100 works in the same data base and ca. 13,100 in animals. Thus, insignificant information exists nowadays in the proteome involvement in algal species, independent of the taxonomic status, ecological importance and economic value of this group of organisms.

2-DE in algal species has allowed the identification of several pathways involved in tolerance mechanisms, associated principally to different abiotic factors. For example, under copper stress the identification of proteins such as peroxiredoxine, enzyme involved in the detoxification of hydrogen peroxide and fatty acid hydroperoxide has Allowed to understand the differential degree of tolerance between Copper tolerant and sensitive species. In fact, using the proteomic protocol described in these species, which uses phenol in the protein extraction, a differential proteome profile in algal individuals between desiccation stress and normal hydration was founded. In this context, new tolerance mechanisms will be revealed using this approximation in order to understand the high desiccation tolerance that exists in this species in comparison with many others, including that from the same phylum. Thus, 2-DE approximation is an important tool that can be interconnected with those obtained to ecological level in order to understand mechanisms of stress tolerance, and therefore explanation of distribution patterns at local and latitudinal scale.

5. Appendix A: Microalgae methodology

Chlamydomonas reinhardtii **(Plantae, Chlorophyta)** (Förster et al., 2006). The figure correspond to the species mentioned.

This protocol is an adaption of the one described by Mathesius and co-workers (Grotewold, 2003) that are suited for plant material.

1. Collect *Chlamydomonas* cells by centrifugation at 5,000 x g for 5 min at 20° C. Determine fresh weight of cell pellets. Samples can be stored at -80°C for later use.
2. Grind pelleted *Chlamydomonas* cells to a fine powder in liquid nitrogen using a mortar and pestle after addition of 0.5 g of glass powder per 1 g fresh weight of pelleted cells.
3. Suspend the ground material in -20°C cold acetone containing 10% w/v TCA and 0.007% w/v DTT. Sonicate this suspension on ice six times for 10 s each with intermittent 1-min breaks using an ultrasonicator. Centrifuge samples at 35,000 x g for 15 min at 4°C.
4. Wash the pellet twice by resuspension in -20°C acetone containing 0.07% w/v DTT, placing it at -20°C for 30 min and centrifuging at 12,000 x g for 15 min at 4°C.
5. Lyophilize the pellet for 3 min and resuspend in sample buffer containing 9 M urea, 4% w/v CHAPS, 1% w/v DTT, 0.8% v/v ampholytes (ones suited for the desired 2D-gel), 25 mM Tris base, 1 mM PMSF and 5 mM EDTA.
6. Sonicate samples twice in a sonic bath in an ice-water mixture for 5 min and centrifuge them at 19,000 x g for 15 min at 20°C.
7. Determine protein concentration of the sample (e.g. with a Bradford assay or a BCA assay) and keep at –80°C until used for isoelectric focusing.

Haematococcus pluvialis **(Plantae, Chlorophyta)** (Wang et al., 2003). The figure correspond to the species mentioned.

In this study a number of key chemical reagents were evaluated, the protocol that yielded the best 2-DE results is detailed below.

1. Collect *H. pluvialis* cells by centrifugation at 3,000 x *g* for 5 min and wash the pellet three times with cold deionized water.
2. Resuspend cell pellet in one volume of 50 mM Tris-HCl buffer pH 8.0, 3 mM DTT, 5 mM MgCl2, 10% glycerol, 0.5% PVP, 5 mM Na2-EDTA, 1 mM PMSF, 5 mM benzamidin, 5 mM acoproic acid and 1% v/v plant protease inhibitor cocktail.
3. Disrupt cells by one passage through a pre-cooled French Press Cell at a pressure of 20,000 psi. Centrifuge cell lysate at 3,000 *g* for 10 min to pellet cell debris.
4. Collect the supernatant and centrifuge at 100,000 *g* for 1 h.
5. Dialyze the supernatant from the previous centrifugation step against 250 mL of 85% w/v sucrose solution at 4°C for 2 h. Precipitate the dialysate with 9 volumes of ice-cold 10% w/v TCA in acetone containing 0.07% w/v β-mercaptoethanol at -20°C overnight.
6. Centrifuge samples at 15,000 x *g*. Discard supernatant and wash the pellet with acetone containing 0.07% w/v β-mercaptoethanol to remove TCA.
7. Then, remove residual acetone by air-drying.
8. Resuspend pellet in solubilization buffer containing 2 M thiourea, 8 M urea, 4% CHAPS, 2 mM TBP, and 0.2% ampholytes (ones suited for the desired 2-DE gel).
9. Determine protein concentration of the sample (e.g. with a Bradford assay or a BCA assay) and keep at –80°C until used for isoelectric focusing.

Nannochloropsis oculata **(Chromista, Ochrophyta)** (Kim et al., 2005). The figure correspond to the species mentioned.

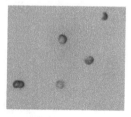

In this study only one method for protein extraction was performed, the details are shown below.

1. Collect *N. oculata* cells by centrifugation at 12,000 x *g* for 10 min. Suspend cell pellet in PBS buffer pH 7.2.

2. Mix suspension with the same volume of sample buffer containing 0.3% w/v SDS, 1% w/v β-mercaptoethanol and 0.05 M Tris-HCl pH 8.0.
3. Denature solution at 100°C for 3 min, cool on ice and treat with DNase/RNase. Precipitate proteins with 10% TCA in 100% Acetone at -70°C for 3 h.
4. Wash the pellet with 100% acetone several times and then air-dry it at room temperature for 5 min.
5. Determine protein concentration of the sample (e.g. with a Bradford assay or a BCA assay) and keep at –80°C until used for isoelectric focusing.

6. Appendix B: Macroalgae methodology

Gracilaria changii **(Plantae, Rhodophyta)** (Wong et al., 2006). The figure correspond to the species mentioned.

In this study two protein extraction methods were analysed in two-dimensional gels. The best results were yielded by the phenol/chloroform method, which is detailed below.

1. Grind frozen seaweeds at -70°C into a fine powder with a mortar and pestle in liquid nitrogen. Put approximately 100 mg of the resulting powder into a 1.5 mL tube for a single extraction.
2. Add 1 mL of TRI reagent (containing phenol and guanidine-isothiocyanate) to 100 mg of seaweed powder and homogenize the mixture.
3. Store the homogenate for 5 min at room temperature to clarify phases. Reserve the phenolic phase and add 200 µL of chloroform per 1 mL of TRI reagent.
4. Cover the samples and shake vigorously for 15 seconds. Store the resulting mixture at room temperature for 2-15 min. Centrifuge mixture at 12,000 g for 15 min at 4°C.
5. Discard upper aqueous phase containing RNA, and retain interphase and lower red phenol-chloroform phase containing DNA and proteins.
6. Add ethanol to the reserved phases in order to precipitate DNA.
7. Retain phenol/ethanol supernatant and add 3 volumes of acetone to precipitate proteins, mix by inversion for 10-15 sec to obtain a homogenous solution.
8. Store sample for 10 min at room temperature and sediment the protein precipitate at 12,000 g for 10 min at 4° C.
9. Discard the phenol/ethanol supernatant and disperse the protein pellet in 0.5 mL of 0.3 M guanidine hydrochloride in 95% ethanol + 2.5% v/v glycerol.
10. Add another 0.5 mL aliquot of the guanidine hydrochloride/ethanol/glycerol solution to the sample and store for 10 min at room temperature. Centrifuge the proteins at 8,000 g for 5 min.

11. Discard the wash solution and perform two more washes in 1 mL each of the guanidine Hydrochloride/ethanol/glycerol wash solution.
12. Perform a final wash in 1 mL of ethanol containing 2.5% glycerol v/v. At the end of the 10 min of ethanol wash, centrifuge the proteins at 8,000 g for 5 min.
13. Discard the alcohol and air-dry the pellet for 7-10 min at room temperature.
14. Resolubilize protein pellet in 40 mM Tris buffer pH 8.8 containing 8 M urea, 4% CHAPS and 2 mM TBP.
15. Determine protein concentration of the sample (e.g., with a Bradford assay or a BCA assay) and keep at –80°C until used for isoelectric focusing.

Scytosiphon gracilis and *Ectocarpus siliculosus* (**Chromista, Ochrophyta**) (Contreras et al., 2008). The figures correspond to the species mentioned.

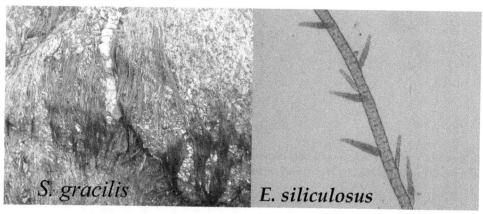

In this method, a major extraction of proteins was obtained in comparison with pervious macroalgae methods described.

1. Remove the excess salt by rinsing in Milli Q water and 50 mM Tris-HCl pH 8.8.
2. Freeze seaweed material at -80°C before pulverization.
3. Homogenize seaweed material using a mortar-driven homogenizer in sample lysis solution composed of 1.5% PVP, 0.7 M sucrose, 0.1 M KCl, 0.5 M Tris-HCl pH 7.5, 250 mM EDTA, protease inhibitor cocktail, 2% v/v β-mercaptoethanol and 0.5% w/v CHAPS
4. Equal volume of Tris-HCl pH 7.5-satured phenol is added and the mixture homogenized at 4°C. Then, by centrifugation the upper phase is removed and the lower phase is re-extracted using the same volume of phenol.
5. The proteins in the phenol phase are precipitated by means of ammonium acetate (0.1 M in methanol). The protein pellet obtained by centrifugation is washed in 80% ice-cold acetone and cold acetone containing 20 mM DTT.
6. Determine protein concentration of the sample and keep at –20°C until used for isoelectric focusing.

Bostrychia radicans / B. moritziana (**Plantae, Rhodophyta**) (Kim et al., 2008). The figure correspond to *B. moritziana*.

In this study a very simple method for protein extraction was performed, the details are shown below.

1. Freeze seaweed material at -80°C before pulverization.
2. Homogenize seaweed material using a mortar-driven homogenizer in sample lysis solution composed of 7 M urea, 2 M thiourea, containing 4% w/v CHAPS, 1% w/v DTT, 2% v/v ampholytes and 1 mM benzamidine.
3. Perform freezing and thawing steps for five times for 1 day*.
4. Extract proteins for 1 h at room temperature with vortexing.
5. Centrifuge mixture at 15,000 g for 1 h at 18°C.
6. Retain soluble fraction and discard insoluble material.
7. Determine protein concentration of the sample (e.g., with a Bradford assay or a BCA assay) and keep at –80°C until used for isoelectric focusing.

* If cell lysing is found to be difficult, used a bead beater in order to facilitate the process.

Saccharina japonica and *Ecklonia cava* (**Chromista, Ochrophyta**) (Yotsukura et al., 2010; 2012). The figures correspond to the species mentioned.

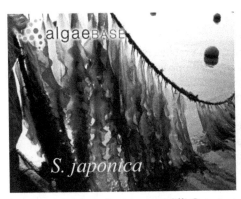

1. Remove the salt excess in Milli Q water.
2. Freeze seaweed material at -80°C before pulverization. Homogenize the tissue in 99.5% cold ethanol and centrifuged.
3. The protein pellet is rinsed in 99.5% ethanol and 100% acetone and resuspended in 0.1 M Tris-HCl buffer pH 8.0, 30% sucrose, 2% SDS, 5% β-mercaptoethanol and phenol.

4. The solution is vortexed and centrifuged at room temperature and the upper phase collected. Agitate solution in 0.1 M ammonium acetate and kept at -20°C.

5. The protein pellet obtained by centrifugation is rinsed in 0.1 M ammonium acetate in methanol and 80% acetone, subsequently dried in evaporator and preserved at –80°C until protein quantification.

Porphyra columbina **(Plantae, Rhodophyta).** The figure correspond to the species mentioned.

This method is an adaptation of the method performed by Contreras et al. 2008, where the details of it are shown below*.

1. Pulverize 3-5 g of frozen seaweeds at -80°C to a fine powder with a mortar and pestle in liquid nitrogen.

2. Resuspend pulverized tissue in 5-10 mL of buffer lysis containing 0.5 M Tris-HCl pH 7.5, 0.7 M sucrose, 0.5 M KCl, 250 mM EDTA, 1.5% w/v PVP, 0.5% w/v CHAPS and 2% v/v β-mercaptoethanol and homogenize for 15 min at 4°C.

3. Add an equal volume of Tris–HCl pH 7.5-saturated phenol and homogenize for 15 min at 4° C. Centrifuge the homogenate at 2,000 g for 30 min.

4. Retain only upper phenol phase containing proteins being careful not to remove the interphase.

5. Add ½ volumes of Tris–HCl pH 7.5-saturated phenol and mix well by inversion.

6. Centrifuge the homogenate at 2,000 g for 20 min.

7. Retain newly only upper phenol phase containing proteins being careful not to remove the interphase and mix with the previously retained upper phase.

8. Add 5 volumes of 0.1 M ammonium acetate on methanol ice-cold.

9. Shake vigorously to mix the solution and leave to precipitate for 4 h at -20°C.

10. Centrifuge at 2,000 g for 40 min. Discard supernatant and wash pellet with 8 volumes of 0.1 M ammonium acetate on methanol ice-cold.

11. Shake vigorously to mix the solution and leave to precipitate for 30 min at -20°C.

12. Centrifuge at 2,000 g for 30 min.

13. The proteins pellet is washed in 80% ice-cold acetone and cold acetone containing 20 mM DTT.

14. The protein pellet is then washed in ice-cold acetone 80% and ice-acetone 60% in methanol to remove the majority of contaminants.

15. Determine protein concentration of the sample and keep at -20°C until used for isoelectric focusing.

* Method not yet published.

7. Acknowledgment

This work was supported by FONDECYT 11085019 to LC. Additional funding cames from FONDAP 1501-0001 (CONICYT) to the Center for Advanced Studies in Ecology and Biodiversity (CASEB) Program 7. Finally, we are especially grateful to Nicole Ehrenfeld and Javier Tapia for image acquisition and Daniela Thomas, Alejandra Nuñez and Aaron Mann for text editing.

8. References

Abe, S., Kurashima, A., Yokohama, Y. & Tanaka, K. (2001). The cellular ability of desiccation tolerance in Japanese intertidal seaweeds. *Botanica Marina*, Vol. 44, No.4, pp.125-131, ISSN 0006-8055

Aguilera, J., Bischof, K., Karsten, U., Hanelt, D. & Wiencke, C. (2002). Seasonal variation in ecophysiological patterns in macroalgae from an Arctic fjord. II. Pigment accumulation and biochemical defence systems against high light stress. *Marine Biology*, Vol. 140, pp. 1087-1095, ISSN: 0025-3162

Alpert, P. (2006). Constraints of tolerance: why are desiccation-tolerant organisms so small or rare?. *The Journal of Experimental Biology*, Vol. 209, 1575-1584, ISSN: 1477-9145

Andrade, S., Contreras, L., Moffett, J.W. & Correa, J.A. (2006). Kinetics of copper accumulation in *Lessonia nigrescens* (Phaeophyceae) under conditions of environmental oxidative stress. *Aquatic Toxicology*, Vol. 78, pp. 398-401, ISSN: 0166-445X

Apte, S.C. & Day, G.M. (1998). Dissolved metal concentrations in the Torres Strait and Gulf of Papua. *Marine Pollution Bulletin*, Vol. 30, pp. 298-304, ISSN: 0025-326X

Baier, M. & Dietz, J. (1997). The plant 2-Cys peroxiredoxin BAS1 is a nuclear-encoded chloroplast protein, its expressional regulation, phylogenetic origin, and implications for its specific physiological function in plants. *The Plant Journal* Vol. 12, pp. 179–190, ISSN: 1365-313X

Baker, C.J. & Orlandi, E.W. (1995). Active oxygen in plant pathogenesis. *Annu. Rev. Phytopathology*, Vol. 33, pp. 299-321, ISSN: 0066-4286

Batley, G.E. (1995). Heavy metals and tributyltin in Australian coastal and estuarine waters. In Zann, L.P. & Sutton ,D. [Eds.] Pollution State of the Marine Environment Report for Australia. Technical Annex 2, Great Barrier Reef Marine Park Authority, Canberra, pp. 63–72.

Bilgrami, K.S. & Kumar, S. (1997). Effects of copper, lead and zinc on phytoplankton growth. *Biologia Plantarum*, Vol. 39, pp. 315-317, ISSN: 0006-3134

Blokhina, O. & Fagerstedt, K.W. (2010). Reactive oxygen species and nitric oxide in plant mitochondria: origin and redundant regulatory systems. *Physiologia Plantarum*, Vol. 138, pp. 447-462, ISSN:1399-3054

Bryan, N.L. & Langston, W.J. (1992) Bioavailability, accumulation and effects of heavy metals in sediments with special reference to United Kingdom estuaries: a review. *Environmental Pollution*, Vol. 76, pp. 89-131, ISSN: 0269-7491

Burritt, D. J., Larkindale, J. & Hurd, C.L. (2002). Antioxidant metabolism in the intertidal red seaweed *Stictosiphonia arbuscula* following desiccation. *Planta*, Vol. 215, pp. 829-838, ISSN: 0032-0935

Cabello-Pasini, A., Diaz-Martín, M.A., Muñiz-Salazar, R., Zertuche-Gonzalez, J.A. & Pacheco-Ruiz, I. (2000). Effect of temperature and desiccation on the photosynthetic performance of *Porphyra perforata*. *Journal of Phycology*, Vol. 36, pp. 10-14, ISSN: 1529-8817

Chinnasamy, G. & Rampitsch, C. (2006). Efficient solubilization buffers for two-dimensional gel electrophoresis of acidic and basic proteins extracted from wheat seeds. *Biochimica et Biophysica Acta - Proteins and Proteomics*, Vol. 1764, pp. 641–4, ISSN: 1570-9639

Clegg, J.S. (2005). Desiccation Tolerance in Encysted Embryos of the Animal Extremophile, Artemia. *Integrative and Comparative Biology*, Vol. 45, pp. 715–724, ISSN 1557-7023

Contreras, L., Moenne, A. & Correa, J.A. (2005). Antioxidant responses in *Scytosiphon lomentaria* (Phaeophyceae) inhabiting copper-enriched coastal environments. *Journal of Phycology*, Vol. 41, pp. 1184-1195, ISSN: 1529-8817

Contreras, L., Medina, M.H., Andrade, S., Oppliger, V. & Correa J.A. (2007a). Effects of copper on early developmental stages of *Lessonia nigrescens* Bory (Phaeophyceae). *Environmental Pollution*, Vol. 145, pp. 75-83, ISSN: 0269-7491

Contreras, L., Dennett, G., Moenne, A., Palma, E. & Correa, J.A. (2007b). Molecular and morphologically distinct *Scytosiphon* species (Scytosiphonales, Phaeophyceae) display similar antioxidant capacities. *Journal of Phycology*, Vol. 43, pp. 1320-1328, ISSN: 1529-8817

Contreras, L., Ritter, A., Denett, G., Boehmwald, F., Guitton, N., Pineau, C., Moenne, A., Potin, P., & Correa, JA. (2008). Two-dimensional gel electrophoresis analysis of brown algal protein extracts. *Journal of Phycology*, Vol. 44, pp. 1315-1321, ISSN: 1529-8817

Contreras, L., Mella, D., Moenne, A. & Correa, J.A. (2009). Differential responses to copper-induced oxidative stress in the marine macroalgae *Lessonia nigrescens* and *Scytosiphon lomentaria* (Phaeophyceae). *Aquatic Toxicology*, Vol. 94, pp. 94-102, ISSN: 0166-445X

Contreras, L., Moenne, A., Gaillard, F., Potin, P. & Correa, J.A. (2010) Proteomic analysis and identification of copper stress-regulated proteins in the marine alga *Scytosiphon gracilis* (Phaeophyceae). *Aquatic Toxicology*, Vol. 96, pp. 85-89, ISSN: 0166-445X

Contreras-Porcia, L., Denett, G., González, A., Vergara, E., Medina, C., Correa, J.A. & Moenne, A. (2011a). Identification of copper-induces genes in the marine alga *Ulva compressa* (Chlorophyta). *Marine Biotechnology*, Vol. 13, pp. 544-556, ISSN: 1436-2228

Contreras-Porcia, L., Thomas, D., Flores, V. & Correa, J.A. (2011b). Tolerance to oxidative stress induced by desiccation in *Porphya columbina* (Bangiales, Rhodophyta). *Journal of Experimental Botany*, Vol. 62, pp. 1815–1829, ISSN 0022-0957

Contreras-Porcia, L., Callejas, S., Thomas, D., Sordet, C., Pohnert, G., Contreras, A., Lafuente, A., Flores-Molina, M.R. & Correa, J.A. (2012). Seaweeds early development: detrimental effects of desiccation and attenuation by algal extracts. *Planta*, Vol. 235, pp. 337-348, ISSN: 0032-0935

Correa, J.A., Castilla, J.C., Ramírez, M., Varas, M., Lagos, N., Vergara, S., Moenne, A., Román, D. & Brown, M.T. (1999). Copper, copper-mine tailings and their effect on marine algae in northern Chile. *Journal of Applied Phycology*, Vol. 11, pp. 57-67, ISSN: 0921-8971

Cremer, F. & Van de Walle, C. (1985). Method for extraction of proteins from green plant tissues for two-dimensional polyacrylamide gel electrophoresis. *Analytical Biochemistry*, Vol. 147, pp. 22–6, ISSN: 0003-2697

Davison, R.I. & Pearson, G.A. (1996). Stress tolerance in intertidal seaweeds. *Journal of Phycology*, Vol. 32, pp. 197-211, ISSN: 1529-8817

Dietz, K.J. (2003). Plant peroxiredoxins. *Annual Review of Plant Biology*, Vol. 54, 93-107, ISSN: 1543-5008

Dietz, K.J, Jacob, S., Oelze, M.L., Laxa, M., Tognetti, V., Nunes de Miranda, S.M., Baire, M. & Finkenmeier, I. (2006). The function of peroxiredoxins in plant organelle redox metabolism. *Journal of Experimental Botany*, Vol. 57, pp. 1697–1709, ISSN 0022-0957

Farrant, J.M. (2000). A comparison of mechanisms of desiccation tolerance among three angiosperm resurrection plant species. *Plant Ecology*, Vol. 151, pp. 29-39, ISSN: 1385-0237

Flengsrub, R. & Kobro, G. (1989). A method for two-dimensional electrophoresis of proteins from green plant tissues. *Analytical Biochemestry*, Vol. 177, pp. 33–36, ISSN: 0003-2697

Förster, B., Mathesius, U. & Pogson, B.J. (2006). Comparative proteomics of high light stress in the model alga *Chlamydomonas reinhardtii*. *Proteomics*, Vol. 6, pp. 4309-4320, ISSN: 1615-9861

Gillet, S., Decottignies, P., Chardonnet, S. & Le Maréchal, P. (2006). Cadmium response and redoxin targets in *Chlamydomonas reinhardtii*: a proteomic approach. *Photosynthesis Research*, Vol. 89, pp. 201-211, ISSN: 0166-8595

Gledhill, M., Nimmo, M., Hill, S.J. & Brown, M.T. (1997). The toxicity of copper II species to marine algae, with particular reference to macroalgae. *Journal of Phycology*, Vol. 33, pp. 2-11, ISSN: 1529-8817

Goyer, A., Haslekas, C., Miginiac-Maslow, M., Klein, U., Le Maréchal, P., Jacquot, J.P. & Decottignies, P. (2002). Isolation and characterization of a thioredoxin-dependent peroxidase from *Chlamydomonas reinhardtii*. *European Journal Biochemistry*, Vol. 269, pp.272–282, ISSN: 1742-4658

Götz, H. & Arnold, C.G. (1980). Comparative electrophoretic study on ribosomal proteins from algae. *Planta*, Vol. 149, pp. 19-26, ISSN: 0032-0935

Grotewold, E. (2003). Methods in Molecular Biology. Plant Functional Genomics. Humana Press, New York.

Grout, J.A. & Levings, C.D. (2001). Effects of acid mine drainage from an abandoned copper mine, Britannia Mines. Howe Sound, British Columbia, Canada, on transplanted blue mussels (*Mytilus edulis*). *Marine Environmental Research*, Vol. 51, pp. 265-288, ISSN: 0141-1136

Hanson, M.R., Davidson, J.N., Mets, L.J. & Bogorad, L. (1974). Characterization of chloroplast and cytoplasmic ribosomal proteins of *Chlamydomonas reinhardtii* by two-dimensional gel electrophoresis. *Molecular and General Genetics*, Vol. 132, pp. 105-118, ISSN: 1432-1874

Hippler, M., Klein, J., Fink, A, Allinger, T. & Hoerth, P. (2001). Towards functional proteomics of membrane protein complexes: analysis of thylakoid membranes from *Chlamydomonas reinhardtii*. *The Plant Journal*, Vol. 28, pp. 595- 606, ISSN: 1365-313X

Hodgkiss, L.J. & Lu, S. (2004). The effects of nutrients and their ratios on phytoplankton abundance in Junk Bay, Hong Kong. *Hydrobiologia*, Vol. 512, pp. 215-229, ISSN: 0018-8158

Hoffmann, A.J. & Santelices, B. (1997). Flora marina de Chile central. Ediciones Universidad Católica de Chile, 434 pp.

Holm-Hansen, O., Naganobu, M., Kawaguchi, S., Kameda, T., Krasovski, I., Tchernyshkov, P., Priddle, J., Korb, R., Brandon, M., Demer, D., Hewitt, R.P., Kahru, M. & Hewes, C.D. (2004). Factors influencing the distribution, biomass, and productivity of phytoplankton in the Scotia Sea and adjoining waters. *Deep-Sea Research II*, Vol. 51, pp. 1333-1350, ISSN: 0967-0645

Ji, Y. & Tanaka, J. (2002). Effect of desiccation on the photosynthesis of seaweeds from the intertidal zone in Honshu, Japan. *Phycological Research*, Vol. 50, pp. 145-153, ISSN: 1440-1835

Kaltschmidt, E. & Wittmann, H.G. (1972). Ribosomal proteins: XXXII: comparison of several extraction methods for proteins from *Escherichia coli* ribosomes. *Biochimie*, Vol. 54, pp. 167-175, ISSN: 0300-9084

Kim, H.K., Shim, J.B., Klochkova, T.A., West, J.A. & Zuccarello, G.C. (2008). The utility of proteomics in algal taxonomy: *Bostrychia radicans/B. moritziana* (Rhodomelaceae, Rhodophyta) as a model study. *Journal of Phycology*, Vol. 44, pp. 1519-1528, ISSN: 1529-8817

Kim, Y.K., Yoo, W.I., Lee, S.H. & Lee, M.Y. (2005). Proteomic analysis of cadmium-induced protein profile alterations from marine alga *Nannochloropsis oculata*. *Ecotoxicology*, Vol. 14, pp. 589-596, ISSN: 0963-9292

Kobayashi, M., Kurimura, Y. & Tsuji Y. (1997). Light-independent, astaxanthin production by the Green microalga *Haematococcus pluvialis* under salt stress. *Biotechnology letter*, Vol. 19, pp. 507-509, ISSN: 0141-5492

Kumar, M., Kumari, P., Gupta, V., Reddy, C.R.K. & Jha, B. (2010). Biochemical responses of red alga *Gracilaria corticata* (Gracilariales, Rhodophyta) to salinity induced oxidative stress. *Journal of Experimental Marine Biology and Ecology*, Vol. 391, pp. 27-34, ISSN: 0022-0981

Lee, M.Y. & Shin, H.W. (2003). Cadmium-induced changes in antioxidant enzymes from the marine alga *Nannochloropsis oculata*. *Journal of Applied Phycology*, Vol. 15, pp. 13-19, ISSN: 0921-8971

Lesser, M.P. (1996). Elevated temperatures and ultraviolet radiation cause oxidative stress and inhibit photosynthesis in symbiotic dinoflagellates. *Limnology and oceanography*, Vol. 41, pp. 271-283, ISSN: 0024-3590

Lesser, M.P. (1997). Oxidative stress causes coral bleaching during exposure to elevated temperatures. *Coral reefs*, Vol. 16, pp. 187-192, ISSN: 1432-0975

Liu, W., Au, D.W.T., Anderson, D.M., Lam, P.K.S. & Wu, R.S.S. (2007). Effects of nutrients, salinity, pH and light:dark cycle on the production of reactive oxygen species in the alga *Chattonella marina*. *Journal of Experimental Marine Biology and Ecology*, Vol. 346, 76-86, ISSN: 0022-0981

Lobban , C.S. & Harrison, P.J. (1994). Seaweed Ecology and Physiology. Cambridge University Press, New York.

Mahong, B., Roytrakul, S., Phaonaklop, N., Wongratana, J. & Yokthongwattana, K. (2012). Proteomic analysis of a model unicellular green alga, *Chlamydomonas reinhardtii*, during short-term exposure to irradiance stress reveals significant down regulation of several heat-shock proteins. *Planta* Vol. 235, pp. 499-511, ISSN: 1432-2048

Marsden, A.D. & DeWreede, R.E. (2000). Marine macroalgal community, structure, metal content and reproductive function near an acid mine outflow. *Environmental Pollution*, Vol. 110, pp. 431-440, ISSN: 0269-7491

Mathesius, U., Keijzers, G., Natera, S.H.A., Weinman, J.J., Djordjevic, M.A. & Rolfe, B.G. (2001). Establishment of a root proteome reference map for the model legume *Medicago truncatula* using the expressed sequence tag database for peptide mass fingerprinting. *Proteomics*, Vol. 1, pp. 1424-1440, ISSN: 1615-9861

Medina, M., Andrade, S., Faugeron, S., Lagos, N., Mella, D. & Correa, J.A. (2005). Biodiversity of rocky intertidal benthic communities associated with copper mine tailing discharges in northern Chile. *Marine Pollution Bulletin*, Vol. 50, pp. 396-409, ISSN: 0025-326X

Mechin, V., Consoli, L., Guilloux, M. L. & Damerval, C. (2003). An efficient solubilization buffer for plant proteins focused in IPGs. *Proteomics*, Vol. 3, pp. 1299-302, ISSN: 1615-9861

Mets, L. & Bogorad, L. (1972). Altered chloroplast ribosomal proteins associated with erythromycin-resistant mutant in two genetic systems of *Chlamydomonas reinhardtii*. *Proceedings of the National Academy of Sciences of the USA*, Vol. 69, pp. 3779-3783, ISSN: 1091-6490

Mishra, A. & Jha, B. (2011). Antioxidant response of the microalga *Dunaliella salina* under salt stress. *Botanica Marina*, Vol. 54, pp. 195-199, ISSN: 1437-4323

Ojeda, F.P. & Santelices, B. (1984) Ecological dominance of *Lessonia nigrescens* (Phaeophyta) in central Chile. *Marine Ecology Progress Series*, Vol. 19, pp. 83-89. ISSN: 0171-8630

Oliveira, R. (1985). Phytoplankton communities response to a mine effluent rich in copper. *Hydrobiologia*, Vol. 128, pp. 61-69, ISSN: 0018-8158

Pearson, G.A., Serrao, E.A. & Cancela, M.L. (2001). Suppression subtractive hybridization for studying gene expression during aerial exposure and desiccation in fucoid algae. *European Journal of Phycology*, Vol. 36, pp. 359-366, ISSN: 0967-0262

Pearson, G.A., Hoarau, G., Lago-Leston, A., Coyer, J.A., Kube, M., Reinhardt ,R., Henckel, K., Serrao, E.T.A., Corre, E. & Olsen, J.L. (2010). An expressed sequence tag analysis of the intertidal brown seaweeds *Fucus serratus* (L.) and *F. vesiculosus* (L.) (Heterokontophyta, Phaeophyceae) in response to abiotic stressors. *Marine Biotechnology*, Vol. 12, pp. 195-213, ISSN: 1436-2228

Pinto, E., Sigaud-Kutner, T.C.S., Leitai, M.A., Okamoto, O.K., Morse, D., & Colepicolo, P. (2003). Heavy metal-induced oxidative stress in algae. *Journal of Phycology*, Vol. 39, pp. 1008-1018, ISSN: 1529-8817

Piperno, G., Huang, B. & Luck, D.J.L. (1977). Two-dimensional analysis of flagellar proteins from wild-type and paralyzed mutants of *Chlamydomonas reinhardtii*. *Proceedings of the National Academy of Sciences of the USA*, Vol. 74, pp. 1600-1604, ISSN: 1091-6490

Ratkevicius, N., Correa, J.A. & Moenne, A. (2003). Copper accumulation, synthesis of ascorbate and activation of ascorbate peroxidase in *Enteromorpha compressa* (L.) Grev. (Chlorophyta) from heavy-metal enriched environments in northern Chile. *Plant Cell and Environment*, Vol. 26, pp. 1599-1608, ISSN: 1365-3040

Reynolds, C.S. (2006) Ecology of Phytoplankton. Cambridge University Press, New York.

Rhee, S.G. (2006). Cell signaling. H_2O_2, a necessary evil for cell signaling. *Science*, Vol. 312, pp. 1882-1883, ISSN: 1095-9203

Rijstenbil, J.W. (2001). Effects of periodic, low UVA radiation on cell characteristics and oxidative stress in the marine planktonic diatom *Ditylum brightwellii*. *European Journal of Phycology*, Vol. 36, pp.1-8, ISSN: 0967-0262

Ritter, A., Ubertini, M., Romac, S., Gaillard, F., Delage, L., Mann, A., Cock, J.M., Tonon, T., Correa J.A. & Potin P. (2010). Copper stress proteomics highlights local adaptation of two strains of the model brown alga *Ectocarpus siliculosus*. *Proteomics*, Vol. 10, pp. 2074-2088, ISSN: 1615-9861

Sabatini, S.E., Juárez, A. B., Eppis, M.R., Bianchi, L., Luquet, C.M. & Ríos de Molina, M.C. (2009). Oxidative stress and antioxidant defenses in two green microalgae exposed to copper. *Ecotoxicology and Environmental Safety*, Vol. 72, pp. 1200-1206, ISSN: 0147-6513

Santelices, B. (1989). Algas marinas de Chile. Distribución Ecológica. Utilización y Diversidad. Pontificia Universidad Católica de Chile.

Smith, C.M., Satoh, K. & Fork, D.C. (1986). The effects of osmotic tissue dehydration and air drying on morphology and energy transfer in two species of *Porphyra*. *Plant Physiology*, Vol. 80, pp. 843-847, ISSN: 0032-0889

Stauber,J.L., Benning,R.J., Hales,L.T. ,Eriksen,R. & Nowak,B. (2001) Copper bioavailability and amelioration of toxicity in Macquarie Harbour, Tasmania, Australia. *Marine & Freshwater Research*, Vol. 51, pp. 1-10 , ISSN: 1323-1650

Sunda, W.G. (1989). Trace metal interactions with marine phytoplankton. *Biological Oceanography*, Vol. 6, pp. 411-442, ISSN: 0196-5581

Takagi, M., Karseno & Yoshida T. (2006). Effect of salt concentration on intracelular accumulation of lipids and triacylglyceride in marine microalgae *Dunaliella* cells. *Journal of Bioscience and Bioengineering*, Vol. 101, pp. 223-226, ISSN: 1389-1723

Tran, N.P., Park, J.K., Hong, S.J. & Lee, C.G. (2009). Proteomics of proteins associated with astaxanthin accumulation in the green algae *Haematococcus lacustris* under the influence of sodium orthovanadate. *Biotechnology Letters*, Vol. 31, pp. 1917-1922, ISSN: 0141-5492 van Tamelen, P.G. (1996). Algal zonation in tidepools, experimental evaluation of the roles of physical disturbance, herbivory and competition. *Journal of Experimental Marine Biology and Ecology*, Vol. 201, pp. 197-231, ISSN: 0022-0981

Varela, D.A., Santelices, B., Correa, J.A. & Arroyo, M.K. (2006). Spatial and temporal variation of photosynthesis in intertidal *Mazzaella laminarioides* (Bory) Fredericq (Rhodophyta, Gigartinales). *Journal of Applied Phycology*, Vol. 18, pp. 827-838, ISSN: 0921-8971

Vega, J.M., Garbayo, I., Domínguez, M.J. & Vigara, J. (2006). Effect of abiotic stress on photosynthesis and respiration in *Chlamydomonas reinhardtii* induction of oxidative stress. *Enzyme and Microbial Technology*, Vol. 40, pp. 163-167, ISSN: 0141-0229

Véliz, K., Edding, M., Tala, F. & Gómez, I. (2006). Effects of ultraviolet radiation on different life cycle stages of the south Pacific Kelps, *Lessonia nigrescens* and *Lessonia trabeculata* (Laminariales, Phaeophyceae). *Marine Biology*, Vol. 149, pp.1015-1024, ISSN: 0025-3162

Vermeer, K. & Castilla, J.C. (1991). High cadmium residues observed during a pilot study in shorebirds and their prey downstream from the El Salvador copper mine, Chile. *Bulletin of Environmental Contamination and Toxicology*, Vol. 46, pp. 242-248, ISSN: 0007-4861

Vranová, E., Inzé, D. & van Breusegen, F. (2002). Signal transduction during oxidative stress. *Journal of Experimental Botany*, Vol. 53, pp. 1227–1236, ISSN 0022-0957

Wang, S.B., Hu, Q., Sommerfeld, M. & Chen, F. (2003). An optimized protocol for isolation of soluble proteins from microalgae for two-dimensional gel electrophoresis analysis. *Journal of Applied Phycology*, Vol. 15, pp. 485-496, ISSN: 0921-8971

Wang, S.B., Chen , F., Sommerfeld, M. & Hu, S. (2004). Proteomic analysis of molecular response to oxidative stress by the green alga *Haematococcus pluvialis* (Chlorophyceae). *Planta*, Vol. 220, pp. 17-29, ISSN: 0032-0935

Wang, Z., Qi, Y., Chen, J., Xu, N. & Yang, Y. (2006). Phytoplankton abundance, community structure and nutrients in cultural areas of Daya Bay, South China Sea. *Journal of Marine Systems*, Vol. 62, pp. 85-94, ISSN: 0924-7963

Wood, Z.A., Poole, L.B. & Karplus, B.A. (2003). Peroxiredoxin evolution and the regulation of hydrogen peroxide signaling. *Science*, Vol. 300, pp. 650–653, ISSN: 1095-9203

Wong, P.F., Tan, L.J., Nawi, H. & AbuBakar, S. (2006). Proteomics of the red alga, *Gracilaria changii* (Gracilariales, Rhodophyta). *Journal of Phycology*, Vol. 42, pp. 113-120, ISSN: 1529-8817

Yotsukura, N., Nagai, K., Kimura, H. & Morimoto, K. (2010). Seasonal changes in proteomic profiles of Japanese kelp: *Saccharina japonica* (Laminariales, Phaeophyceae). *Journal of Applied Phycology*, Vol. 22, pp. 443-451, ISSN: 0921-8971

Yotsukura, N., Nagai, K., Tanaka, T., Kimura, H. & Morimoto, K. (2011). Temperature stress-induced changes in the proteomic profiles of *Ecklonia cava* (Laminariales, Phaeophyceae). *Journal of Applied Phycology*. Vol. 24, pp. 163-171. ISSN: 1573-5176

Zou, D. & Gao, K. (2002). Effects of desiccation and CO_2 concentration on emersed photosynthesis in *Porphyra haitanensis* (Bangiales, Rhodophyta), a species farmed in China. *European Journal of Phycology*, Vol. 37, pp. 587-592, ISSN: 0967-0262

Zuppini, A., Gerotto, C. & Baldan, B. (2010). Programmed cell death and adaptation: two different types of abiotic stress response in a unicellular chlorophyte. *Plant and Cell Physiology*, Vol. 51, pp. 884-895, ISSN 0032-0781

Zurek, R. & Bucka, H. (2004). Horizontal distribution of phytoplankton and zooplankton from the littoral towards open waters under wind stress. *Oceanological and Hydrobiological Studies*, Vol. 33, pp.69-81, ISSN: 1730-413X

2

Application of Gel Electrophoresis Techniques to the Study of Wine Yeast and to Improve Winemaking

María Esther Rodríguez, Laureana Rebordinos, Eugenia Muñoz-Bernal,
Francisco Javier Fernández-Acero and Jesús Manuel Cantoral
Microbiology Laboratory, Faculty of Marine and Environmental Sciences
University of Cadiz, Puerto Real
Spain

1. Introduction

Yeasts are unicellular fungi that are frequently used as a model and tools in basic science studies. This is the case of the laboratory yeasts *Saccharomyces cerevisiae*, which were introduced in the laboratory for genetics and molecular studies in about 1935. There is, however, a second type of yeast comprising those used in industrial processes, for example, in brewing, baking and winemaking. Wine yeast and its properties have been known to humans for as long as civilizations have existed, and the earliest evidence of this yeast has been dated to Neolithic times (Mortimer, 2000).

Most wine yeast strains are diploid and have a low frequency of sporulation. Another important characteristic of wine yeasts, and those used in other industries, is their highly polymorphic chromosomes: their genetic constitution is affected by the frequent and extensive mutation they undergo. These effects include (i) aneuploidy, (ii) polyploidy, (iii) amplification and deletion of chromosomal region or single gene, and (iv) the presence of hybrid chromosomes. The chromosomal polymorphism obtained by applying the technique known as pulsed field gel electrophoresis (PFGE) has been used to characterize and to classify strains that belong to the same species.

In the wine industry, knowledge of the yeast species responsible for the alcoholic fermentation is important because these yeasts with their metabolism contribute significantly to the organoleptic characteristics of the finished wine (Fleet, 2008). The diverse range of yeasts associated with the vinification process can be classified in two groups. The first group is formed principally by the genera *Hanseniaspora, Torulaspora, Metschnikowia, Candida, Zygosaccharomyces*, etc. These yeasts initiate spontaneous alcoholic fermentation of the must, but they are soon replaced by the second group, formed by *Saccharomyces* yeasts, which are present during the subsequent phases of the fermentation until it is completed. Within the genus *Saccharomyces* the species most relevant for the fermentation process are *S. cerevisiae* and *S. bayanus* var. *uvarum*; this is because they have become of interest for their biotechnological properties. However, there is currently increasing interest in the non-*Saccharomyces* yeasts for the development of innovative new styles of wine (Viana et al.,

2009). In the industry, knowledge of specific strains of these microorganism species is important for (i) their selection; (ii) their use as starter cultures; and (iii) improving the fermentation process.

During the 1990's the development of molecular techniques has enabled the identification and characterization of different strains belonging to the same species of yeast, and it has been possible to establish the ecology of spontaneous fermentations in many of the world's winemaking regions (Fleet, 2008). These techniques also constitute a powerful tool not only for the selection of the most suitable yeast, since they tell us which yeasts are the most representative in the fermentation process, but also for obtaining information on the addition to the must of particular strains of yeast in the case of inoculated fermentations (Rodríguez et al., 2010).

Two of the approaches most often used for the molecular characterization of industrial yeast are analysis of the electrophoretic karyotypes by pulsed-field gel electrophoresis (PFGE) and analysis of the restriction fragment length polymorphism of the mitochondrial DNA (mtDNA-RFLP). We have used PFGE in winemaking to analyse the diversity of wild yeasts in spontaneous fermentation of a white wine produced in a winery in SW Spain with the object of selecting the most suitable autochthonous starter yeast; and from the results of the inoculation, we were able to make decisions for improving the efficiency of the process and to establish procedures for the proper performance of the inoculation (Rodríguez et al., 2010). We have also applied the analysis of the karyotypes to characterize natural yeasts in biodynamic red wines in another region of Spain. In this chapter we also evaluate the use of the mtDNA-RFLP technique for quick monitoring of the dominance of inoculated strains in industrial fermentation, without any need for the prior isolation of yeast colonies (Rodríguez et al., 2011).

Another electrophoretic technique has been used to show substantial changes in protein levels in selected wine yeasts under specific growth conditions. It has recently been stated that the proteome is "the relevant level of analysis to understand the adaptations of wine yeasts for fermentation" (Rossignol et al., 2009). Following this, in-depth studies are now being made of the proteome of wine yeast strains and the relationship between the proteome and wine quality and winery processes. We are now exploring more generally the relevance of proteomics to wine improvement. In this chapter, we will summarize the efforts being made by the proteomics research community to obtain the knowledge needed on proteins in the post-genomics era

2. Pulsed-field gel electrophoresis (PFGE) for the study of yeast population

PFGE as a system encompasses a series of techniques in which the intact chromosomes of microorganisms like yeasts and filamentous fungi are submitted to the action of a pulsing electric field in two orientations that is changing direction, in a matrix of agarose. The best-known PFGE modality is the CHEF system (Contour-clamped Homogeneous Electric Fields); this consists of a hexagon of 24 electrodes surrounding the gel that produce a homogeneous electric field alternating between two directions orientated at 120° with respect to each other. Using this system, Chu et al. (1986) resolved the electrophoretic karyotype of *Saccharomyces cerevisiae* in 15 bands in a size range of 200-2200 kb. Before performing the electrophoresis the yeast cells must be suitably treated, avoiding the direct manipulation of the genetic material to prevent possible rupture of the chromosomes. The cells are then embedded in blocks of agarose which are subsequently treated with a reducing agent and K proteinase to destabilize the wall and cytoplasmatic membranes, respectively (Figure 1), thus facilitating the release of the DNA when submitted to the action of an electric field.

This methodology for correctly obtaining the karyotype of *S. cerevisiae* is based on the procedure described by Carle & Olson (1985) and optimized by Rodríguez et al. (2010). It also depends on the concentration of the agarose gel (1%), buffer (0.5 x TBE), initial and final switch (60-120 seconds respectively), run time (24 hours), voltage (6 V/cm) and buffer temperature (14 °C).

The analytical results given by this technique are the number and size of the yeast chromosomes, and it allows specific strains of *Saccharomyces* to be differentiated because their karyotypes show distinct bands running below the 500-kb marker. It also allows the differentiation between *S. cerevisiae* and *Saccharomyces bayanus* var. *uvarum* (*S. uvarum*) species (Naumov et al., 2000, 2002).

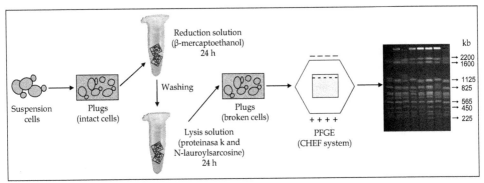

Fig. 1. Methodology for characterizing yeast strains, using pulsed field gel electrophoresis to obtain the karyotype.

In previous research the PFGE technique has been used to analyse the dynamics of the yeast population during the spontaneous fermentations of wine (Demuyter et al., 2004; Martínez et al., 2004; Naumov et al., 2002; Raspor et al., 2002; Rodríguez et al., 2010), and it has also been used to characterize other industrial yeasts including baker's and brewer's yeast (Codón et al., 1998). Another relevant application of PFGE has been to characterize the yeast population which is present in the *flor velum* that grows on the surface of fino-type sherry wines in the barrel, during their biological ageing process (Mesa et al., 1999, 2000). The results revealed an interesting correlation between the yeast genotypes and the different blending stages.

One disadvantage of the technique is that it is laborious, expensive and requires specialized personnel; increasingly, therefore, analysts are resorting to other simpler and faster techniques to discriminate between yeast clones, like, for example, interdelta analysis of sequences or microsatellite analysis (Cordero-Bueso et al., 2011; Le Jeune et al., 2007; Schuller et al., 2007). However, the methodology proposed in Figure 1 enables a large number of yeast isolates to be processed, and PFGE is considered a most suitable technique for discriminating between yeast clones (Schuller et al., 2004).

In our laboratory, this technique has been used to characterize the wine yeast population responsible for the spontaneous fermentation of a white wine produced in a winery in SW Spain (Rodríguez et al., 2010). Analyses of industrial-scale fermentations (in 400 000-l fermentation vessels) were carried out during two consecutive vintages. In 1999 and 2000 a total of 211 and 228 yeast colonies, respectively, from different vessels, were characterised by karyotyping. The degree of polymorphism observed was high, and 17 different karyotypic patterns were detected in 1999, and 21 patterns in 2000. In the two campaigns, we also found patterns belonging to non-*Saccharomyces* yeasts, the karyotypes of which did not show the four bands running below 500-kb. During the fermentation, this population was displaced by *S. cerevisiae* strains; patterns I, II, III and V were predominant during entire fermentation process in 1999, whereas in 2000 patterns II and V were predominant (see Figure 2). Those were the yeast strains selected for inoculating the industrial fermentations, as will be explained bellow.

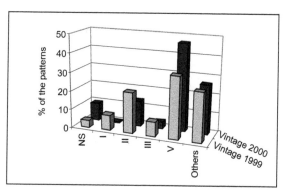

Fig. 2. Frequencies of the majority karyotype patterns (%) obtained in the spontaneous fermentations of 1999 and 2000. NS corresponds to non-*Saccharomyces* yeasts.

The results of the characterization of the yeasts also showed that the different strains changed their proportion, and there was a sequential substitution of strains during the fermentation; this gave a valuable indication of the dynamics of the yeasts population throughout the process. Some of these changes were specific to a particular fermentation phase, suggesting that the yeast strains with different electrophoretic karyotypes also differ in their adaptation to the evolving environment at different phases of the fermentation process.

Although the diversity of wild yeast can contribute to high-quality and unique flavour in the finished wine, spontaneous fermentation is often unpredictable and might introduce less desirable traits to the product, sometimes even spoiling a production batch. Other risks associated with spontaneous fermentation include either slow or arrested fermentation. To avoid these problems, winemakers often add cultures of selected yeasts, in the form of active dried yeasts or autochthonous yeasts. Nevertheless, in some cases, these yeasts used as starters are not able to displace the wild yeasts present in the must, since the wild yeasts can be very competitive (Esteve-Zarzoso et al., 2000; Lopes et al., 2007).

In our work on the analysis of the karyotype, it has been possible to monitor the yeast population under industrial conditions for several years when fermentations of the white wine were inoculated with selected autochthonous yeast strains. This has allowed inoculation strategies to be designed for the correct development of inoculated yeast while retaining the unique regional character of the finished wine (Rodríguez et al., 2010). The strains with patterns I, II, III and V were the most representative during the spontaneous fermentation process (Figure 2) and they could be isolated at the late fermentation phase. These autochthonous yeasts show valuable traits of enological interest, such as high fermentative capacity, ethanol tolerance, and they had a killer phenotype. The capacity of each strain to compete within the mixed population was also tested under semi-industrial conditions by PFGE. The results show that the strains with karyotypes II and V were the most vigorous competitors, followed by the strain with patterns III and I, which were detected in lower proportions (Rodríguez et al., 2010). Therefore, the strains with patterns II, III and V were used to inoculate the fermentations in the year 2001; strains with patterns II and V were used in 2002, 2003 and 2004; and from the year 2005 until the present (2011), only the strain with pattern V has been used.

The inoculation of industrial vessels of the winery of this study presented several peculiarities. For example: (i) in each vintage year several vessels with a total capacity of 400 000-l were inoculated; (ii) the inoculums comprising the selected autochthonous yeast strains were prepared from fresh YEPD plates (1% yeast extract, 2% glucose, 2% peptone and 2% agar) by preparing a starter in which each scaling-up round was performed when the °bé reached a value between 1-2, and in each round, the fermentation volume was increased tenfold to give high initial levels of inoculum (> 60 x 10^6 viable cells/ml) and ensure the correct development of the inoculated strain; (iii) once the starter cultures were scaled-up and added to a 400 000-l container, partial volumes were withdrawn and used for the inoculation of other 400 000-l vessels of the winery; and (iv) the 400 000-l vessels received random additions of fresh must until reaching the final volume. The frequency and timing of these additions depended on the production yield of fresh must during the vintage campaign.

In our results (Table 1) only the strain with pattern V, called P5, was dominant under industrial conditions and, for this reason, the number of the starter strains were reduced over the years and currently only the strain with karyotype V is used for the inoculation of the industrial fermentations. In spite of this strain's good capacity for achieving dominance,, in some years a high degree of polymorphism was detected in the fermentations; and the cause of the unexpected predominance of wild yeast karyotype was linked to several factors, including: a sudden decrease in the temperature of one of the vessels during the scaling-up process of the inoculation in 2002; the method of inoculation and the scaling-up process, which were changed in 2003, whereby the inoculums of pure culture did not represent 10% and each scaling-up round was performed when the inoculums had a high sugar content (around 3-5 °Bé); and the storage of must in the vessels in which spontaneous fermentation had occurred.

When karyotype V was dominant in the fermentations, the wine obtained had fruity characteristics, with well-balanced acidity, that satisfied the wine producer. Although we did not obtain a comprehensive aromatic characterization of the wine, the panel of wine tasters (Figure 3) considered the wine produced in these fermentations better than the wine obtained either in the spontaneous fermentations of 1999 and 2000 or in the vintage years of 2002 and 2003, when the karyotype V (strain P5) was not detected in high proportion in the yeast population.

Patterns	2001	2002	2003	2004	2005	2007	2008
II	7.6	9.2	3.6	0.4	0	10.7	0
III	8.5	0	0	0	0	0	0
V	83	0.6	19.3	99.6	79.8	50	100
(NS)	0.2	0	0.4	0	0.4	0	0
Others	0.7	90.2	76.7	0	19.8	39.3	0
Total yeast analysed	423	174	223	235	278	142	240

Table 1. Frequencies (%) of the inoculated strains in the industrial fermentations for seven vintage years. This analysis was not performed in the vintage year of 2006. NS: non-*Saccharomyces*.

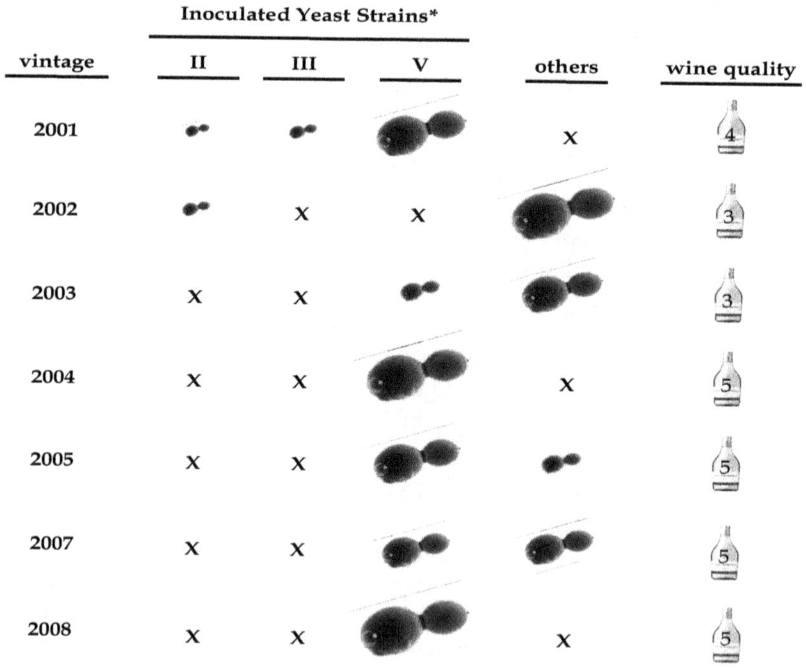

	Inoculated Yeast Strains*				
vintage	II	III	V	others	wine quality
2001	••	••	⬤⬤	X	4
2002	••	X	X	⬤⬤	3
2003	X	X	••	⬤⬤	3
2004	X	X	⬤⬤	X	5
2005	X	X	⬤⬤	••	5
2007	X	X	⬤⬤	⬤⬤	5
2008	X	X	⬤⬤	X	5

*Strain group II was inoculated in the years 2001 to 2004; the strain with pattern III was inoculated only in 2001; and strain group V was inoculated in all vintages

Fig. 3. Composition of the wine yeast population in each vintage, and its relationship with the quality of the final product. For each year the size of the yeast cell shown is proportional to the contribution of each strain to the total wine yeast population of the winery. The x symbol indicates that the proportion of the strain(s) within the population was below 5%. Wine quality was evaluated by a panel of expert wine-testers from the winery, who graded the final product on a scale from 1 to 5 based on fruity wine with well-balanced acidity desired by the producer (5 indicates highest quality and 1 lowest quality). Predominance of the strain with pattern V corresponded to a better quality of the wine.

By using PFGE to study the yeast population of the inoculated fermentations, the producer was able to make informed decisions for improving the process; the common factors in the vintages of 2001, 2004, 2005, 2007 and 2008, in which the inoculated strain was dominant, can be highlighted. These factors were the following: (i) the culture was not scaled-up to the next volume until the yeast had fully depleted the sugar to less than 1 °Bé (one degree is equivalent to 18 g/l of fermentable sugars in the must); therefore all cultures reached a high alcohol content before the addition of fresh must; (ii) the inoculum was always diluted less than 10-fold in each scaling-up round; (iii) the temperature of the fermentation was kept at 17 °C.

We think that these criteria favoured the adaptation of the inoculums to the conditions of the must obtained in each vintage and to the final conditions within the 400 000-l industrial vessels. In addition, these criteria favoured the predominance of the inoculated strain with pattern V.

In another study with biodynamic red wines, carried out in the Ribera del Duero D. O. Region (Valladolid, Spain), spontaneous fermentations were also analysed applying PFGE. We studied seven fermentations in three phases during the fermentation process: initial (IF), middle (MF) and final (EF), and 20 isolated strains per sample were characterized by applying PFGE (417 strains in 2008, and 412 strains in 2009). The results for two consecutive vintages studied showed the presence of different types of the yeast during the fermentations that were grouped in three populations. The first population was formed by non-*Saccharomyces* yeast, whose strains showed patterns with the absence of bands running below the region of 500 kb, which are specific to *S. cerevisiae* strains as reported above. The second population comprised *Saccharomyces bayanus* var. *uvarum* (*S. uvarum*); and the third population included *Saccharomyces cerevisiae* yeast. The strains of *S. uvarum* were differentiated from the *S. cerevisiae* strains by the presence of two small chromosomes in the region of 245-370 kb, instead of three as for S. *Cerevisiae*, as reported by Naumov et al. (2000, 2002). Non-*Saccharomyces* (NS) yeasts were dominant in the initial phase of fermentation but were displaced in the subsequent and final phases of the process by another population of yeasts. *S. uvarum* yeasts were present mainly in the phase mid-way through the fermentation; then the population of *S. cerevisiae* yeasts displaced the NS and *S. uvarum* yeasts, and remained dominant until the end of the fermentation, in the majority of the deposits analysed. The low frequency of detection of *S. uvarum* at the end of the fermentation could be indicative of its lower ethanol tolerance compared to *S. cerevisiae*. Within each population yeast strains were also found with different karyotyping patterns, and the distribution (by %) of these varied in the seven deposits analysed during the two consecutive years studied. Thus, for *S. uvarum*, considerable variability of strains and a total of 12 different electrophoretic patterns were detected (Figure 4): uI-uVII for vintage 2008; and uI-uIII, uV, uVIII-uXII in 2009. The strains with patterns uI, uII, uIII and uV, followed by uIV (in 2008) and uIX (in 2009) were the most representative in two years studied. Within the population of the *S. cerevisiae* yeasts, the variability of the patterns was higher than in *S. uvarum* ; 29 (cI-cXXIX) and 27 (cI-cVII, cX-cXII, cXV, cXVII, cXIX, cXXII, cXXIV, cXXX-cXLI) electrophoretic karyotype patterns were detected for 2008 and 2009 respectively. The *S. cerevisiae* yeast strains most representative of the fermentation process in these years were those that showed the karyotypes cIII, cVI, cXI and cXII.

The yeast population dynamics presented in this biodynamic red wine were different from those observed in other studies of white wines in which *S. uvarum* was dominant during spontaneous alcoholic fermentation (Demuyter et al., 2004).

Although *S. uvarum* has been found in other producing regions of the world, such as Alsace (Demuyter et al., 2004), at the moment there are no studies about the population dynamics of *S. uvarum* in Ribera del Duero, Spain.

The use of the PFGE technique allows analysts to detect a high degree of polymorphism in the population of the yeast and to monitor the dynamics of yeast ecology during the fermentation; this is because it is able to show the occurrence of gross chromosomal rearrangements, which is the phenomenon that mainly accounts for the rapid evolution of yeast clones subjected to industrial conditions (Infante et al., 2003). The technique also shows the most representative yeast strains in the industrial winemaking process, which are partially but significantly responsible for the finished wine's quality. Knowledge of these main strains can be used as a criterion for making a first selection of the autochthonous yeast. Later, with these previously selected strains, PFGE can be applied to study the features of these strains that are of enological interest, as described in recent years, which fall into three main categories: (i) properties that affect the performance of the fermentation process; (ii) properties that determine the quality of the wine; and (iii) properties associated with the commercial production of yeast (reviewed in Fleet, 2008). The yeasts selected by these means can then be used for inoculation in the fermentation process, thus improving winemaking.

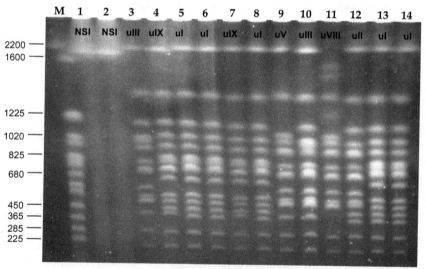

Fig. 4. Electrophoretic karyotype of 14 colonies isolated from the sample taken in 2009 from a vessel in the phase mid-way through the fermentation process. Colonies 3-14 correspond to different karyotype patterns (uI-uIX) of yeast strains found in the *S. uvarum* species. Isolates 1-2 correspond to non-*Saccharomyces* strains which show the same pattern (NSI), with absence of bands running below 1225 kb. The chromosomes of the *S. cerevisiae* YNN295 strain were used as reference (M).

3. Application of mtDNA-RFLP as a rapid method for monitoring the inoculated yeast strains in wine fermentations

Although PFGE has been reported to be the most efficient in discriminating between different strains of *S. cerevisiae*, the mtDNA-RFLP technique is frequently used to differentiate between yeast isolates of the same species (González et al., 2007) because it enables a larger number of strains to be analyzed in a shorter time; it is a fast, simple, reliable and economic method, which does not require sophisticated material or specialized personnel (Fernández-Espinar et al., 2006). For these reasons, it is a very suitable technique for use by industry.

Most of the mitochondrial DNA in yeasts does not code proteins, and contains a high proportion of AT bases. Analysts can take advantage of this characteristic to characterize yeasts; it involves measuring variation in sequences in the mtDNA affecting the restriction sites of several endonucleases. Endonucleases such as *Alu*I, *Hinf*I or *Rsa*I, recognise the very frequent restrictions in the chromosomal DNA but not in the mitochondrial DNA, leading to a total cleavage of the chromosomal DNA in small pieces. These pieces can be easily differentiated from the mitochondrial fragments, which appear as bands with an electrophoretic mobility corresponding to molecules greater than 2 kb, generating polymorphisms that allow the characterization between yeast strains.

In previous studies applying this technique it has been demonstrated that the population of a fermentation vessel is "taken over" by wild yeasts, which displace the inoculated yeast strain, reducing it to a minority presence (Esteve-Zarzoso et al., 2000; Lopes et al., 2007; Raspor et al., 2002). In our research, when we have analyzed the inoculated fermentation of white wine as described above, we have found several examples of real situations that led to a significant decrease in the proportion of the inoculated strain (pattern V) and, in consequence, the quality of the wine was reduced. In order to minimize the impact of unwanted yeasts, wineries need a simple method for rapid diagnosis of the degree of dominance of inoculated strains, a method that could be performed routinely during the fermentation process (Ambrona et al., 2006, López et al., 2003). With this object we have used RFLP analysis of mtDNA for the rapid monitoring of the dominance, or otherwise, of inoculated yeast strains in industrial fermentations of white and red wines in a winery in southern Spain (Rodriguez et al., 2011).

We apply this technique directly to samples of fermenting wine without previously isolating yeast colonies. For white wine fermentations, a rapid assay is performed consisting of taking a sample of fermenting must, purifying the DNA from harvested cells, and obtaining the restriction patterns by digestion with endonuclease *Hinf*I. The same protocol is applied to red wine fermentation, but an overnight cultivation step is added before purification of the DNA (Figure 5).

The criterion for considering the result of the rapid test to be positive was obtaining restriction patterns of mtDNA that were identical for the total cells and the inoculated strain; when this is the case, the starter yeast can be taken as being dominant in the fermentation. The result was considered negative when additional bands, or absence of bands, were observed in the patterns; in this case neither the dominance, nor even the presence, of the inoculated yeast strain can be assured.

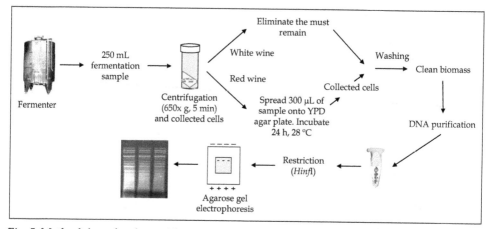

Fig. 5. Methodology for the rapid test assays by restriction analysis of the mtDNA.

This methodology of the RFLP test has been applied since 2005 in a winery of SW Spain and the results were obtained 11 and 23 hours after sampling, for white and red wine respectively (Rodríguez et al., 2011). If the wine-producer knows whether or not the presence of the inoculated yeasts has suffered a sudden decrease, and the inoculated strain is no longer dominant, in any phase of the fermentation process, a rapid intervention can be made. Since this year the winery has only used the selected autochthonous strain P5 for the inoculation of its white wine fermentations; this strain was in a clear majority in spontaneous fermentations, as reported above, and it shows karyotype pattern V. The peculiarities of the inoculation of industrial vessels for this wine have been described above (section 2). After applying the RFLP test, the correct course of the fermentation of all starters was assured before the inoculation of the industrial vessels. Generally, all the fermentations are tested in at least two different phases of the process: first after refills with fresh must and again when the fermentations are finished.

In addition, the results were checked using PFGE to validate the previous results obtained from the RFLP test. For this validation, 34 samples tested by RFLP were analyzed by electrophoretic karyotype of 323 colonies for white wine. The results indicated that when RFLP test was positive, the inoculated strain was present in the fermentation at 64% or more. When the RFLP tests were negative it was confirmed by PFGE that the starter yeast was present at only 60% or less.

The red wine of the winery was fermented in stainless steel vessels of 27.000 l and no refills of must were carried out. The fermentations were inoculated with several commercial active dried wine yeasts (ADWY) by hydration following manufacturer's instructions. For this wine, for 331 colonies analyzed by PFGE, the results of the RFLP test correctly predicted the results obtained later by applying PFGE. However, in this case, the limit found for white wine cannot be established because, from the positive results obtained by applying the RFLP test, the presence of the inoculated strain was greater than 75%, and all the negative results were at 55% or less (Rodríguez et al., 2011). Nevertheless, further experiments will be necessary to confirm these correlations because the RFLP test shows qualitative results and the actual percentage implantation of the starter yeast cannot be known when the results are positive or negative. Figure 6 shows examples of positive and negative results for the last two years (2009 and 2010) for the dominance and non-dominance respectively of the inoculated strain P5 in white wine of the same winery described above.

For the 2009 vintage, we tested several different vessels at the initial phase of the fermentation, for white wine (Panel A). The results were positive for all cases after applying the rapid test, i.e. the restriction patterns of the samples and the inoculated strain were identical, and the strain P5 was responsible at the beginning of the fermentation process displacing other wild yeasts.

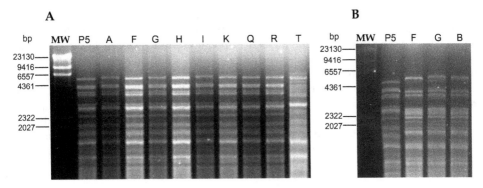

Fig. 6. Rapid test based on mtDNA-RFLP with *Hinf*I of samples from white wine fermentations which were inoculated with autochthonous yeast strain P5. Panel A shows results of samples taken in 2009 from vessels A, F, G, H, I, K, Q, R and T at the initial fermentation phase (11-7 °Bé). Panel B shows results of samples taken in the 2010 vintage from vessels F, G and B during the main phase of the fermentation (5-3 °Bé). MW is the lambda-*Hind*III molecular marker.

Only the fermentation vessel reference T was considered negative for the RFLP test at the beginning of the process. In this vessel evidence was observed of spontaneous fermentation before the inoculation, due to the conditions in which the must was stored. When this must was inoculated, the strain P5 did not implant successfully and it was concluded that another wild yeast population was dominant.

In situations like this, the winemaker can take the decision not to use the fermenting must for the inoculation of other vessels.

Panel B (Figure 6) shows examples of the negative results of the RFLP test in three vessels mid-way during the fermentation process. In these deposits the population was perhaps similar because the restriction pattern of the total cells in each sample was similar. After the vintage, 20 colonies were isolated from the same sample previously analyzed by rapid test in the vessel G (Figure 6, panel B) in order to confirm the composition of the yeast population by karyotype. Surprisingly, all the clones show the same pattern as a commercial yeast strain used for the inoculation in the fermentation of another type of white wine in the previous year. It is assumed that this commercial yeast was also dominant in the fermentations sampled in vessels F and B. In previous studies researchers have reported the risks in using commercial yeast, because they can become part of the microbiota of the winery, effectively creating their own ecosystem, and can subsequently be predominant in the fermentations (Santamaría et al., 2005). In our study, we think that this commercial yeast was present in equipment which was not properly cleaned. When the wine-producer was more careful in the next vintage, there were no problems of contaminations by commercial yeast, and the dominant yeast in the fermentations was the inoculated autochthonous yeast P5 (data not shown). Therefore, it was confirmed that the commercial yeast had not acquired an ecological niche because it presumably did not adapt well to the ecosystem of a properly-cleaned winery.

As stated, the results of the RFLP can be obtained 11 and 23 hours after taking the sample for white and red wine respectively. However, this time can be shortened further, because it depends on the method used to rupture cells, on the number of samples analyzed per day, and on whether the samples contain a greater amount of must residues. In the case of red wine, there was another problem in shortening the test time, because the residues were difficult to clear by centrifugation; we think that some compounds remaining in the digested DNA samples were inhibitory for the endonuclease. Therefore, a step has been added in the protocol of the rapid test (Figure 5) in which the sample of the red must is plated on YPD-agar and incubated overnight at 28 °C. Nevertheless, we think that, for red wine, the time taken to obtain the results could also be shortened further, like that for white wine, if the clean biomass can be separated from the must residues in a few minutes. To achieve this, further experiments will need to be carried out.

4. Relevance of proteomic analysis in the winemaking process

In brief, proteomics can be described as a set of techniques for unravelling complex mixtures of proteins. In spite of it being a relatively recent technique, most of the systems used are widely known by the research community. However, the crucial work for its final "take-off" as a viable technique has been the modifications made to the mass spectrometry system, to allow the analysis of peptides and proteins. The exponential growth in the number of entries for genes and/or proteins in the databases now makes protein analysis and identification much easier, as well. This, combined with the use of powerful methods of fractionation and separation of peptides and proteins, such as 2D-PAGE (two dimensional polyacrylamide electrophoresis) and high resolution liquid chromatography, proteomics has been consolidated since the mid-90's, as the science for massive protein analysis; it is now the main methodology for unravelling biological processes, leading some authors to describe the current period as the "post-genomic era".

Proteomics has been defined as the set of techniques for studying the complex mixture of proteins, named the Proteome, that exists in any specific cell, microorganism, tissue, etc, used in specific experimental conditions, culture, sampling, etc. It is a highly dynamic system, and is more complex than genomics because, while the genome of an organism is more or less constant, the number of proteomes obtained from a specific genome is infinite. It depends on the assayed cell, tissue, culture conditions, etc., because each change produces a modification in the observed proteome. An additional factor of complexity derives from the fact that there are changes that occur in proteome that are not encoded in the genome. These changes mainly originate from two sources: (i) the editing of the mRNA; and (ii) post-translational modifications (PTMs) that normally serve to modify or modulate the activity, function or location of a protein in different physiological or metabolic contexts. More than

200 different PTMs have been described (including phosphorylation, methylation, acetylation, etc.) that transform each single gene into tens or hundreds of different biological functions. Before the advances made in proteomics, the differential analysis of the genes that were expressed in different cell types and tissues in different physiological contexts was done mainly through analysis of mRNA. However, for wine yeast, it has been proved that there is no direct correlation between mRNA transcripts and protein content (Rossignol et al., 2009). It is known that mRNA is not always translated into protein, and the amount of protein produced by a given amount of mRNA depends on the physiological state of the cell. Proteomics confirms the presence of the protein and provides a direct measure of its abundance and diversity.

In terms of methodology, proteomics approaches are classified in two groups: (i) gel-free systems based on the use of various chromatography methods; and (ii) gel-based methods that use mainly two-dimensional polyacrylamide gel electrophoresis (2DE). This latter approach will form the focus of our discussion here, given the subject matter of this book. As a succinct summary, the typical workflow of a proteomic experiment begins with the experimental design. This must be studied in depth, and it will delimit the conclusion obtained, even more so when comparisons are made between two strains, cultures or physiological stages, among others. As an optimum, only one factor among the various different assayed conditions must change (Fernandez-Acero et al., 2007). Several biological replicates, usually from 3 to 5, will be required depending on the strategy adopted. The next key step is to obtain a protein extract of high enough quality to separate complex mixtures of proteins. Usually, the protein extraction is done in sequential steps. First the tissue, cells, etc. are ruptured using mechanical or chemical techniques. Then, proteins are precipitated and cleaned. Most existing protocols use acetone and trichloroacetic acid. In the next step the proteome is defined and visualized using electrophoretic techniques. 2DE has been widely used for this purpose. Using this technique, proteins are separated using two different parameters. In the first dimension, proteins from the purified extract are separated by their iso-electric point using an iso-electrofocusing (IEF) device. Then, the focused strips are loaded in a polyacrylamide gel where the proteins are separated by their molecular weight. This system allows the separation of hundreds of proteins from one complex mixture. The gels are visualized with unspecific (Comassie, Sypro, etc.) or specific (e.g. Phospho ProQ diamond) protein stains. The gels are digitalized and analyzed with specific software to reveal the significant spots. These spots are identified using mass spectrometry; commonly, for 2DE approaches, MALDI TOF/TOF is used. The huge list of identified proteins obtained is studied to discover the biological relevance of each identification.

In spite of the many achievements of proteomics, only a few proteomic studies have been carried out on wine yeast, whereas mRNA expression has been widely used to study a broad range of industrial conditions. However, Rossignol et al. (2009) show that substantial changes in protein levels during alcoholic fermentation are not directly associated with changes in the transcriptome; this suggests that the mRNA is selectively processed, degraded and/or translated. This conclusion is important: it is the proteome, not the genome nor the transcriptome, that is the relevant level of analysis for understanding the adaptations of wine yeasts during alcoholic fermentation, since these are responsible for the phenotype.

The usual strategy for wine production is the inoculation of selected yeast strains into the must, decreasing the lag phase, a quick and complete fermentation of the must, and a high degree of reproducibility of the final product. The development of global analysis methodology has allowed a detailed analysis to be made of changes in gene expression and protein levels at various time-points during vinification. Zuzuarregui et al. (2006) presented a comparison between the mRNA and protein profiles of two yeast strains with different fermentation behaviours, which correlates with divergence in the fermentation profiles. The results indicate changes in the mRNA and protein levels and, probably, post-translational modifications of several proteins, some of them involved in stress response and metabolism.

Another proteomic approach was aimed at studying the adaptation of a wild-type wine yeast strain, isolated from a natural grape must, to physiological stresses during spontaneous fermentation (Trabalzini et al., 2003). Using 2DE, changes in the yeast proteome were monitored during glucose exhaustion, before the cells begin their stationary phase. The proteome adaptation of *S. cerevisiae* seems to be directed or caused by the effects of ethanol, leading to both hyperosmolarity and oxidative responses. Through the use of a wild-type *S. cerevisiae* strain and PMSF, which is a specific inhibitor of vacuolar proteinase B, it was also possible to distinguish the specific contributions of the vacuole and the proteasome autoproteolytic process. This is the first study that follows the adaptation of a physiologically wild wine yeast strain progressively to the exhaustion of an essential nutrient, glucose.

To monitor yeast stress Salvadó et al. (2008), using ADWY (active dried wine yeast) inoculated into the must, have observed its behaviour in different stress situations, i.e. high sugar concentration or low pH. The main responses after inoculation in a fermentable medium were the activation of several genes of the fermentation pathway and the monoxidative branch of the pentose pathway, and the induction of a huge cluster of genes related to ribosomal biogenesis and protein synthesis. The changes that occur during the lag phase are characterized by an overall change in the protein synthesis and reflect the physiological conditions of the yeast, which affects the fermentative capacity and fermentation performance. Certain enological practices increase these stressful conditions for ADWY. This is the case of low-temperature fermentation, which improves taste by restructuring flavour profiles, with potential enological applications. This study focuses on changes that occur in ADWY after inoculation in a synthetic wine. These changes reflect adaptation to a new medium.

Previous reports have shown that proteomic analysis of wine yeast is the most relevant tool for understanding the physiological changes involved in winery processes. The information obtained may improve the quality of the final product. Our group has been a pioneer in fungal proteomic approaches (Fernandez-Acero et al., 2007, 2011; Garrido et al., 2010), and in line with this, we are now exploring the relevance of proteomics in wine improvement (Muñoz-Bernal et al., 2011). Our group has developed new protocols for obtaining the proteome and subproteomes of yeast, and the results to date suggest that there is a lot of biological information to be studied and analyzed from the proteomic perspective. The relevance of this achievement for winery processes could be significant.

5. Conclusions

Application of the PFGE technique allows the yeast population in the wine fermentation process to be characterized. The technique has been reported to be the most efficient for discriminating between *S. cerevisiae* yeast clones (Schuller et al., 2004), and it differentiates these from the specie *S. bayanus* var. *uvarum* (Naumov et al., 2002). It is also able to reveal the occurrence of gross chromosomal rearrangements, which account for the rapid evolution shown by yeast in industrial environments (Infante et al., 2003). Using PFGE, we have detected a high degree of polymorphism in the population of spontaneous fermentations of different types of wine produced in different regions of Spain, and it was observed that there were yeast strains that were specific to a particular phase of the fermentation process. This suggests that yeast strains with different karyotypes also differ in their adaptation to the evolving environment at different phases of the fermentation process. Studies for the molecular characterization of wine yeast represent a first step for selecting autochthonous yeast strains which are better adapted to specific conditions of a particular wine-making region. Moreover, such knowledge in respect of yeast populations may lead to the identification of a new natural source of wine yeast that could be used by the industry in the future as a new commercial starter (Fleet, 2008).

Studies by PFGE of the yeast population in inoculated fermentations also allow producers to understand and make informed decisions for improving their processes. Our results suggest that the success of the inoculation protocol is highly dependent on adequate preparation of the inoculums, which must facilitate the adaptation of the inoculated strains to the final conditions of the fermentation.

The RFLP test designed to monitor and confirm that the population of the inoculated yeast has reached and maintained predominance, in white or red wines, is proposed as a response to one of the major challenges for microbiological control in the wine industry. In our results real situations are shown taking place during actual wine fermentations; for example spontaneous fermentations sometimes occur before the inoculation. We offer a test which the winemaker can use to obtain a reliable indication of whether or not wild yeasts are displacing the inoculated strains. If the strategy presented is followed, the wine producer would be able to identify and correct in time the unwanted evolution of the yeast population - usually by re-inoculating the selected strains and/or correcting a deviation in temperature or change in some other parameter of the vessel that might have caused the unwanted situation.

Our studies are among the first examples carried out at the industrial scale showing how molecular techniques can be successfully applied to improve quality and efficiency in the winemaking process.

Despite the achievements already made, we are also exploring the potential use of the latest molecular proteomics techniques to unravel the biological component of the complex winemaking processes. Proteomics data collected to date strongly suggest that these techniques are potentially very useful for controlling the fermentation process and for assuring the quality of the finished wine; they offer excellent prospects for improving these processes in the near future.

6. Acknowledgements

This work was supported by grants PETRI 95-0855 OP from the DGICYT of the Ministry of Science and Innovation, and OT 054/174/015/020/114/136/104 from Bodegas Barbadillo S.L. of Sanlúcar de Barrameda, Spain, and CDTI-IDI-20101408.

7. References

Ambrona, J., Vinagre, A., Maqueda, M., Álvarez, M. L., Ramirez, M. (2006). Rhodamine-pink as a genetic marker for yeast populations in wine fermentation. *Journal of Agricultural and Food Chemistry*, Vol. 54, pp. 2977-2984, ISSN 0021-8561

Carle, G. F., Olson, M. V. (1985). An electrophoretic karyotype for yeast. *Proceedings of the National Academy of Sciences of the United States of America*, Vol. 82, pp. 3756-3760, ISSN 0027-8424

Chu, G., Vollarath, D., Davis, R. W. (1986). Separation of large DNA molecules by contour-clamped homogeneous electric field. *Science*, Vol. 234, pp. 1582-1585, ISSN 0036-8075

Codón, J. M., Benítez, T., Korhola, M. (1998). Chromosomal polymorphism and adaptation to specific industrial environments of *Saccharomyces* strains. *Applied Microbiology and Biotechnology*, Vol. 49, pp. 154-163, ISSN 0175-7598

Cordero-Bueso, G., Arroyo, T., Serrano, A., Tello, J., Aporta, I., Vélez, M. D., Valero, E. (2011). Influence of the farming system and vine variety on yeast communities associated with grape berries. *International Journal of Food Microbiology*, Vol. 145, pp. 132-139, ISSN 0168-1605

Demuyter, C., Lollier, M., Legras, J. L., Le Jeune, C. (2004). Predominance of *Saccharomyces uvarum* during spontaneous alcoholic fermentation, for three consecutive years, in an Alsatian winery. *Journal of Applied Microbiology*, Vol. 97, pp. 1140-1148, ISSN 1364-5072

Esteve-Zarzoso, B., Gostíncar, A., Bobet, R., Uruburu, F., Querol, A. (2000). Selection and molecular characterization of wine yeast isolated from the `El Penedès´ area (Spain). *Food Microbiology*, Vol. 17, pp. 553-562, ISSN 0740-0020

Fernández-Acero, F. J., Carbú, M., Garrido, C., Vallejo, I., Cantoral, J. M. (2007). Proteomic Advances in Phytopathogenic Fungi. *Current Proteomics*, Vol. 4, No. 2, pp. 79-88, ISSN 1570-1646

Fernández-Acero, F. J., Carbú, M., El-Akhal, M. R., Garrido, C., González-Rodríguez, V. E., Cantoral, J. M. (2011). Development of Proteomics-Based Fungicides: New Strategies for Environmentally-Friendly Control of Fungal Plant Diseases. *International Journal of Molecular Sciences*, Vol. 12, pp. 795-816, ISSN 1422-0067

Fernández-Espinar, M. T., Martorell, P., De Llanos, R., Querol, A. (2006). Molecular methods to identify and characterize yeasts in foods and beverages. In: Querol, A., Fleet, G. H. (Eds), The Yeast Handbook: Yeast in Food and Beverages, Springer, pp. 55-82

Fleet, G.H. (2008). Wine yeasts for the future. *FEMS Yeast Research*, Vol. 8, pp. 979-995, ISSN 1567-1356

Garrido, C., Cantoral, J. M., Carbú, M., González-Rodríguez, V. E., Fernández-Acero, F. J. (2010). New Proteomic Approaches to Plant Pathogenic Fungi. *Current Proteomics*, Vol. 7, pp. 306-315, ISSN 1570-1646

González, S. S., Barrio, E., Querol, A. (2007). Molecular identification and characterization of wine yeast isolated from Tenerife (Canary Island, Spain). *Journal of Applied Microbiology*, Vol. 102, pp. 10185-1025, ISSN 1364-5072

Infante, J.J., Dombek, K.M., Rebordinos, L., Cantoral, J.M. & Young, E.T. (2003). Genome-wide caused chromosomal rearrangements play a major role in the adaptive evolution of natural yeast. *Genetics*, Vol. 165, pp. 1745-1759, ISSN 0016-6731

Le Jeune, C., Lollier, M., Demuyter, C., Erny, C., Legras, J. L., Aigle, M., Masneuf-Pomarède, I. (2007). Characterization of natural hybrids of *Saccharomyces cerevisiae* and *Saccharomyces bayanus* var. *uvarum*. FEMS Yeast Research, Vol. 7, pp. 540-549, ISSN 1567-1356

Lopes, C. A., Rodríguez, M. E., Sangorrín, M., Querol, A., Caballero, A. C. (2007). Patagonian wines: implantation of an indigenous strain of Saccharomyces cerevisiae in fermentations conducted in traditional and modern cellars. *Journal of Industrial Microbiology & Biotechnology*, Vol. 34, pp. 139-149, ISSN 1367-5435

Lopez, V., Frenández-Espinar, M. T., Barrio, E., Ramón, D., Querol, A. (2003). A new PCR-based method for monitoring inoculated wine fermentations. *International Journal of Food Microbiology*, Vol. 81, pp. 63-71, ISSN 0168-1605

Martínez, M., Gac, S., Lavin, A., Ganga, M. (2004). Genomic characterization of *Saccharomyces cerevisiae* Straits isolated from wine-producing areas in South America. Journal of Applied Microbiology, Vol. 96, pp. 1161-1168, ISSN 1364-5072

Mesa, J. J., Infante, J. J., Rebordinos, L., Cantoral, J. M. (1999). Characterization of yeast involved in the biological ageing of Sherry wines. *LWT Food Science and Technology*, Vol. 32, pp. 114-120, ISSN 0023-6438

Mesa, J. J., Infante, J. J., Rebordinos, L., Sánchez, J. A., Cantoral, J. M. (2000). Influence of the yeast genotypes on enological characteristics of Sherry wines. American Journal of Enology and Viticulture, Vol. 51, 1, pp. 15-21, ISSN 0002-9254

Mortimer, R.K. (2000). Evolution and variation of the yeast (*Saccharomyces*) genome. *Genome Research*, Vol. 10, pp. 403-409, ISSN 1088-9051

Muñoz-Bernal, E., Fernández-Acero, F. J., Rodríguez, M. E., Cantoral, J. M. (2011). Variación en el perfil 2DE en *Saccharomyces bayanus* var. *uvarum* inducido por temperatura. Análisis diferencial de proteoma. XXIII Congreso Nacional de Microbiología, Salamanca (Spain), 11-14 July 2011

Naumov, G. I., Masneuf, I., Naumova, E. S., Aigle, M., Dubourdieu, D. (2000). Association of *Saccharomyces bayanus* var. *uvarum* with some French wines: genetics analysis of yeast populations. *Research in Microbiology*, Vol. 151, pp. 683-691, ISSN 0923-2508

Naumov, G. I., Naumova, E. S., Antunovics, Z., Sipiczki, M. (2002). *Saccharomyces bayanus* var. *uvarum* in Tokaj wine-making of Slovakia and Hungary. *Applied Microbiology and Biotechnology*, Vol. 59, pp. 727-730, ISSN 0175-7598

Raspor, P., Cus, F., Povhe Jemec, K., Zagorc, T., Cadez, N., Nemanic, J. (2002). Yeast population dynamics in spontaneous and inoculated alcoholic fermentations of *Zametovka* must. *Food Technology and Biotechnology*, Vol. 40, pp. 95-102, ISSN 1330-9862

Rodríguez, M.E., Infante, J.J., Molina, M., Domínguez, M., Rebordinos, L. & Cantoral, J.M. (2010). Genomic characterization and selection of wine yeast to conduct industrial fermentations of a white wine produced in a SW Spain winery. *Journal of Applied Microbiology*, Vol. 108, pp. 1292-1302, ISSN 1364-5072

Rodríguez, M.E., Infante, J.J., Molina, M., Rebordinos, L. & Cantoral J.M. (2011). Using RFLP-mtDNA for the rapid monitoring of the dominant inoculated yeast strain in industrial wine fermentations. *International Journal of Food Microbiology*, Vol. 145, pp. 331-335, ISSN 0168-1605

Rossignol, T., Kobi, D., Jacquet-Gutfreund, L. & Blondin, B. (2009). The proteome of a wine yeast strain during fermentation: correlation with the transcriptome. *Journal of Applied Microbiology*, Vol. 107, pp. 47-55, ISSN 1364-5072

Salvadó, Z., Chiva, R., Rodríguez-Vargas, S., Rández-Gil, F., Mas, A., Guillamón, J. M. (2008). Proteomic evolution of a wine yeast during the first hours of fermentation. *FEMS Yeast Research*, Vol. 8, pp. 1137-1146, ISSN 1567-1356

Santamaría, P., Garijo, P., López, R., Tenorio, C., Gutiérrez, A. R. (2005). Analysis of yeast population during spontaneous alcoholic fermentation: Effect of the age of the cellar and the practice of inoculation. *International Journal of Food Microbiology*, Vol. 103, pp. 49-56, ISSN 0168-1605

Schuller, D., Valero, E., Dequin, S. & Casal, M. (2004). Survey of molecular methods for the typing of wine yeast strains. *FEMS Microbiology Letters*, Vol. 231, pp. 19-26, ISSN 1567-1356.

Schuller, D., Casal, M. (2007). The genetic structure of fermentative vineyard-associated *Saccharomyces cerevisiae* populations revealed by microsatellite analysis. *Antonie van Leeuwenhoek*, Vol. 91, pp. 137-150, ISSN 0003-6072

Trabalzini, L., Paffetti, A., Scaloni, A., Talamo, F., Ferro, E., Coratz, G., Bovalini, P., Santucci, A. (2003). Proteomic response to physiological fermentation stresses in a wild-type wine strain of *Saccharomyces cerevisiae*. *Biochemistry Journal*, Vol. 370, pp. 35-46

Viana, F., Gil, J. V., Vallés, S., Manzanares, P. (2009). Increasing the levels of 2-phenyl acetate in wine through the use of a mixed culture of *Hanseniaspora osmophila* and *Saccharomyces cerevisiae*. *International Journal of Food Microbiology*, Vol. 135, pp. 68-74, ISSN 0168-1605

Zuzuarregui, A., Monteoliva, L., Gil, C., del Olmo, M. (2006). Transcriptomic and proteomic approach for understanding the molecular basis of adaptation of *Saccharomyces cerevisiae* to wine fermentation. *Applied and Environmental Microbiology*, Vol. 72, No. 1, pp. 836-847, ISSN 0099-2240

3

Gel Electrophoresis Based Genetic Fingerprinting Techniques on Environmental Ecology

Zeynep Cetecioglu[1], Orhan Ince[1] and Bahar Ince[2]
[1]Istanbul Technical University, Environmental Engineering Department, Maslak, Istanbul
[2]Bogazici University, Institute of Environmental Science, Rumelihisarustu-Bebek, Istanbul
Turkey

1. Introduction

Molecular tools in environmental microbiology have been applied extensively in last decades because of the limitations in culture-dependent methods (Amann *et al.*, 1995; Muyzer *et al.*, 1996; Head *et al.*, 1998). Despite isolation techniques are provided detailed knowledge about the single species in terms of morphology, biochemistry, and also genetic (Bitton, 2005), they have important drawbacks. The first one is to find the selective media favoring the desired microbial group. Additionally, isolated species cannot reflect their behaviors in the natural environment. Until today, 19.000 microbial species have been isolated (DSMZ, 2011; http://www.dsmz.de), however it is accepted that this number is only a small portion of real diversity (Amann *et al.*, 1995). Besides, using the molecular tools in natural and engineering systems, we can find the answer to the questions such as 'which species do exist?', 'which species are active?', 'how many microorganisms are there?', which species do utilize the specific compounds?'.

Microbial ecology studies need identification of species based on a comprehensive classification system that perfectly reflects the evolutionary relations between the microorganisms (Pace, 1996). Zuckerkandl and Pauling (1965) indicated that nucleic acids could document evolutionary history. Due to the pioneering studies, nucleic acids, especially 16S rRNA, are the ultimate biomarkers and hereditary molecules probably because of their essential role in protein synthesis, making them one of the earliest evolutionary functions in all cellular life-forms (Olsen *et al.*, 1986; Pace *et al.*, 1986; Woese, 1987; Stahl *et al.*, 1988). In particular, 16S rRNA and 16S rDNA have been used in phylogenetic analysis and accepted as ideal evolutionary chronometer.

Genetic fingerprinting techniques are one of the most applied molecular tools based on 16S rRNA in microbial ecology studies. These techniques such as denaturing gradient gel electrophoresis (DGGE), temperature gradient gel electrophoresis (TGGE), amplified ribosomal DNA restriction (ARDRA) or restriction fragment length polymorphism (RFLP), terminal restriction fragment length polymorphism (T-RFLP), and single strand conformation polymorphism (SSCP), have been developed for estimation of diversity in ecosystems, screening clone libraries, following the diversity changes with respect to time and location and also identification of species (Hofman-Bang, 2003). This approach comes into prominence because of fast, less labor-intensive features (Muyzer and Smalla, 1998).

These methods have been used to characterize the microbial diversity in different environments such as activated sludge (Liu *et al.*, 1997; Curtis and Craine, 1998), anaerobic reactors (Leclerc *et al.*, 2004), sediments (Muyzer and De Wall, 1993, Cetecioglu *et al.*, 2009), lake water (Ovreas *et al.*, 1997), hot springs (Santegoeds *et al.*, 1996), biofilm (Santegoeds *et al.*, 1998). The method can be used for as both qualitative and semi-quantitative approaches on biodiversity estimations.

In this chapter, these genetic fingerprinting techniques based on gel electrophoresis are discussed. Also exemplarily applications are presented.

2. Microbial ecology and characterization of microbial community via molecular tools

Biochemical conversions occurred in environment are determined by black box model because of limitations to identify microbial communities which are responsible of these (un)known processes (Amann *et al.*, 1995). Acquisition of pure cultures is necessary to obtain an insight into the physiology, biochemistry and genetics of isolated microorganisms. In spite of developments on cultivation methods everyday, still a small portion of the microbial species within the nature can be isolated by culture-dependent techniques (Giovannoni *et al.*, 1990). Another problem in microbial ecology is the complications on identification and classification of the species based on their morphological features. Since the morphological features of the microorganisms cannot give the detailed information about their evaluation relationships. In order to determine the role of microbial diversity in natural or engineered systems, the questions about microbial population including 'Who is there? How many microorganisms are there them? Where are they located? What are they doing? How do populations respond to changes in environmental conditions? What is the relationship between diversity and community stability?' have to be answered. Accordingly, culture independent methods, which give information about microbial ecosystem in terms of diversity, function, etc., are more reliable (Muyzer *et al.*, 1998; Head *et al.*, 1998).

To increase our knowledge about microbial communities and our understanding of their composition, dynamics and interactions within microbial ecosystems, nucleic acid analysis give a wide range opportunity nowadays. Molecular phylogeny not only employs nucleic acid documentation and evolutionary history but also provides a motivation for identification and quantification of microbial species (Olsen and Woese, 1993). The phylogenetic tree of all living organisms is represented in Figure 1. Ribosomal RNA and its gene are the main biomarkers and hereditary molecules for prokaryotes because of their essential role in protein synthesis making them one of the earliest evolutionary functions in all cellular life-forms (Woese, 1987). Therefore prokaryotes can be detected, identified and enumareted by the analysis of 16S rRNA and 16S rDNA.

16S rRNAs and 16S rDNAs, which encode them, are ideal biomarker because they exist in all prokaryotes, they have conserved and their variable regions give the opportunity to identify species even strains as seen in Figure 2. While the conserved regions of 16S rRNA make this molecule as an evolutionary clock instead of their selectively neutral mutational changes (Woese, 1987, Amann *et al.*, 1995), their variable regions allow phylogenetic determination on different taxonomic level (Amann *et al.*, 1995; Head *et al.*, 1998).

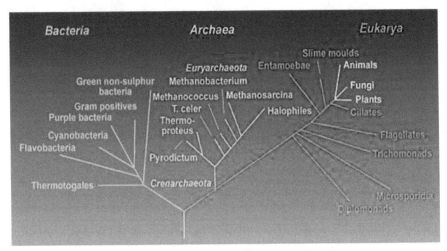

Fig. 1. The rRNA phylogenetic tree of life (Madigan *et al.*, 2009).

As a result, to design general or specific primers and probes for 16S rDNAs and 16S rRNAs provides study options about identification and evolution of microorganisms because this molecule is fairly large ($\approx$1500 nucleotides) including sufficient sequence information. Also the abundance is high within most cells (10^3 to 10^5 copies) and they can be detected easily (Amann *et al.*, 1995). While even secondary structure of 16S rRNA molecule is highly conserved, many variable regions randomly change during evolution. This differential variation explains the relationship between microorganisms evolutionarily. Data obtained from this analysis are adequate to compare statistically significant phylogenetic relations (Olsen *et al.*, 1986). Therefore 16S rRNA and its encoding gene have been widely used to investigate community diversity. The rapidly growing 16S rDNA sequence data bank, accessible (http://www.ebi.ac.uk/) provides the opportunity to get information about 16S rDNA sequences of the determined cultured and uncultured species (Dahllöf, 2002).

In spite of the advantages of using 16S rRNA molecule for phylogenetic analysis, the main limitations are that the heterogeneity between multiple copies of this molecule in one species interferes pattern analysis, confuses the explanation of diversity obtained from clone libraries and sequences retrieved from banding patterns (Dahllöf, 2002).

3. Fingerprinting techniques and their application areas

Fingerprinting techniques provide a separation in microbial community according to their genetic pattern or profile (Muyzer, 1998). A variety of fingerprinting techniques such as *denaturing/temperature gradient gel electrophoresis, amplified ribosomal DNA restriction analysis, terminal restriction fragment length polymorphism, and single strand confirmation polymorphism* has been developed to assess diversity and dynamics in the ecosystem (Hofman-Bang, 2003). The first fingerprinting technique was used in 1980's, which based on the electrophoretic separation in high-resolution polyacrylamide gels of 5S rRNA and tRNA

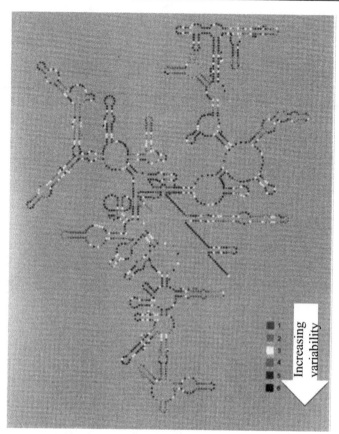

Fig. 2. Secondary structure of the 16S rRNA of E. coli, showing conserved and variable regions (Van de Peer *et al.*, 1996).

obtained from natural samples (Hofle, 1988 and 1990). In 1993, Muyzer *et al.* introduced a new fingerprinting technique to apply on microbial ecology, *denaturing gradient gel electrophoresis* (DGGE). In this method, PCR amplified DNA fragments can be separated according to their nucleic acid pattern. This method has become widespread in a short time. Then another similar technique has been developed, *temperature gradient gel electrophoresis* (TGGE). These methods provide not only analysis of the structure and species composition of microbial communities but also identification of several uncultured microorganisms (Heuer *et al.*, 1997 and Cetecioglu *et al.*, 2009).

3.1 Denaturing/Temperature Gradient Gel Electrophoresis (DGGE/TGGE)

DGGE is a gel electrophoresis technique to separate same length-DNA fragments based on their base sequence differences. In theory, it is sensitive to observe even one base difference on sequence because of melting patterns of the fragments (Muyzer *et al.*, 1993). This method provides a fast, and labor-intensive approach to determine the diversity and the microbial community within an ecosystem, to monitor the changes on dynamics and also to screen the clone libraries (Muyzer and Smalla, 1998). Furthermore, DGGE can be used as qualitative and semi-quantitative approach for biodiversity estimations.

3.1.1 Principles of the experiment

The optimal gradient is the main concern for DGGE/TGGE experiments since the main purpose is separation of DNA fragments according to their melting behaviours. Perpendicular polyacrylamide gels are used according to incremental gradients of denaturants or temperature. The sample including same-length DNA fragment mixtures is loaded to gel for running by electrophoresis. After completing electrophoresis, the gel is stained by a dye such as ethidium bromide, SYBR gold, SYBR green, etc. for obtaining sample pattern. While linear gradient is created by chemical denaturants as urea and formamide for DGGE, temporal temperature gradient is used to separate the DNA fragments in TGGE. Melting pattern of double strand DNA fragments is based on their hydrogen bond content: GC rich DNA fragments melts at higher denaturant/temperature region of the gradient. Complete separation of the double strand DNA is prevented by using GC-clamp primer during the amplification of target DNA region (Dorigo et al., 2005). The schematic explanation of DGGE is given in Figure 3.

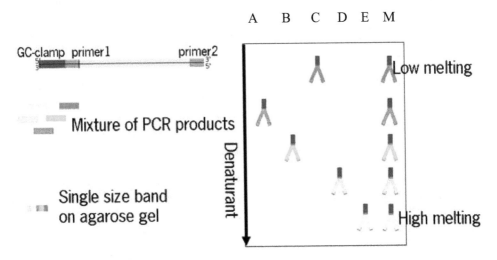

Fig. 3. Principle of DGGE (A: organism a, B: organism b, C: organism c, D: organism d, E: organism E, M: mix sample) (Plant Research International, 2011).

The main difficulties and limitations of the DGGE/TGGE can be listed as:

1. Proper primer selection to represent whole community
2. Optimization of electrophoresis conditions (Muyzer et al., 1993)
3. Limitations on sensitivity for detection of rare community members (Vallaeys et al., 1997)
4. Separation of only small DNA fragments up to 500 bp (Muyzer and Smalla, 1998)
5. Biases coming from PCR amplification such as chimeric products or fidelity errors
6. Heteroduplex formations, multiple bands or due to resolution of the gel, or different fragments resulting from existence of several rRNA coding regions, (Curtis and Craine, 1998).

3.1.2 Application area

DGGE/TGGE is used for several purposes in microbial ecology. The first and the most common application is to reveal and to compare community complex of the microbial diversity within different environments. Curtis and Craine (1998) used this technique to show the bacterial complexity of different activated sludge samples. Connaughton et al. (2006) used PCR-DGGE method to find out bacterial and archaeal community structure in a high-rate anaerobic reactor operated at 18 °C. This technique was used to reveal the microbial community in a lab-scale thermophilic trickling biofilter producing hydrogen (Ahn et al., 2005). Another biofilm study showed the bacterial diversity in a river by 16S rDNA PCR-DGGE method (Lyautey et al., 2005). In another study, the authors showed that the different bacterial and archaeal profiles within the highly polluted anoxic marine sediments in the different locations from the Marmara Sea (Cetecioglu et al., 2009). Ye et al. (2011) showed the temporal variability of cyanobacteria in the water and sediment of a lake.

Furthermore the scientists use these techniques, mostly DGGE, to analyse the community changes over time. Santagoeds et al. (1998) used PCR-DGGE method to monitor the changes in sulphate reducing bacteria in biofilm. Ferris and Ward (1997) also performed similar approach to reveal seasonal changes in bacterial community from hot spring microbial mat. Kolukirik et al. (2011) used 16S rDNA PCR-DGGE technique to represent the local and seasonal bacterial and archaeal shifts in hydrocarbon polluted anoxic marine sediments.

These fingerprinting techniques are widely used to monitor simple communities instead of complex environments. It is one of the detection methods to analyse the cultivation/ isolation approaches and to determine the enrichment cultures (Santagoeds et al., 1996; Ward et al., 1996; Teske et al., 1996; Muyzer, 1997; Bucholz-Cleven et al., 1997).

Also DGGE/TGGE are commonly chosen for comparison of the efficiency of the DNA extraction protocols (Heuer and Smalla, 1997; Lieasack et al., 1997) and the screening of the clone libraries (Heuer and Smalla, 1997; Lieasack et al., 1997, Kolukirik et al., 2011) because rapid and reliable results are caused to perform less time (Kowalchuk et al., 1997).

3.2 Amplified Ribosomal DNA Restriction Analysis (ARDRA)

Recognition site of restriction enzymes are changed for different microbial species. The principle of amplified ribosomal DNA restriction analysis (ARDRA), also called as restriction fragment length polymorphism (RFLP), is based on this knowledge. The combination of PCR and restriction can, for example, be used for enhanced amplification of minor DNA templates (Green and Minz, 2005).

In the first step of this technique, ribosomal DNA is amplified by PCR to avoid undesired and/or dominant DNA templates. Then, the 16S rDNA PCR products are digested into specific DNA fragments by restriction enzymes. At the final step, the fragments are loaded to high-resolution gel for electrophoresis. The schematic representation of the principle of ARDRA is given in Figure 4. The main advantage of this technique is to provide rapid comparison of rRNA genes (Moyer et al., 1994).

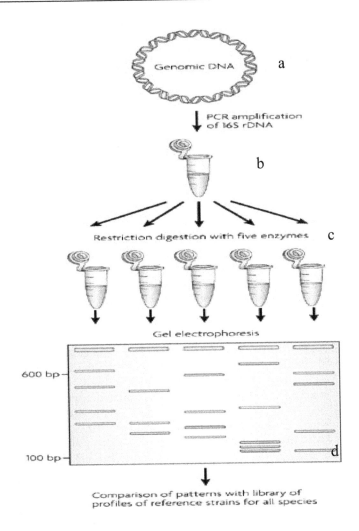

Fig. 4. Steps of ARDRA (a: Genomic DNA extraction, b: PCR reaction for specific region, c: restriction digestion, d: gel electrophoresis) (Dijkshoorn *et al.*, 2007).

The application areas of this technique are also similar to DGGE. It is varied from detection isolates or clones to determination of whole community in an environment. For these different purposes, different gel types can be used. While agarose gel is sufficient to detect isolates or clones, polyacrylamide gels are necessary for better resolution in the community analysis (Martinez-Murcia *et al.*, 1995).

In the literature, there are different studies performed by ARDRA. Lagace *et al.* (2004) identified the bacterial community of maple trees. A wide variety of the organisms were detected from different groups. Barbeiro and Fani used this technique to investigate more specific bacterial group, Acinetobacteria, within 3 sewage treatment plants (1998). In 1995, Vaneechoutte and his colleagues performed similar study for Acinetobacter strains. They showed that this technique is less prone to contamination problems for detection. In another study, ARDRA was used to screen bacterial and archaeal clone libraries to detect the microbial community within an anaerobic reactor to treat fodder beta silage (Klocke et al., 2007). Also there are some studies to investigate the microbial community in soil (Smith et al., 1997; Viti and Giovannetti, 2005).

3.3 Terminal Restriction Length Polymorphism (T-RFLP)

Terminal Restriction Fragment Length Polymorphism (T-RFLP) is another fingerprinting technique to obtain profiles of microbial communities. The principle of this method is to separate the genes according to position of their restriction site closest to a labelled end of an amplified gene (Figure 5). The main difference from ARDRA is that the restriction enzymes using in T-RFLP only detect terminal restriction fragments (T-RF). Also this method is used qualitative and quantitative analysis like DGGE (Liu et al., 1997).

The method is carried out in a series of steps including PCR, restriction enzyme digestion, gel electrophoresis and recognition of labelled fragments. Like most other fingerprinting techniques, PCR amplification of a target gene is the first step of T-RFLP.

After DNA extraction, target gene amplification is carried out using one or both the primers having their 5′ end labelled with a fluorescent molecule. Then amplicons are digested by restriction enzymes. Following the restriction reaction, the digested DNA fragments are separated using either capillary or polyacrylamide gel electrophoresis in a DNA sequencer with a fluorescence detector so that only the fluorescently labelled terminal restriction fragments (TRFs) are visualized. At the final step, electropherom is obtained as a result of T-RFLP profiling. Using this graph, electropherom, only target restricted DNA fragments are detected and also satisfactorily quantified by automated electrophoresis systems. Quantification analysis gives an opportunity to make various statistical methods, such as similarity indices, hierarchical clustering algorithms, ordination methods, and self-organizing maps (Liu et al., 1997).

In the literature, T-RFLP was carried out for different purposes like other fingerprinting techniques. In 1997, while Liu et al. used this technique to characterize microbial diversity in different environments such as activated sludge, enriched sludge from lab-scale bioreactor, aquifer sand, termite, Moeseneder and his colleagues (1999) optimized T-RFLP to determine marine bacterioplankton communities and to compare this technique to DGGE. In 2000, Horz and his colleagues reported major sub-groups of ammonia oxidizing bacteria by using amoA functional gene. Methane-oxidizing bacteria from landfill site cover soil were detected by T-RFLP combined with RNA dot-blot hybridization (Stralis-Pavese et al., 2006). Also in the same study, RFLP method is used to screen clone libraries. Lueders and Friedrich tried to determine PCR amplification bias by T-RFLP in 2003. Blackwood and his colleagues used T-RFLP for quantitative comparison of microbial communities from different environments such as soil and bioreactors (2003). Additionally this technique was used to screen clone libraries (Moeseneder et al., 2001). Liu et al. (2011) performed T-RFLP to determine the microbial shift during bioremediation of petroleum hydrocarbon contaminated soil.

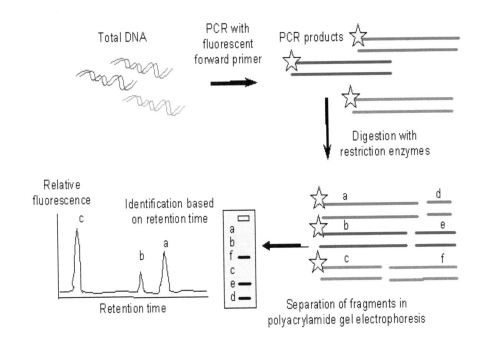

Fig. 5. Steps of T-RFLP (Kaksonen, 2011).

3.4 Single Strand Conformation Polymorphism (SSCP)

Single Strand Conformation Polymorphism (SSCP) is also a fingerprinting technique to separate same-length DNA fragments according to their differences in mobility caused by the secondary structure. The principle of this technique is represented in Figure 6. None of denaturant is used in this method to detect the mobility of the secondary structure of DNA fragments. Each band on SSCP gel corresponds to a distinct microbial sequence, indicating the presence of a microbial strain or species retrieved from the sample (Leclerc *et al.*, 2001; Lee *et al.*, 1996). The main limitation of SSCP, which is similar to DGGE/TGGE, is that one single strand DNA sequence can form more than one stable conformation and this fragment can be represented by multiple bands (Tiedje *et al.*, 1999). The advantage of this technique compared to other fingerprinting methods is that it does not require GC-clamp and gradient gel. SSCP is easier and more straightforward.

SSCP is mostly performed to determine the microbial community profile in different environments such as bioreactor and natural ecosystems. Firstly Lee *et al.* (1996) applied this method to obtain genetic profile of microbial communities. Then Schwieger and Tebbe (1998) used SSCP to determine the community profile including up to 10 bacterial strains. In another study, this method was combined with colony PCR to determine population levels of single and multiple species within plant and environmental samples (Kong *et al.*, 2005). Schmalenberger *et al.* (2008) investigated bacterial communities in an acidic fen by SSCP following by sequencing analysis. In this study, each representative band was cut, then cloned and sequenced to identify species. Also SSCP was carried out to determine the bacterial profile in an aerobic continuous stirred tank reactor (CSTR) treating textile wastewater (Khelifi *et al.*, 2009). Also this technique was applied for determination of *Clostiridum* sp. based on difference their [Fe-Fe]-hydrogenase gene (Quemeneur *et al.*, 2010).

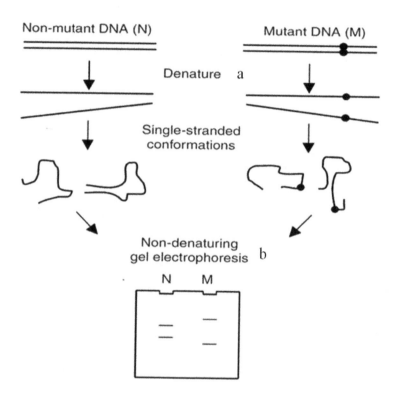

Fig. 6. Steps of SSCP (a: denaturation of ds DNA, b: electrophoresis) (Gasser *et al.*, 2007).

4. Conclusion

The principles of all fingerprinting techniques are similar. DGGE/TGGE, ARDRA, T-RFLP and (SSCP) have been developed to screen clone libraries, to estimate the level of diversity in environmental samples, to follow changes in community structure, to compare diversity and community characteristics in various samples and simply to identify differences between communities. While some of the scientists have showed that sensibilities and resolution of all these techniques are similar, DGGE is still more common application compared to other mentioned techniques. The main reasons of it are that the application of

DGGE is easier and more effective and also less equipment is necessary for it.

5. References

Ahn, Y., Park, E-J., Oh, Y-K., Park, S., Webster, G., Weightman, A.J. 2005. Biofilm microbial community of a thermophilic trickling biofilter used for continuous biohydrogen production. *FEMS Microbiology Letters*, 249: 31-38.

Amann, R.I., Ludwig, W., and Schleifer, K.H., 1995: Phylogenetic identification and in situ detection of individual microbial cells without cultivation. *Microbiol Rev*, 59: 143–169.

Barberio, C., fani, R., 1998. Biodiversity of an Acinetobacter population isolated from activated sludge. *Res. Microbiol.*, 149: 665-673.

Bitton, G., 2005. *Wastewater Microbiology*. 2nd ed., New York, Wiley-Liss.

Blackwood, C.B., Marsh, T., Kim, S-H., Paul, E.A., 2003. Terminal Restriction Fragment Length Polymorphism Data Analysis for Quantitative Comparison of Microbial Communities. *Appl. Environ. Microb.*, 69(2): 926-932.

Buchholz-Cleven, B.E.E., Rattunde, B., Straub, K.L., 1997. Screening for genetic diversity of isolates of anaerobic Fe(II)oxidizing bacteria using DGGE and wholecell hybridization. *Systematic and Applied Microbiology*, 20, 301–309.

Cetecioglu, Z., Ince, B. K., Kolukirik, M., and Ince, O. (2009). Biogeographical Distribution and Diversity of Bacterial and Archaeal Communities within Highly Polluted Anoxic Marine Sediments from the Marmara Sea. *Marine Pollution Bulletin*, 3(58): 384-395.

Connaughton, S., Collins, G., and O'Flaherty, V. 2006. Development of microbial community structure and activity in a high-rate anaerobic bioreactor at 18 ^{0}C. *Water Research*, 40: 1009 – 1017

Curtis, T.P., Craine, N.G., 1998. The comparison of the bacterial diversity of activated sludge plants. *Water Science and Technology*, 37, 71-78.

Dahllöf, I., 2002. Molecular community analysis of microbial diversity. *Current Opinion in Biotechnology*, 13, 213–217.

Dijkshoorn, L., Nemec, A., Seifert, H., 2007. An increasing threat in hospitals: multidrug-resistant Acinetobacter baumannii. *Nature Reviews Microbiology*, 5, 939–951

Dorigo,U., Volatier, L., Humbert, J. F., 2005. Molecular approaches to the assessment of biodiversity in aquatic microbial communities. *Water Research*, 39, 2207-2218.

Ferris, M.J., Ward, D.M., 1997. Seasonal distributions of dominant 16S rRNA defined populations in a hot spring microbial mat examined by denaturing gradient gel electrophoresis. *Applied and Environmental Microbiology*, 63, 1375–1381.

Giovannoni, S.J., Britschgi, T.B., Moyer, C.L., Field, K.G. 1990. Genetic diversity in Sargasso Sea bacterioplankton, *Nature*, 345, 60–63.

Green SJ & Minz D (2005) Suicide polymerase endonuclease restriction, a novel technique for enhancing PCR amplification of minor DNA templates. *Appl. Environ. Microbiol.*, 71: 4721-4727

Head, I.M., Saunders, J.R., Pickup R.W., 1998. Microbial evolution, diversity, and ecology: A decade of ribosomal RNA analysis of uncultivated microorganisms. *Microbiology and Ecology*, 35, 1-21.

Heuer, H., Smalla, K., 1997. Application of denaturing gradient gel electrophoresis (DGGE) and temperature gradient gel electrophoresis (TGGE) for studying soil microbial communities. In: van Elsas JD, Trevors JT & Wellington EMH (Eds) *Modern Soil Microbiology*, Marcel Dekker, New York. 353-373.

Höfle MG (1988) Identifcation of bacteria by low molecular weight RNA profiles: a new chemotaxonomic approach. *J. Microbiol. Meth.*, 8: 235-248

HöfleMG (1990) Transfer RNAs as genotypic fingerprints of eubacteria. *Arch. Microbiol.*, 153: 299-304

Hofman-Bang, J., Zheng, D., Westermann, P., Ahring, B. K., Raskin, L., 2003. Molecular ecology of anaerobic reactor systems. *Advances in Biochemical Engineering and Biotechnology*, 81, 151-203.

Horz, H-P., Rotthauwe, J-H., Lukow, T., Liesack, W., 2000. Identification of major subgroups of ammonia-oxidizing bacteria in environmental samples by T-RFLP analysis of amoA PCR products. *Journal of Microbiological Methods*, 39: 197-204.

Kaksonen, A. Molecular approaches for microbial community analysis. 08/11/2011. http://wiki.biomine.skelleftea.se/biomine/molecular/index_12.htm

Khelifi, E., Bouallagui, H., Touhami, Y., Godon, J-J., Hamdi, M., 2009. Bacterial monitoring by molecular tools of a continuous stirred tank reactor treating textile wastewater. *Bioresource Technology*, 100: 629-633.

Klocke, M., Mahnert, P., Mundt, K., Souidi, K., Linke, B., 2007. Microbial community analysis of a biogas-producing completely stirred tank reactor fed continuously with fodder beet silage as mono-substrate. *System. Appl. Microb.*, 30: 139-151.

Kolukirik, M., Ince, O., Cetecioglu, Z., Celikkol, S., Ince, B.K., 2011: Local and Seasonal Changes in Microbial Diversity of the Marmara Sea Sediments, *Marine Pollution Bulletin*, 62 (11): 2384-2394.

Kong, P., Richardson, P.A., Hong, C., 2005. Direct colony PCR-SSCP for detection of multiple pythiaceous oomycetes in environmental samples. *Journal of Microbiological Methods*, 61: 25-32.

Kowalchuk, G.A., Stephen, J.R., de Boer, W., Prosser, J.I., Embley, T.M., Woldendorp, J.W., 1997. Analysis of ammoniaoxidizing bacteria of the subdivision of the class Proteobacteria in coastal sand dunes by denaturing gradient gel electrophoresis and sequencing of PCR-amplified 16S ribosomal DNA fragments. *Applied and Environmental Microbiology*, 63, 1489-1497.

Lagace, L., Pitre, M., Jacques, M., Roy, D., 2004. Identification of the Bacterial Community of Maple Sap by Using Amplified Ribosomal DNA (rDNA) Restriction Analysis and rDNA Sequencing. *Appl. Environ. Microb.*, 70(4): 2052-2060.

Leclerc, M., Delbes, C., Moletta, R. and Godon J-J., 2001. Single strand conformation polymorphism monitoring of 16S rDNA Archaea during start-up of an anaerobic digester. *FEMS Microbiology Ecology*, 34, 213–20.

Leclerc, M., Delgenes, J.P., Godon, J.J., 2004. Diversity of the Archaeal community in 44 anaerobic digesters as determined by single strand conformation polymorphism analysis and 16S rDNA sequencing. *Environmental Microbiology*, 6, 809–819.

Lee, D-H., Zo, Y-G., Kim, S-J., 1996. Nonradioactive Method To Study Genetic Profiles of Natural Bacterial Communities by PCR–Single Strand-Conformation Polymorphism. *Appl. Microb. Environ.*, 62(9): 3112-3120.

Liesack, W., Janssen, P.H., Rainey, F.A., WardRainey, N.L., Stackebrandt, E. 1997. Microbial diversity in soil: The need for a combined approach using molecular and cultivation techniques. In: van Elsas JD, Trevors JT & Wellington EMH (Eds) *Modern Soil Microbiology*. Marcel Dekker, New York. pp. 375–439

Liu, P-W.G., Chang, T.C., Whang, L-M., Kao, C-H., Pan, P-T, Cheng, S-S., 2011. Bioremediation of petroleum hydrocarbon contaminated soil: Effects of strategies and microbial community shift. *International Biodeterioration&Biodegradation*, 65: 1119-1127.

Liu, W.T., Marsh, T.L., Cheung, H., Forney, L.J., 1997. Characterization of microbial diversity by determining terminal restriction fragment length polymorphisms of genes encoding 16S rRNA. *Applied and Environmental Microbiology*, 63, 4516–4522.

Lueders, T. and Friedrich, M.W., 2003. Evaluation of PCR Amplification Bias by Terminal Restriction Fragment Length Polymorphism Analysis of Small-Subunit rRNA and mcrA Genes by Using Defined Template Mixtures of Methanogenic Pure Cultures and Soil DNA Extracts. *Appl. Environ. Microb.*, 69(1): 320-326.

Lyautey, E., Lacoste, B., Ten-Hage, L., Rols, J-L., Garabetian, F. 2005. Analysis of bacterial diversity in river biofilms using 16S rDNA PCR-DGGE: methodological settings and fingerprints interpretation. *Water Research*, 39: 380-388.

Madigan, M.T., Martinko, J.M., Parker, J., 2009. *Brock Biology of Microorganisms*, (12th edition), Prentice Hall, Inc., New Jersey.

Martinez-Murcia, A.J., Acinas, S.G., Rodriguez Valera, F., 1995. Evaluation of prokaryotic diversity by restrictase digestion of 16S rDNA directly amplified from hypersaline environments. *FEMS Microbiology Ecology*, 17, 247–256.

Moeseneder, M.M., Arrieta, J.M., Muyzer, G., Winter, C., Herndl, G.J., 1999. Optimization of Terminal-Restriction Fragment Length Polymorphism Analysis for Complex Marine Bacterioplankton Communities and Comparison with Denaturing Gradient Gel Electrophoresis. *Appl. Environ. Microb.*, 65(8): 3518-3525.

Moeseneder, M.M., Winter, C., Arrieta, J.M., Herndl, G.J., 2001. Terminal-restriction fragment length polymorphism _T-RFLP/ screening of a marine archaeal clone library to determine the different phylotypes. *Journal of Microbiological Methods*, 44: 159-172.

Moyer, C., Dobbs, F. C., Karl, D.M. 1994. Estimation of diversity and community structure through restriction fragment length polymorphism distribution analysis of bacterial 16S rRNA genes from a microbial mat at an active, hydrothermal vent system, Loihi Seamount, Hawaii. *Applied and Environmental Microbiology*, 60, 871-879.

Muyzer, G., Brinkhoff, T., Nübel, U., Santegoeds, C., Schafer, H., Wawer, C., 1997. Denaturing gradient gel electrophoresis (DGGE) in microbial ecology. In: Akkermans, A.D.L, van Elsas, J.D., de Bruijn, F.J., (Eds) *Molecular Microbial Ecology Manual*, 3.4.4, 1-27. Kluwer Academic Publishers, Dordrecht, The Netherlands.

Muyzer, G., De Waal, E.C., and Utterlinden, A.G., 1993: Profiling of complex microbial populations by denaturing Gradient Gel Electrophoresis analysis of polymerase chain reaction-amplified genes coding for 16S rRNA. *Appl Environ Microbiol*, 59: 695-700.

Muyzer, G., Hottentrager, S., Teske, A., Wawer, C., 1996. Denaturing gradient gel electrophoresis of PCR amplified 16S rDNA - A new molecular approach to analyse the genetic diversity of mixed microbial communities. In: Akkermans ADL, van Elsas JD & de Bruijn FJ (Eds) *Molecular Microbial Ecology Manual*, 3.4.4, 1-23. Kluwer Academic Publishers, Dordrecht, The Netherlands.

Muyzer, G., Smalla, K., 1998. Application of denaturing gradient gel electrophoresis (DGGE) and temperature gradient gel electrophoresis (TGGE) in microbial ecology. *Antonie Van Leeuwenhoek*, 73, 127-141.

Olsen, G.J., D.J. Lane, S.J. Giovannoni, D.A. Stahl and N.R. Pace. (1986). Microbial ecology and evolution: a ribosomal RNA approach. *Ann. Rev. Microbiol.*, 40:337-365.

Olsen, G.J., Woese, C.R., 1993. Ribosomal RNA: a key to phylogeny. *The FASE B Journal*, 7, 113-123.

Øvreas, L., Forney, L., Daae, F.L., Torsvik, V., 1997: Distribution of bacterioplankton in meromictic Lake Saelevanet, as determined by denaturant gradient gel electrophoresis of PCR-amplified gene fragments coding for 16S rRNA. *Appl. Environ. Microbiol.*, 63, 3367-3373.

Pace, N. R., 1996. New perspective on the natural microbial world: Molecular microbial ecology. *ASM News*, 62:463-470

Pace, N.R., G.J. Olsen, and C.R. Woese. (1986). Ribosomal RNA phylogeny and the primary lines of evolutionary descent. *Cell*, 45:325-326

Plant Research International, Wageningen University, Characterization identification and detection, 12.12.2011, Available from: <http://documents.plant.wur.nl/pri/biointeractions/images/slide14.jpg, 2011>

Quemeneur, M., Hamelin, J., Latrille, E., Steyer, J-P., Trably, E., 2010. Development and application of a functional CE-SSCP fingerprinting method based on [FeeFe]-hydrogenase genes for monitoring hydrogen-producing Clostridium in mixed cultures. *Intern. Jour. Hydrogen Energy.*, 35: 13158-13167.

Santegoeds, C.M., Ferdelman, T.G., Muyzer, G., de Beer, D., 1998. Structural and functional dynamics of sulfate-reducing populations in bacterial biofilms. *Applied and Environmental Microbiology*, 64, 3731-3739.

Santegoeds, C.M., Nold, S.C., Ward, D.M., 1996. Denaturing gradient gel electrophoresis used to monitor the enrichment culture of aerobic chemoorganotrophic bacteria from a hot spring cyanobacterial mat. *Applied and Environmental Microbiology*, 62, 3922–3928.

Schmalenberger, A., Tebbe, C.C., Kertesz, M.A., Drake, H.L., Küsel, K., 2008. Two-dimensional single strand conformation polymorphism (SSCP) of 16S rRNA gene fragments reveals highly dissimilar bacterial communities in an acidic fen. *European Journal of Soil Biology*, 44: 495-500.

Schwieger, F and Tebbe, C.C., 1998. A New Approach To Utilize PCR–Single-Strand-Conformation Polymorphism for 16S rRNA Gene-Based Microbial Community Analysis. *Appl. Microb. Environ.*, 64(12): 4870-4876.

Smit, E. Leeflang, P., Wernars, K., 1997. Detection of shifts in microbial community structure and diversity in soil caused by copper contamination using amplified ribosomal DNA restriction analysis. *FEMS Microbiology Ecology*, 23: 249-261.

Stahl, D.A., Flesher, B., Mansfield, H.R., Montgomery, L., 1988: Use of phylogenetically based hybridization probes for studies of ruminal microbial ecology. *Applied and Environmental Microbiology*, 54, 1079–1084.

Stralis-Pavese, N., Bodrossy, L., Reichenauer, T.G., Weilharter, A., Sessitsch, A., 2006. 16S rRNA based T-RFLP analysis of methane oxidizing bacteria—Assessment, critical evaluation of methodology performance and application for landfill site cover soils. *Applied Soil Microbiology*, 31: 251-266.

Teske, A., Wawer, C., Muyzer, G., Ramsing, N. B., 1996. Distribution of sulfate-reducing bacteria in a stratified fjord (Mariager Fjord, Denmark) as evaluated by most-probable-number counts and denaturing gradient gel electrophoresis of PCR-amplified ribosomal DNA fragments. *Applied and Environmental Microbiology*, 62, 1405–1415.

Tiedje, J.M., Asuming-Brempong, S., Nusslein, K., Marsh, T.L., Flynn, S.J., 1999. Opening the black box of soil microbial diversity. *Applied Soil Ecology*, 13, 109-122.

Vallaeys, T., Topp, E., Muyzer, G., Macheret, V., Laguerre, G., Soulas, G., 1997. Evaluation of denaturing gradient gel electrophoresis in the detection of 16S rDNA sequence variation in rhizobia and methanotrophs. *FEMS Microbiology Ecology*, 24, 279–285.

Van de Peer, Y., van der Auwera G., de Wachter R., 1996. The evolution of stramenopiles and alveolates as derived by "substitution rate calibration" of small ribosomal subunit RNA. *Journal of molecular evolution*, 42, 201-210.

Vaneechoutte, M., Dijkshoorn, L., Tjernberg, I., Elaichouni, A., De Vos, P., Claeys, G., Verschraegen, G., 1995. Identification of Acinetobacter Genomic Species by Amplified Ribosomal DNA Restriction Analysis. *Jour. Of Clinical Microb.*, 33(1): 11-15.

Vanhoutte, T., Huys, G., Cranenbrouck, S., 2005: Exploring microbial ecosystems with denaturing gardient gel electrophoresis (DGGE). *BCCM Newletter*: http://bccm.belspo.be/newsletter/17-05/bccm02.htm

Viti, C. and Giovannetti, L., 2005. Characterization of cultivable heterotrophic bacterial communities in Cr-polluted and unpolluted soils using Biolog and ARDRA approaches. *Applied Soil Ecology*, 28: 101-102.

Ward, D.M., Santegoeds, C.M., Nold, S.C., Ramsing, N.B., Ferris, M.J., Bateson, M.M., 1996. Biodiversity within hot spring microbial mat communities: molecular monitoring of enrichment cultures. *Antonie van Leeuwenhoek*, 71, 143–150.

Woese, C.R., 1987. Bacterial evolution. Microbiology Reviews, 51, 221–271.

Ye, W., Tan, J., Liu, X., Lin, S., Pan, J., Li, D., Yang, H., 2011. Temporal variability of cyanobacterial populations in the water and sediment samples of Lake Taihu as determined by DGGE and real-time PCR. *Harmful Algae*, 10: 472-479.

4

Gel Electrophoresis of Grapevine (*Vitis vinifera* L.) Isozymes – A Review

Gizella Jahnke, János Májer and János Remete
University of Pannonia Centre of Agricultural Sciences
Research Institute for Viticulture and Enology
Badacsony
Hungary

1. Introduction

Several articles were written from the beginning of the fifties about the presence of plant enzymes in multiple forms. The major discussion was questioning whether these forms are artifacts that rose during the purification or not. To show that these forms are not artifacts in 1952 Jermyn divided his original peroxidase juice into two parts by acidic precipitation. The precipitate contained the A and B, while the supernatant the C and D points. Two components were found in the purified peroxidase solution; one migrated to the anode, the other to the cathode (Jermyn and Thomas, 1954).

The first major step for the starting up of isozyme analysis was the development of starch gel electrophoresis by Smithies (1955). The second major step was the demonstration of the direct visualization of isozymes in the stach gel by specific histochemical stains by Hunter and Markert in 1957 (McMillin in Tanksley and Orton, 1983).

The term isozyme was formed by Market and Moller (1959), using this word for different molecular forms of enzymes with the same substrate specifity.

Proteins - as the primary products of structural genes - are very alluring for the direct genetic studies. Variation in the DNA coding sequences frequently (but not all the cases) causes variation in the primary conformation of the proteins. In un-natural environments the detection of this variation is very difficult, because in such conditions the base of the separation is only the size of the protein (molecular weight). In natural environments the change of a single amino acid can detectably modify the migration. The extraction from a single tissue can contain a lot of proteins, which - in the case of non-specific (e.g. Comassie blue) staining - can result in a complex pattern, that makes it difficult to identify the homolog (allelic) and non-homolog enzymes. This problem can be solved by the application of enzyme-specific staining after the electrophoresis (Shields et al. in Tanksley and Orton, 1983).

The analysis of isozymes and their functions is the subject of functional genomics. The study of the gene expression in the level of RNA and proteins can give answers to a lot of open questions (Bernardi, 2004).

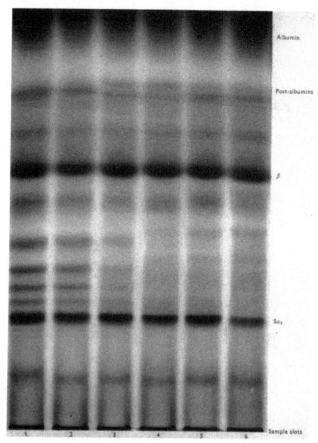

Fig. 1. A photograph of the result obtained by vertical starch-gel electrophoresis (approx. 19 hr. at 5v/cm.) with serum samples from six healthy individuals. Only the section of gel from the sample slots to the albumin is included in the photograph. Samples 1 and 2 are from female identical twins, 45 years old. Samples 3-5 are from 9-year-old female non-identical quadruplets and sample 6 is from their mother. (Smithies, 1958)

2. Grouping of isozymes

Isozymes are divided into three categories depending on the way they are biosynthesized:

1. isoenzymes (multilocus isozymes) arise from multiple gene loci, which code for structurally distinct polypeptide chains of the enzyme;
2. allozymes (or alleloenzymes), are structurally distinct variants of a particular polypeptide chain, coded by multiple alleles at a single locus;
3. secondary isozymes result from post-translational modifications of the enzyme structure;

The distinction between multiple alleles and multiple gene loci as causes of isozyme formation is that multiple alleles are the result of differences between individual members of a certain species, whereas multiple loci are common to all members of a species (Markert, 1975.).

The most probable reason for the presence of multilocus enzyme forms is the gene duplication. The gene duplication – the multiplication of genes in the genom –can come into existence by, for example, not equal crossing over. The frequency of mutation of various structural genes can be different, as a result of which some genes only rarely present in different allelic variant, as more alleles present in the population of other isoforms. This difference can be accepted as the evidence of a separate locus.

For the formation of multilocus isoforms a different evolutionary way can be imagined. It is probable, that the variation of the structural genes of originally different enzymes can cause the formation of similar catalytic functions. (H. Nagy, 1999).

Enzymes with variable substrates generally show higher variability itseves (catechol oxidase, acid phosphatase, peroxidase, esterase), but the amount of allozymic polymorphism is an increasing function of environmental variation. "Observations on natural pupulatians are cited which substantiate the claim that allozymic polymorphism is primarily due to selection acting on environmental variation in gene function. ...a large portion of the observed allozymic variation is due to a rather specific type of phenomenon: substrate variability" (Gillespie and Langley, 1974).

Enzymes with a single, special substrate show lower variability (glucose phosphate isomerase, phosphoglucomutase, glutamate-oxalacetate transaminase, glucose-6-phosphate transaminase etc.), but the banding patterns are less affected by the environment (Gillespie and Kojima, 1968; Gillespie and Langley, 1974).

3. Separation of isozymes by gel electrophoresis

Isoenzymes can be separated by electrophoresis or isoelectric focusing. The isozymes – under given proper circumstances – show peculiar patterns in the gel, which are called zymogramm (Hunter and Markert 1957).

Electrophoresis is a type of chromathography. The power for the separation of proteins is the difference in voltage between the two ends of the gel. The movement of proteins in the electric field is effected by their weight, shape and charge (Smith, 1960; Bálint and Bíró in Bíró, 1989).

The gel for the separation can be made from starch, agarose or acrilamide (Fig 2.). A standardized method of starch gel electrophoresis is used by UPOV (1996) for the analysis of identity of plant cultivars by isozyme analysis (Baum, 1986).

Advantages of the starch gel are that it is non-toxic, and more isozymes can be analysed by the slicing of a thick gel. More recently polyacrylamide gels are used because of their larger resolution. The porous structure of poliacrilamide gel is formed through a process of polymerization of acrylamide ($CH_2=CH-CO-NH_2$) and bis-acrylamide ($CH_2=CH-CO-NH-CH_2-NH-CO-CH=CH_2$). As a result of polimerisation a colourless, diaphanous, flexible, and consistent gel arises, which is resistant to scalding or chilling. The density, viscosity and size of poles are determined by the concentration of acrylamide and bis-acrylamide. (Hajósné Novák és Stefanovitsné Bányai in Hajósné Novák, 1999).

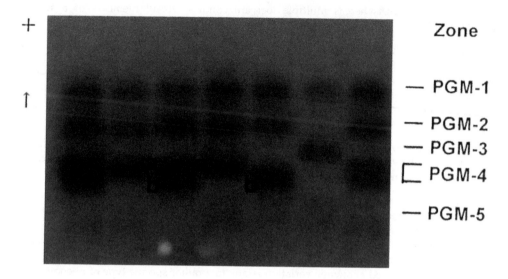

Fig. 2. Segregation of isozyme bands in PGM zone 4. Each column shows representative individuals of a selfed population of Cherry Bell-4. (Nomura et al., 1999.)

4. Separation of isozymes by isoelectric focusing

The charge of proteins is determined by the ratio of acidous and alkaline molecule parts and the rate of their dissociation. The rate of dissociation is determined by the pH of the surroundings of the molecule. The isoelectric point (pI) of the protein is the pH at which the acidous and alkaline molecule parts equally dissociate, the protein's net charge being zero. In a surrounding where the pH is lower than the isoelectric point, the net charge of the protein will be positive, in turn when the pH is higher than the IP, the net charge will be negative (Hames, 1990).

Isoelectric focusing of proteins can be carried out in a gel, in which a pH gradient is generated. Under voltage the proteins migrate to the point of the gel, where their net charge is equal to zero (pI). For this method thin poliacrilamide or agarose gels are used (Fig. 3).

The advantages of isoelectric focusing in opposition of gel electrophoresis are that the isozyme variants can be identified based on their isoelectric points, which results more accurate determination of isoforms, than the identification based on Rf values. On the other hand, the used gels in isoelectric focusing are thinner, so the separation is faster (Patterson and Payne, 1989).

Previously isoelectric focusing had disadvantages, as it required practice and the staining of the gels, because of the wide pH gradient, was difficult (Patterson és Payne, 1989), but nowadays these cause no problem. With the use of ampholites, the preparation of pH gradient gels needs even less practice, and you can even purchase ones. Neither causes problems with the staining of this so-called immobilized pH gradient gels, because the pH gradient can be removed by the washing of the gel.

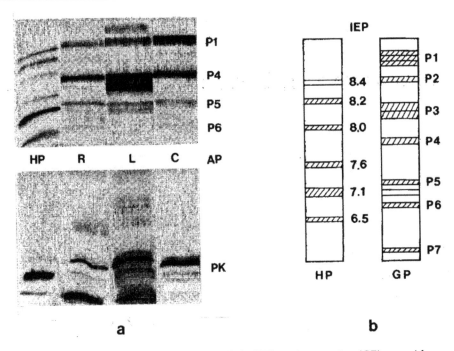

Fig. 3. Isoelectric focusing patterns of horseradish (HP), and grapevine (GP) peroxidases. IEP isoelectric points, P1-P7 names of isoenzyme bands, AP application zone, PK cathodic peroxidase. -a) anodic and cathodic pattern of peroxidases obtained from different organs of *Vitis riparia*. L Leaf, R root, C callus. The bands of horseradish peroxidases are distorted due to a margin effect on the slab. -b) Shematic diagram of the anodic part of peroxidase patterns. (Bachmann and Blaich, 1988)

5. Advantages and disadvantages of isozyme analyses

"The utility of isozymes as genetic markers is generally attributed to their frequent polymorphism, codominance, single gene-Mendelian inheritance, rapid, simple and relatively inexpensive assay and their ubiquity in plant tissues and organs (even in embrios and pollen). Although the selective neutrality of isozymes has been debated, it seems highly probable that they are adaptive under certain circumstances." (Bretting and Widrlechner 1995a).

Other advantages of isozyme analysis are the rapid analyses of samples, a small amount of plant material is sufficient. Young plants can be tested and selected based on their genotypes for features, which morphologically appear later. These can mean significant temporal and financial savings in the case of the breeding of annual plants. Now the best cost-efficient markers are the isozymes (Bretting and Widrlechner, 1995b).

Disadvantages of isozyme analysis as against the DNA markers, that they are organ-, tissue- and developmental stage-specific (Fig 4.). They often go through post-transcriptional modifications, which limit their usage (Staub et al., 1996).

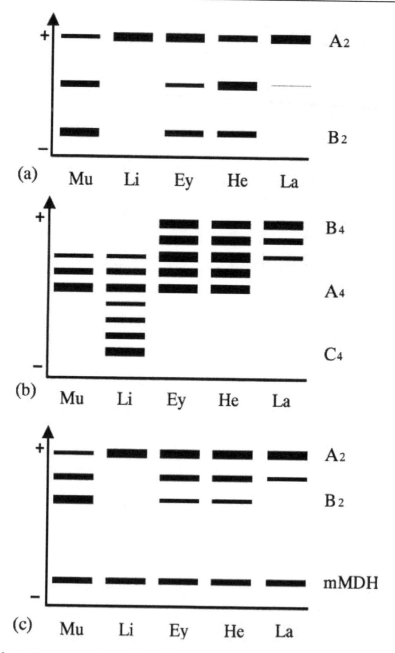

Fig. 4. Schematic representation of the differential expression of GPI (a), LDH (b) and MDH (c) isozymes in adult tissues (Mu, white muscle; Li, liver; Ey, eye; He, heart) and larvae (La) of L. cephalus. The egg pattern is identical to the larval one. Differences in line thickness refer to different staining intensities. (Manaresi et al. 1998)

6. Isozyme analysis in grape

Enzyme banding patterns for over 60 varieties of wine and table grapes were determined by gel electrophoresis by Wolfe (1976). Enzymes were extracted from ripe berries of each variety and separated by electrophoresis in a starch gel. Enzyme bands were detected by developing the gels in a buffered solution that produced an insoluble dye at the site of enzyme activity. The varieties were assayed for leucine aminopeptidase, indophenol oxidase, acid phosphatase, catechol oxidase, alcohol dehydrogenase, esterase, and peroxidase. The first four enzymes listed were found the most useful for distinguishing varieties.

Enzyme-banding patterns of catechol oxidase, acid phosphatase, esterase, alcohol dehydrogenase, indophenol oxidase, and leucine aminopeptidase obtained by enzyme staining of starch gel electropherograms allow the distinction of berries of the grape cultivars Perlette, Thompson Seedless, Superior Seedless and an early ripening sport of Superior Seedless (Schwennesen et al., 1982).

Twenty-seven cultivars and feral accessions from four Vitis species were examined by SUBDEN et al. (1987) for 12 isozyme systems exhibiting polymorphism. Using extracts from woody tissue and a protocol to avoid isozyme inactivation by polyphenolics and other materials, 27 of 29 strains exhibited unique isozyme banding patterns for glucose-6-phosphate isomerase, peptidase, and acid phosphatase. Implications for genetic homogeneity screening of nursery stock or identifying unknown samples are discussed.

German researchers analysed the isozymes of peroxidase by isoelectric focusing. Purified internodal phloem extracts from dormant wood were used. In the 6-11 pH range 8 bands were found, 71 Vitis species and varieties were identified (Bachmann and Blaich, 1988).

Genetic analysis of 11 allozyme polymorphisms was performed by Weeden (1988) on the progeny of 'Cayuga White' x 'Aurore', two complex interspecific grape (Vitis) hybrids. Segregation for most of the polymorphisms closely approximated monogenic Mendelian ratios, and eight new isozyme loci were defined for grape. Joint segregation analysis among the isozyme loci revealed three multilocus linkage groups (ACP-1 – PGM-c; ACP-2 – AAT-c; GPI-c – LAP-1). These results demonstrate that sufficient allozyme polymorphism exists in grape to establish many multilocus linkage groups and that this genetic analysis can be accomplished using extant progeny or progeny readily produced from highly heterozygous clones.

The pattern of the systems PER and ACP from 8 vines of Vitis vinifera L. has been studied in 1988 by Royo et al. (1989). A method to differentiate and characterize 6 clones of Vitis vinifera L. has been established by gaining the variability of the PER pattern from the band pattern constantly present in any vine, and the total band pattern from another vine (not only amongst the vines but also along the vegetative cycle). In the vines investigated no difference has been found for the ACP system.

Three enzymes in 5 cultivars of Vitis vinifera L. are analyzed by PAGE in young leaves. With acid phosphatase, arylesterase and glutamat-oxalat transaminase more or less different isoenzyme patterns of the different cultivars were obtained. There were no interclonal differences. The most polymorph enzyme was the arylesterase. The best results were obtained with young leaves from sprouting buds (Eiras-Dias et al., 1989).

Starch-gel electrophoresis was used by Walters et al (1989) for the analysis of *Vitis vinifera* L. cultivars, interspecific *Vitis* hybrids and wild individuals of *Vitis riparia* Michx. They suggest a simple and inexpensive procedure for the extraction of active enzymes from grape, which is rapid and efficient. Starch-gel electrophoresis with different optimized gel-electrode buffer systems is used for 40 different isoenzymes, 14 of which were consistently resolvable and showed variation among different cultivars.

Isozyme analysis is one of the means suitable to characterize clonally propagated cultivars. Isoelectric focusing was used to reveal differences in isozyme patterns between tissue-cultured plants and mother plants, for the cultivars Barbera, Queen of the Vineyards, Dolcetto and Delight. In cultivar Barbera both 2n and 4n plants were considered. Leaf samples were collected from shoots grown on cuttings under controlled environmental conditions and from plants obtained by tissue culture. The buds used for tissue culture were taken from the same shooted cuttings. Leaf extracts were analyzed by isoelectric focusing considering the following isozymes: AcPH (acid phosphatase), GPI (glucose phosphate isomerase) and PGM (phosphoglucomutase). The banding patterns of GPI and PGM showed differences among the cultivars, while for AcPH there seemed to be no differences among them in the pH range considered. There were no differences between isozyme patterns of the Barbera 2n and Barbera 4n. The main difference between in vitro plants and mother plants was the amount of isozyme evaluated by densitometric measurements. In all the cultivars, the amount of isozymes for AcPH was higher in mother plants than in in vitro ones, while for PGM and GPI it was the opposite. This can be due to the different environmental conditions affecting cellular metabolism (Botta et al., 1990).

The idea of using woody stems during the resting period instead of leaves for the isozyme analysis arose in 1990. Kozma et al. (1990) analysed the esterase isozymes of varieties from different convarietas and interspecific hybrid families by poliacrilamide gel electrophoresis and isoelectric focusing. Based on their results they established, that the phloem extracts from woody stems collected in the resting period or shoots colleting in spring give good reproducible patterns, but the leaf extracts give irreproducible patterns.

Tests were carried out on different types of calli and somatic embryos of V. rupestris using 2-D electrophoresis. The investigation carried out by Martinelli et al. (1993) is focused on the isozyme patterns of AcP (acid phosphatase), ADH (alcohol dehydrogenase), EST (esterase), G6PDH (gluconate-6-phosphate dehydrogenase) and PGM (phosphoglucomutase). A typical variation of isozyme pattern could be observed during the different steps of somatic embryogenesis. De-differentiated callus showed other types of isoenzyme pattern compared to those obtained during the development of somatic embryos.

Similarly to Kozma et al. Bachmann (1994) used extracts from phloem of dormant canes for the isozyme analysis. This comprehensive study has analysed the peroxidase isozyme banding patterns of 313 different cultivars and species of *Vitis* using isoelectric focusing on polyacrylamid gels. The author reports that acidic peroxidases were characteristic for *Vitis vinifera* L. cultivars with only a 5 % frequency of occurrence in other *Vitis* species. Variation in neutral to basic peroxidases could be used to group together similar cultivars independent of berry colour, e.g. Pinot noir, Pinot gris, Pinot blanc and Pinot meunier. However, other examples of colour variants, e.g. Merlot noir and blanc were clearly different using peroxidase banding.

Shiraishi et al. (1994) used GPI and PGM isozyme banding patterns for the detection of hybrid origin of seedlings during their triploide grape breeding. First they analysed 99 diploide cultivars, 20 diploide plants from 8 wild *Vitis* species and populations from the crosses between them . In the GPI-2 locus 13 in the GPM locus 11 allels was found. Data showd high genetical differences between Vitis species . After that, the GPI-2 and PGM-2 genotype of 6 diploide and 4 tetraploide cultivars (used for the crosses) were determined. 15 diploide x tertraploide crosses were made. Trisomy gene expression was detected in 92 out of 98 seedlings, as 6 showed diploide patterns.

Seed proteins and enzymes (AcP, ADH, EST, G-6-PDH, MDH, PGM, POD) from several cultivars and wild ecotypes of *Vitis vinifera* L. have been used to evaluate taxonomic differences between *V. vinifera* sspp. sativa and sylvestris (Scienza et al., 1994). Only total proteins in the pH range of 4.0-5.5 and AcP, EST and G-6-PDH were useful for genotype differentiation. The cluster analysis (UPGMA), based on Jaccard genetic distance and determined on the presence/absence of electrophoretic profiles, reveals 2 distinct groups, supporting the hypothesis of the authors that *V. sativa* and *V. silvestris* should be regarded as 2 separate taxa.

Studies on the induction characteristics and the fine structure of grapevine cells cultured in vitro were undertaken with cultivar Monastrell berry samples of different developmental stages between fruit set and veraison (Zapata et al., 1996). Medium composition, electron microscopy application and protoplast isolation procedures are explained. It could be shown that the intensity of cell development and callus induction percentage depended on the berry growth stage; the de-differentiation process is mainly located in meso-carp tissues. Cultured cells showed to be highly vacuolated with their cytoplasm reduced to a very thin peripheral layer (containing golgi sacks).

Ros Barceló et al. (1996) studied the gene expression of isozymes of providase in downy mildew resistant (*Vitis vinifera* x *Vitis rupestris*) x *Vitis riparia* hybrids and in the susceptible *Vitis vinifera* parent. The peroxidase isoenzyme type B3 (PI=8,9) expressed in the phloem and leaves of resistant hybrids was completely absent in the susceptible parent.

To test whether the basic peroxidase isoenzyme B3 may be considered as a molecular marker of disease resistance in *Vitis* species, suspension cell cultures derived from the downy mildew susceptible *V. vinifera* parent species were treated with an elicitor (cellulase Onoztika R-10) from the soil fungus *Trichoderma viride*, a specific and well-known elicitor of disease resistance reactions in grapevines. The results showed that treatment with the elicitor induces, simultaneously with the activation of the disease resistance mechanism, the appearance of B3 in the cell cultures. These results suggest that the basic peroxidase isoenzyme B3 may be considered as a marker of disease resistance in Vitis species.

Isoenzymes from grapevine woody stems and shoots were evaluated for their use in identification of varieties and clones by Royo et al. (1997). Plant extracts were separated by polyacrylamide gel electrophoresis. Isoenzyme analysis was carried out for esterases, peroxidases, catechol oxidase, glutamate oxalacetate transaminase and acid phosphatase. The plant material was grown and sampled at two localities in Spain, with different climatic conditions. Sampling was carried out bimonthly for two consecutive years in order to find out the influence of the environment and time of the year. Each isozyme system had a pattern defined by 'fixed' bands that were always present at both localities and during the resting period of the plant (autumn – winter).

An evaluation of the genetic diversity of 'Albariño' (*Vitis vinifera* L.) was carried out by Vidal et al. (1998). The 73 isozyme and 308 RAPD markers were common in the samples tested. The results show the existence of a genetic homogeneity within 'Albariño'cultivated in Galicia. Minor ampelographic differences among samples could be due to external factors rather than to genetic differences.

DNA and isoenzyme analyses were used to characterize 20 table grape cultivars including Moscato d'Amburgo, Italia, Sultanina, Bicane and some recently released new cultivars (Crespan et al., 1999). GPI and PGM isoenzyme systems were able to separate the cultivars into 9 groups whereas the 8 microsatellite loci that were analysed revealed a higher discriminating power. Parentage analysis confirmed that the cultivar Italia was obtained from the crossing Bicane x Moscato d'Amburgo.

Hungarian researchers used isoelectric focusing for the peroxidase and esterase isozymes of some grapevine cultivars. Samples were gathered at different times of the year. The leaf samples after blooming were found the best for the identification of varieties, but they found the phloem extracts of woody stems also suitable for cultivar identification (Stefanovits-Bányai et al., 1999; Stefanovits-Bányai et al, 2002).

Sixty-four Muscat flavoured grapevine accessions were analysed in the work of Crespan and Milani (2001). An analysis was performed at two isozymes and 25 microsatellite loci. The 64 accessions were reduced to 20, which were easily distinguishable from each other at the molecular level by as few as two microsatellite loci. The remaining 44 were found to be synonyms. Three mutants with red and pink coloured berries were identified in the Moscato bianco group. Moscato nero encompasses at least two, Moscato rosa three different varieties. It seems that only two of the analysed Muscats are the main progenitors of the Muscat family: Moscato bianco and Muscat of Alexandria, which in turn are joined by a direct parent-offspring link.

Sànchez-Escribano et al. (1998) analysed 43 table grape varieties by 6 isoenzyme systems (PER, CO, GOT SOD, EST, AcP). The last 2 enzymes were found unsuitable for identification, by the combination of the zymogram of the other 4 enzymes, they were able to identify 31 cultivars, as the remaining 12 were clustered to 5 groups.

Protein and esterase isozyme patterns of authentic grapes and wines of 13 white wine cultivars were determined by means of isoelectric focussing (range of pI: 2.5-10) by Paar et al (1999). Esterase staining with grapes showed active zones mainly in the alkaline pI-range, with most of the cultivars, however indicating no qualitative, but only quantitative differences. Staining of the protein patterns of grapes and wines with Coomassie Brilliant Blue proved to be well suitable for the differentiation of cultivars. With grapes as well as wines the most predicative bandings focussed in the acid pI-range of 4. With the cultivars Grüner Veltliner, Rotgipfler and Riesling Italico the protein banding patterns were so characteristic, that these cultivars were easily identified, whereas with the other cultivars detailed comparisons of the phenogrammes were necessary.

Isozyme and RAPD markers were used for the characterization of Hungarian grapevine varieties and their parents (HAJÓS-NOVÁK And HAJDÚ 2003). The cathecol-oxidase system was found the most suitable for identification proposes.

Isozyme and SSR analysis were carried out for the differentiation of the grapevine cultivars Kéknyelű and Picolit. The name of the grapevine cultivar 'Kéknyelű' has become inseparable from the name of the Badacsony vine region (Hungary), whose fame is also well known beyond the Hungarian frontier. In the Vitis International Variety Catalogue (http://www.genres.de/idb/vitis/) 'Kéknyelű' is reported, as the synonym of the Italian grapevine cultivar 'Picolit'. Vertical poliacrylamide-gel electrophoresis was used for the investigation isoenzymes of catechol-oxidase (CO) and acid phosphatase (AcP). Microsatellite analyses were carried out at 6 loci (VVS2, VVS16, VrZag79, VVMD7, VMC4A1, VMC4G6). The results of the isoenzymatic and microsatellite analyses confirmed, that this two cultivars are different (Jahnke et al., 2007).

Jahnke et al. (2009) investigated the genetic diversity of Hungarian grapevine cultivars with biochemical and molecular markers (isoenzyme and SSR). The isoenzyme patterns of 4 enzyme systems (catechol-oxidase, glutamate-oxalacetate-transaminase, acid phosphatase and peroxidase) and the microsatellite profile in 6 loci (VVS2, VVS16, VVMD7, VMC4A1, VMC4G6, VrZag79) of 48 grapevine varieties were analysed.

The results with CO, GOT, AcP and PER enzymes were reproducible and the zymograms obtained from the woody stems were independent from the time of sampling during the dormant period of the grape (Fig 5.).

Based on the isoenzyme patterns of these 4 enzymes most of the investigated varieties (40/48) were identified. A correlation was found between the isoenzyme patterns and the classification to convarietas of the varieties.

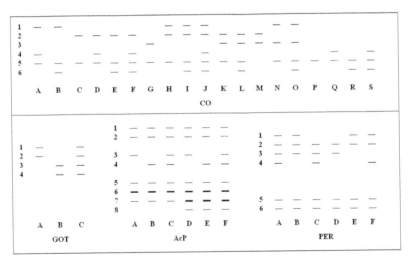

Fig. 5. Characteristic interpretative zymograms observed for CO, GOT, AcP and PER enzymes. The letters mark the different types of isoenzyme patterns, while numbers refer to the number of different isoenzyme bands (Jahnke et al. 2009.)

It was established, that while the varieties of the convarietas pontica differed from those of the convarietas orientalis and occidentalis, the two latter groups could have not been differentiated from each other. Based on the SSR (simple sequence repeat) analyses 46 of the

48 investigated varieties were identified. Even 'Pinot blanc' and 'Pinot gris' cultivars belonging to the same conculta (Pinot) could be differentiated in their VMC4A1 locus.

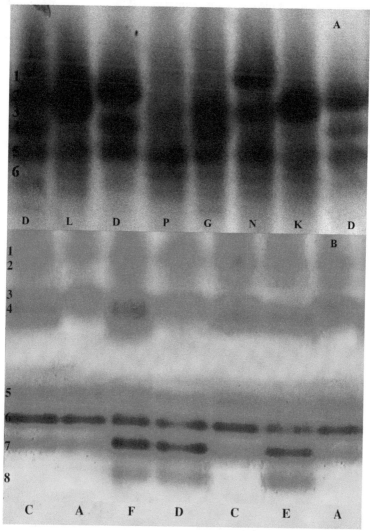

Fig. 6. Isoenzyme gel photos for CO (A) and AcP (B) respectively. The numbers show the band numbers, and the capital letters the banding pattern types shown in Figure 5. (Jahnke et al. 2009)

7. References

Bachmann O. (1994): Peroxidase isoenzyme patterns in Vitaceae. Vitis 33: 151-153.

Bachmann O., Blaich R. (1988): Isoelectric focusing of grapevine peroxidases as a tool for ampelography. Vitis 27: 147-155.

Bálint M., Bíró E. (1989): A fehérjekutatás fizikai-kémiai, preparatív és analitikai módszerei. 129-166. p. In: Bíró E. (Eds.) Biokémia I. Budapest: Tankönyv K.

Baum B. R. (1986): International Registration of Cultivars with Emphasis on Barley: Procedures and Methods of Producing a Register. Acta Horticulturae (182):237-250.

Bernardi G. (2004): Structural and Evolutionary Genomics. Natural Selection in Genome Evolution. Elsevier, Amsterdam

Botta R., Vallania R., Miaja M. L., Vergano G., Me G. (1990): Isozyme pattern comparison between tissue-cultured grapevines and mother plants. Vitis Special Issue: 88-92.

Bretting P. K., Widrechner M. P. (1995b): Genetic Markers and Horticultural Germplasm Management. HortScience 30(7) 1349-1356.

BrettingP. K., Widrlechner M. P (1995a): Genetic markers and plant genetic resource management. Plant Breeding Reviews (13): 11-86.

Crespan M., Botta R., Milani N. (1999): Molecular characterisation of twenty seedless table grape cultivars (Vitis vinifera L.). Vitis 38: 87-92.

Crespan M., Milani N. (2001): The Muscats: A molecular analysis of synonyms, homonyms and genetic relationship within a large family of grapevine cultivars. Vitis 40: 23-30.

Eiras-Dias J. E. J., Sousa B., Cabral F., Carcalho I. (1989): Isozymatic characterisation of portuguese vine varieties of Vitis vinifera L. Rivista di Viticoltura e di Enologia 1: 23-26.

Gillespie J. H., Kojima K. (1968): The degree of polymorphisms in enzymes involved in energy production compared to that in nonspecific enzymes in 2 Drosophila-ananassae populations. Proceedings of the National Academy of Science of the USA 61: 582-585.

Gillespie J. H., Langleyc. H. (1974): A general model to account of enzyme variation of natural populations. Genetics 76: 837-846.

H. Nagy A. (1999): A genetikai variabilitás vizsgálata izoenzimekkel. 25-39 p. In: HAJÓSNÉ NOVÁK M. (Eds.): Genetikai variabilitás a növénynemesítésben. Budapest: Mezőgazda Kiadó, 142 p.

Hajósné Novák M., Stefanovitsné Bányai É. (1999): Az izoenzimek elválasztása gélelektroforézissel. 99-104 p. In: HAJÓSNÉ NOVÁK M. (Szerk.): Genetikai variabilitás a növénynemesítésben. Budapest: Mezőgazda Kiadó, 142 p.

Hajós-Novák M., Hajdu E. (2003): Isozyme and DNA fingerprinting characterization of two hungarian wine grape hybrids and their parents. Acta Horticulturae 603: 211-216.

Hames B. D. (1990): One-dimensional polyacrylamide gel electrophoresis. 1.-139. p. In: Hames B.D., Rickwood D. (Eds.): Gel electrophoresis of proteins. ILR press at Oxford University press, Oxford- New York- Tokyo, Sec. Ed.

Hunter R. L., Markert C. L. (1957): Histochemical Demonstration of Enzymes Separated by Zone Electrophoresis in Starch Gels. Science (125): 1294-1295

Hunter R. L., Markert C. L. (1957): Histochemical demonstration of enzymes separated by zone electrophoresis in starch gels. Science 125: 124-1295.

Jahnke G., Korbuly J., Májer J., Györffyné Molnár Júlia (2007): Discrimination of the grapevine cultivars 'Picolit' and 'Kéknyelű ' with molecular markers. Scientia Horticulturae (114): 71-73.

Jahnke G., Májer J., Lakatos A., Györffyné Molnár J., Deák E., Stevanovits-Bányai É., Varga P. (2009): Isoenzyme and microsatellite analysis of Vitis vinifera L. varieties from the Hungarian grape germplasm. Scientia Horticulturae (120): 213–221.

Jermyn M. A., Thomas R. (1954): Multiple components in horse-radish peroxidase. Biochemical Journal. 56: 631-639.

Kozma P., H. Nagy A., Juhász O. (1990): Inheritence of isoenzymes and soluable proteins in grape vareties and F1 hybrids. Proceedings of the 5th International Symposium on Grape Breeding ; Vitis Special Issue: 134-141

Li, B.; Xu, Y. & Choi, J. (1996). Applying Machine Learning Techniques, Proceedings of ASME 2010 4th International Conference on Energy Sustainability, pp. 14-17, ISBN 842-6508-23-3, Phoenix, Arizona, USA, May 17-22, 2010

Lima, P.; Bonarini, A. & Mataric, M. (2004). Application of Machine Learning, InTech, ISBN 978-953-7619-34-3, Vienna, Austria

Manaresi S., Mantovani B., Zaccanti F. (1998): Comparative Analysis of Glucosephosphate Isomerase, Lactate Dehydrogenase and Malate Dehydrogenase Isozymes in 9 Cyprinid Species from Italy. Zoological Science 15:461-467.

Markert C., Moller L. F. (1959): Multiple forms of enzymes: tissue, ontogenetic, and species specific patterns. Proceedings of the National Academy of Sciences of the USA. 45: 753-763.

Markert C. (Ed.). (1975): Isozymes. Vol 1-4, Academic Press, New York.

Martinelli L., Scienza A., Villa P., De Ponti P., Gianazza E. (1993): Enzyme Markers for Somatic Embriogenezis in Vitis. Journal of Plant Phisiology 141: 476-481. p.

Mcmillin D. E. (1983): Plant isozymes: a historical perspective. 3-13. p. In: TANKSLEY S. D., ORTON T. J. (Szerk.): Isozymes in Plant Genetics and Breeding. Amsterdam: Elsevier Science Publishers B.V.

Nomura K., Yoneda K., Inoue H., Tateishi A., Kanehira T., Isobe K. (1999): Identification and linkage analysis of isozymes loci in open-pollinated varieties of radish (Raphanus sativus L.). Scientia Horticulturae 82: 1-8.

Paar E., Doubek S., Eder R. (1999): Differenzierung von Wiessweinsorten mittels isoelektischer Fokussierung. Mitteilungen Klosterneuburg 49:176-185.

Patterson B. A., Payne L. A. (1989): Zymograms of plant extracts using isoelectric focusing on ultra thin layers. Acta Horticulturae 247:163-169.

Ros Barceló A., Zapata J. M., Calderón A. A. (1996): A Basic Peroxidase Isoenzyme, Marker of Resistance Against Plasmopara viticola in Grapevines, is Induced by an Elicitor from Trichoderma viride in Susceptible Grapevines. Journal of Phytopathology 144: 309-313

Royo B., Gonzalez J., Laquidain M. J., Larumbe M. P. (1989): Caracterization mediante analisi isoenzimatico de clones de la vinifera "Garnacha" (Vitis vinifera L.). Investigación agraria. Producción y protección vegetales. 4: 343-354.

Royo J. B., Cabello F., Miranda S., Gogorcena Y., Gonzalez J., Moreno S., Itoiz R., Ortiz J. M. (1997): The use of isoenzymes in characterisation of grapevines (Vitis vinifera L.). Influence of the environment and time of sampling. Scientia Horticulturae 69: 145-155.

Sànchez-Escribano E., Ortiz J. M., Cenis J. L. (1998): Identification of table grape cultivars (Vitis vinifera L.) by the isozymes from the woody stems. Genetic Resources and Crop Evolution 45: 173-179.

Schwennesen J., Mielke E. A., Wolfe W. H. (1982): Identification if Seedless Table Grape Cultivars and Bud Sport with Berry Isozymes. HortScience 17: 366-368.

Scienza A., Villa P., Tedesco G., Parini L., Ettort C., Magenes S., Gianazza E. (1994): A chemotaxonomic investigation on Vitis vinifera L. II. Comparison among ssp. sativa traditional cultivars and wild biotypes of ssp. silvestris from various Italian regions. Vitis 33: 217-224.

Scienza A., Villa P., Tedesco G., Parini L., Ettort C., Magenes S., Gianazza E. (1994): A chemotaxonomic investigation on Vitis vinifera L. II. Comparison among ssp. sativa traditional cultivars and wild biotypes of ssp. silvestris from various Italian regions. Vitis 33: 217-224.

Shields C. R., Orton T. J., Stuber W. (1983): An outline of general resource needs and procedures for the electrophoretic separation of active enzymes from plant tissue. 443-467 p. In: Tanksley S. D., Orton T. J. (Eds.): Isozymes in Plant Genetics and Breeding. Amsterdam: Elsevier Science Publishers

Shiraishi S, Ohmia Wakama C., Hiramatsu M. (1994): Variation of Gluceosephosphate Isomerase and Phosphoglucomutase Isozymes in Vitis and Their use in Grape Breeding. J. Fac. Agr., Kyushu Univ., 38 (3.4), 255-272.

Siegwart, R. (2001). Indirect Manipulation of a Sphere on a Flat Disk Using Force Information. International Journal of Advanced Robotic Systems, Vol.6, No.4, (December 2009), pp. 12-16, ISSN 1729-8806

Smith I. (1960). Chromatographic and electrophoretic techniques. Vol. II Zone electrophoresis. Heinemann, London.

Smithies O. (1955): Zone Electrophoresis in Starch Gels.: Group Variations in the Serum Proteins of Normal Human Adults. Biochemical Journal 61: 629-641.

Smithies O. (1958): An Improved Procedure for Starch-gel Electrophoresis: Further Variations in the Serum Proteins of Normal Individuals. Biochemical Journal 71: 585-587.

Staub J. E., Serquen F. C., Gupta M. (1996): Genetic Markers, Map Construction, and Their Application in Plant Breeeding. HortScience 31: 729-740.

Stefanovits-Bányai É, Lakatos Zs., Kerepesi I., Balogh I. (1999): Esterase and peroxidase isozyme patterns of some Vitis vinifera L. species. Publicationes Uniersitate Horticulturae Alimentariae LIX: 42-46.

Stefanovits-Bányai É., Lakatos S., Hajós-Novák M., Hajdu E., Balogh I. (2002): Recent developments in biochemical characterisation of Vitis vinifera L. varieties in Hungary. International Journal of Horticultural Science 8: 57-61.

Subden R. E., Krizus A., Lougheed S. C., Carey K. (1987): Isozyme Characterisation of Vitis Species and Some Cultivars. American Journal of Enology and Viticulture 38:176-181.

UPOV (1996): Guidelines for the conduct of tests for distinctness, uniformity and stability. Rape seed (Brassica napus L. oleifera).

UPOV, Geneva, Switzerland TG/36/6

Vidal J. R., Moreno S., Masa A., Ortiz J. M. (1998): Study of the genetic homogeneity of Albariño (Vitis vinifera L.) growing in Galicia (Spain) using isozyme and RAPD markers. Vitis 37: 145-146.

Walters T. W., Posluszny U., Kevan P. G. (1989): Isozyme analysis of the grape (Vitis). I. A practical solution. Canadian Journal of Botany 67: 2894-2899.

Weeden N. F., Reisch B. I., Martens M. H. E. (1988): Genetic Analysis of Isozyme Polimorohism in Grape. Journal of the American Society for Horticultural Sciences 113 (5): 765-769.

Wolfe W. H (1976): Identification of grape varieties by isozyme banding patterns. American Journal of Enology and Viticulture 27: 68-73.

5

Gel Electrophoresis for Investigating Enzymes with Biotechnological Application

Maria de Lourdes T. M. Polizeli[1*], Simone C. Peixoto-Nogueira[1],
Tony M. da Silva[1], Alexandre Maller[2] and Hamilton Cabral[3]
*[1]Biology Department, Faculty of Philosophy
Sciences and Letters of Ribeirão Preto, São Paulo University
[2]Biochemistry and Immunology Department
School of Medicine of Ribeirão Preto, São Paulo, São Paulo University
[3]Science Pharmaceutical Department
School of Pharmaceutical Science of Ribeirão Preto, São Paulo University
Brazil*

1. Introduction

Enzymes are vital biological catalysts with essential action in the metabolism of all living beings. Moreover, enzymes have a very significant role in various industrial sectors, including baking, brewery, detergents, textile, pharmaceutical, animal feed, cellulose pulp biobleaching, biofuels and others (Polizeli et al., 2005; 2009; 2011). Industrial enzymes are especially produced from microorganisms, such as bacteria, yeasts and filamentous fungi.

Here, we describe some basic knowledge about microbial enzymes with potential application in industry, their properties and some biochemical methods for the detection of amylase, pectinase, proteases, and xylanase activities on polyacrylamide gel electrophoresis. These methods are more qualitative procedures than quantitative, once they may be used to confirm the electrophoretic homogeneity of purified enzymes through chromatographic process. In addition, we described the principle of each method approached in this chapter, which grants a better understanding of each procedure.

1.1 Overall considerations about industrial enzymes

Enzymes are optimum biological catalysts present in all living beings, and they, under adequate conditions, catalyze in their active sites the natural substrates from the metabolic route reactions. Quite often, the metabolic enzymes act in a sequence, and the product generated in a reaction becomes the substrate for the following phase, as diagrammed:

$$\text{Substrate 1} \xrightarrow{\text{enzyme 1}} \text{Product 1} + \text{Enzyme 1} \tag{1}$$

$$\text{Substrate 2} (= \text{Product 1}) \xrightarrow{\text{enzyme 1}} \text{Product 2} + \text{Enzyme 2} \tag{2}$$

Some properties of the enzymes make them excellent competitors against traditional chemical catalysts, due to their great catalytic efficacy (kcat), considering that the main objective of any biological transformation is to obtain a short-term high conversion of substrate into product. Besides this characteristic, enzymes present a high specificity and selectivity, according to their metabolic role, acting in optimum conditions of pH and temperature. Also, they do not pose damage to the environment because they are biodegradable.

Applied biological catalysis had its origins in ancient China and Japan, in the manufacturing of foods and alcoholic beverages in which amylases and proteases of vegetal and microbial origins were employed. It dates back to the end of the XIX century with the introduction of standardized preparations of rennilases in the production of cheese. After this period, the implementation of new industrial applications of enzymes was slow, arising intensively in the last 40 years.

The main reasons for the current importance of enzymes in the industrial scenario are due to the development of application processes of proteases in detergents, pectinases in juices and glucoamylase in the production of glucose from starch. Additionally, the employment of recombinant DNA techniques allowed the obtainment of high productivity and the most suitable design of enzymes.

Following, we present the application of enzymes in several technical sectors justifying the development of methods for the visualization of enzymatic activities in electrophoresis gels.

1.2 Some industrial enzymes and their applications

Proteases, amylases, cellulases and lipases are used in the preparation of detergents and that is the greatest industrial market. Enzymes are added to increase wash efficiency. For so, they require lower usage temperatures, reduce wash periods and agitation costs. They act on both proteic and starchy residuals as well as on fats; they enhance clothes softness and restore color brightness. They require, in general, suitable thermostability and activity in alkaline pH.

Animal feed – the addition of cellulases, xylanases, proteases, lipases, ligninases and phytases in ruminants and monogastrics foods leads to the digestibility of grass and forage, reduces pollutants, decreases the release of carbon dioxide, soluble carbon hydrates and phosphorus. β-glucans increase food viscosity decreasing starch digestibility, but the addition of β-glucanases increases food assimilation, resulting in weight gain (Facchini et al., 2011a, 2011b).

Swine and poultry feed needs the addition of phytase, a phosphatase that acts on acid or alkaline media and dephosphorylates phytic acid releasing phosphorus to the environment (Haefner et al., 2005). That can damage the environments where the soil contains plenty of phosphate, as it is observed in Europe.

Xylanases and ligninases may also take part in the biobleaching of the cellulose pulp for the manufacturing of paper, whereas cellulases are used for the modification of the textile fiber properties, giving them the pre-wash effect. Those three enzymatic systems participate in the sugar cane hydrolysis for the bioethanol manufacturing. Such procedure has been widely adopted in many countries like Brazil, which has a number of flexpower vehicles (Betini et al., 2009; Michelin et al., 2009, 2011).

Proteases and lipases also have a role in the dairy industry, acting in the production of cheese (Gupta et al., 2002). Chemokine, extracted from the stomach of calves, acts in the Milk coagulation, leading to the formation of cheese. Lipases give the aroma and hot flavor in cheese (Hasan et al., 2006). Still in the food industry, pectinases and cellulases increase the extraction of oils through pressing (coconut, sunflower, soybean, olive, etc.).

One sector that has been widely economically explored is the application of amylases in processes of starch saccharification (Silva et al., 2009a, 2009b). For so, there is the need of several enzymes such as α-amylase, which forms maltooligosaccharides; glucoamylases and β-amylases, which hydrolyze starch to glucose and maltose and the glucosyltransferases with the production of cyclodextrins. With synergistic action of the amylolytic system, there is the production of maltose syrup used in breweries, as well as the glucose syrup which is preferably converted by glucose isomerase to fructose syrup, due to the high sweetening power. Such compounds may be used in the manufacturing of sauce, child feeding, gums, candies, ice-cream, pharmaceutical products, canned products etc. Amylases also act in the textile industry to remove starch added to cotton to increase resistance (Gupta et al., 2003).

In the pharmaceutical industry, the use of enzymes is increasingly growing, reflecting the *in vivo* catalysis potential. We can highlight the use of pancreatin, obtained from the swine pancreas, which is used as adjuvant in the digestive process of people who have genetic disorders leading to digestive problems or who, due to surgical removal of the pancreas or precocious aging, present digestive problems. Many enzymes are used as therapeutical agents, such as asparaginase and glutaminase, collagenase, hyaluronidase, ribonuclease, streptokinase, uricase and uroquinase (Prakashan, 2008).

Semi synthetic penicillins (ampicillin and amoxicillin) were launched in the market to replace penicillin, given the acquired resistance of some microorganisms. All kinds of penicillin have the same basic structure: 6-aminopenicillanic acid (6-APA), a thiazolidine ring bound to a beta lactam that takes a free amino group. In the synthesis of semi synthetic penicillin there is the enzymatic hydrolysis of G penicillin with the penicillin G acylase. After purified and concentrated, the 6-APA released in the hydrolysis is used as an intermediary in the synthesis of amoxicillin or ampicillin (Cabral et al., 2003).

The enzymes can be used in the cosmetics industry in creams against skin aging and acne, buffing cream, oral hygiene and hair dying.

Enzymes may be used in analytical applications, due to their high specificity, identifying substances in complex mixtures such as blood, urine and other biological fluids. They participate in tests for glucose (glucose oxidase), urea, amino acids, proteins, ethanol, etc (Godfrey & West, 1996).

In fine chemistry, the list of compounds produced by enzymatic biocatalysis is huge, and we highlight, as an example, vitamin C and several L-amino acids. Acrylamide is used as a monomer in the production of polyacrylamide, widely used as flocculating polymer. Acrylamide was initially produced chemically, but the technological disadvantages such as the formation of toxic residuals and the costs in the purification process made the enzymatic way a viable process.

2. Some industrial enzymes and their applications

2.1 Preparation of polyacrylamide gel electrophoresis for activity enzymatic

The detection of enzymatic activities for industrial use, in electrophoresis gel, happens when PAGE (polyacrylamide gel electrophoresis) is employed, which is an electrophoresis performed in non-denaturing conditions. The Figure 1 illustrates some steps used to preparation of the electrophoresis gel.

In this kind of procedure, there is not preferably the addition of the sodium dodecyl sulfate – SDS detergent, β-mercaptoethanol or another reducing agent, such as dithiothreitol - DTT or urea. Also, the protein samples in its native form (not denaturated) are not boiled before their application in the gel because enzymes will lose its activities, if denaturated. The enzymatic activity may also be detected in gels of the SDS-PAGE type, if the samples were not boiled or added by any reducing agent that denatures the protein. Sodium dodecyl sulfate (SDS) is an anionic surfactant whose role is to bestow the proteins with uniform load density. SDS presents a high negative load and a hydrophobic tail that interacts with the polypeptidic chains in an approximated ratio of 1.4 g of SDS for each gram of protein, making them negatively loaded. In the lack of SDS, the proteins with equal mass may migrate differently in the pores of the gel due to the load differential of their tridimensional structures.

PAGE may be performed in a pH 4.5 or 8.9, depending on the isoelectric point - pI, of the sample under study. In order to accomplished zymograms, it is performed SDS-PAGE; however, the samples generally correspond to a crude extract or a partial purified extract which are either not boiled or added by β-mercaptoethanol, DTT or urea.

2.2 Preparation of the sample for application in electrophoresis

The preparation of proteic solutions for the application in electrophoresis is an important phase. It is important to highlight once again that most enzymes used industrially have microbial origin (fungi, yeasts or bacteria) and also that normally, their synthesis is followed by the elimination of a number of primary and secondary metabolites produced by the very microorganisms, as well as other compounds present in the cultivation medium, such as vitamins, salts, carbohydrates, amino acids and peptides.

In order to avoid the interference of such factors in the electrophoresis, especially when there is the application of enzymatic extracts without previous purification, it is necessary to pay close attention to the type of sample that is being prepared.

Below are some measures and precautions that must be adopted:

i.   Dialysis: This procedure aims at the removal of substances with smaller molar mass, such as salts, carbohydrates and amino acids, which may interfere in the electrophoresis quality;

ii. Attention to the concentration of the sample applied in the electrophoresis. In general lines, around 10 µg of proteins is necessary for a good visualization in the electrophoresis gel after the dying phase. For the detection of the enzymatic activity, considerable enzymatic levels are necessary. If the protein solution presents a lower concentration, the application of any procedure for the concentration of proteins is necessary, such as lyophilization, use of filtering membranes with defined molar mass, ammonium sulfate precipitation and even the use of solvents (ice acetone or ethanol). We must bear in mind that those two last processes need an additional dialysis for the removal of the ammonium sulfate or carbohydrates when the precipitation happens through the action of solvents.

iii. For the detection of activity in electrophoresis gel, the run must take place in low temperatures, such as a refrigerator or a cold chamber.

iv. Attention must also be paid to the native load of proteins and the separation must depend only on its molar mass. For so, proteins may be mixed with SDS, becoming negatively loaded, as it has already been described.

2.3 Electrophoresis separation techniques

2.3.1 Electrophoresis in non-denaturing conditions (PAGE)

Electrophoresis is going to be performed in pH (4.5 or 8.9), in a polyacrylamide gel that may range from 5 to 15%, depending on the size and the load of the protein under study. For proteins loaded negatively, the running buffer will consist of Tris-HCl and glycine, pH 8.9. For proteins loaded positively, there is going to be a buffer with β-alanine and glacial acetic acid, pH 4.5. Both procedures must be performed at 4°C. Table 1 indicates necessary volumes to obtain PAGE gels with different concentrations.

2.3.1.1 PAGE for acid proteins (-), (Davis, 1964)

Solution A

Tris-HCl	9.75 g
HCl (1 M)	12 mL
TEMED	0.05 mL
Distilled water	25 mL
pH adjusted for pH 8.7	

Solution C

Acrylamide	9.6 g
Bis-acrylamide	0.32 g
Distilled water	20 mL
pH adjusted for pH 8.9	

Solution G

Ammonium persulfate	0.007 g
Distilled water	5 mL

Dye
Bromophenol blue
Glycerol

Running buffer
Tris-HCl 50mM
glycine 36mM, pH 8.9

Preparation of the samples and markers:

- In a sterile microtube, put 18 μL of proteic sample and 2 μL of running buffer Tris-HCl 50mM and glycine 36mM, pH 8.9;
- Add 2 μL of the dying solution Bromophenol blue 0.1% and 4 μL of glycerol.

2.3.1.2 PAGE alkaline proteins (+), Reisfeld et al. (1962)

Solution A

KOH (1 M)	48 mL
Glacial acetic acid	17.2 mL
TEMED	4 mL
Complete with distilled H_2O	100 mL
pH adjusted for 4.3	

Solution C

Acrylamide	19.2 g
Bis-acrylamide	0.54 g
Complete with distilled H_2O	40 mL
pH adjusted for 4.5	

Solution G

ammonium persulfate	0.28 g
Complete with distilled H_2O	100 mL

Dye
Methyl green
Glycerol

Sample preparation:

- In a sterile microtube put 18 μL of sample and 2 μL of running buffer consisting of 31.2 g of β-alanine, 8 mL of glacial acetic acid and an amount of distilled water sufficient to reach the volume of 1000 mL;
- Add 2 μL of the solution Methyl green 0.1% and 4 μL of glycerol.

Solution	Polyacrylamide concentration (%)							
	4	5	6	7	8	10	12	14
A (mL)	1.05	1.05	1.05	1.05	1.05	1.05	1.05	1.05
C (mL)	0.71	0.876	1.05	1.22	1.49	1.77	2.10	2.45
H_2O (mL)	6.65	6.47	6.30	6.13	5.97	5.60	5.25	4.95

Table 1. Preparation of 8.4 mL of PAGE in different concentrations.

- Set up the electrophoresis bowl;
- In a Becker, add the solutions A, C and H_2O. Mix;

- Add 50 µL of ammonium persulfate 10% and quickly put the mixture in the electrophoresis bowl;
- Place the comb;
- Wait until the gel solidifies and apply the samples (up to 30 µL) in different lanes;
- Connect the energy cables in the respective jacks of the energy source adjusted to 70 mAmps and 120 Volts;
- Turn on the source and wait for the samples to run throughout the gel before turning it off;
- Remove the gel carefully and process it.

2.3.2 Electrophoresis gel SDS-PAGE, Laemmli (1970)

The electrophoresis must be performed in pH 8.9 and in the presence of SDS (sodium dodecyl sulfate), with the gel concentration ranging from 5 to 15%. Table 2 indicates necessary volumes to obtain SDS-PAGE gels with different concentrations.

Solution A

Tris-HCl	36.5	g
TEMED	230	µL
Distilled water	9	mL

Adjust the pH with HCl concentrated for pH 8.9 and store it at 4-6°C.

Solution C

Acrylamide	28	g
Bis-acrylamide	0.74	g
Distilled water	100	mL

Store at 4-6°C in a glass flask with Amberlite resin due to the degradation of acrylamide in acid and ammonia.

Solution E

| SDS | 0.21 | g |
| Distilled H_2O | 100 | mL |

Store at room temperature.

Sample buffer

| Tris-HCl | 0.755 | g |
| Glycerol | 1 | mL |

Dissolve with 17.5 mL of distilled H_2O and adjust the pH to 6.75 with concentrated HCl and add:

SDS	2	g
Bromophenol blue	0.001	g
Distilled water	100	mL

Freeze aliquots for a further use.

Run Buffer

Tris	3.025	g
glycine	14.4	g
SDS	1.0	g

Dissolve in an amount of distilled water sufficient for 1000mL, pH 8.9.

Preparation of samples and markers:

- In a sterile microtube, place a sample buffer and a molar weight marker in the ration of 1:1;
- In a sterile microtube, place the buffer sample and the sample in the ratio of 1:3.

Solution/Reagent	Gel concentration (%)					
	5	6	7	8	10	12
A (mL)	1	1	1	1	1	1
C (mL)	1.4	1.7	2	2.3	2.85	3.45
E (mL)	3.8	3.8	3.8	3.8	3.8	3.8
H_2O (mL)	1.8	1.5	1.2	0.9	0.35	-
Ammonium persulfate(g)	0.00425	0.00425	0.0057	0.006	0.007	0.00708

Table 2. Preparation of SDS-PAGE

- Set up the electrophoresis bowl;
- In a Becker, add the solutions A, C, E and H_2O and mix;
- Add the ammonium persulfate and quickly place the solution in the electrophoresis bowl; Place the comb;
- Wait until the gel solidifies and apply the patterns (0.5-5 µL) and the samples (up to 30 µL) in different lanes;
- Connect the energy cables in the respective jacks of the source regulated for 70 mAmps and 120 Volts;
- Connect the source and wait until the samples run throughout the gel before turning it off;
- Remove the gel carefully and process it.

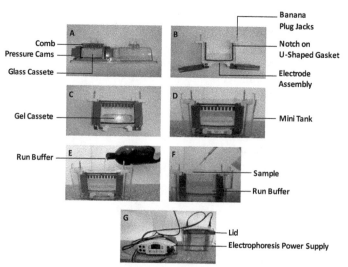

Fig. 1. Preparation and development of vertical polyacrylamide electrophoresis gel using Bio-Rad™ system. (A) and (B) accessories for stacking gel. (C) and (D) preparation of the electrophoresis gel; (E) addition of run buffer; (F) application of the samples; (G) development of the electrophoresis.

3. Methods for specific enzymes detection in electrophoresis gel

3.1 Amylase

The method consists of the conduction of an electrophoresis in polyacrylamide gel polymerized with starch 0.5%. The electrophoretic run is performed in the pH adequated to the amylase isoelectric point. After the end of the run, the gel must be immersed in the suitable temperature and buffer during at least one hour. The gel is going to be revealed with a solution of iodine (I_2 10 mM) and potassium iodide (KI 14 mM) until the appearance of activity bands. Fig. 2A illustrates the activity of α-glucosidase, one of the enzymes of the amylolytic system, which leads to the formation of glucose as end product (Aquino et al., 2001; 2003; Silva et al., 2009a, 2009b).

3.2 Pectinase

Method I – After conducting a PAGE 4.5 or 8.9 (depending on the enzyme pI), the gel containing the enzyme must be incubated with the substrate – a solution containing citric pectin or sodium polypectate 1% in the suitable buffer of the enzyme under study. In Fig. 2B there was the use of 1% of sodium polypectate in a sodium acetate buffer 100mM, pH 4.0 and incubation at 50°C (enzyme optimum temperature), for 2 hours for the dying with 0.02% Ruthenium red [($Ru_3O_2(NH_3)_{14})C_{16}.4H_2O$)], a dye capable of interacting with the pectic substates (Sterling, 1970). Thus, in the region where the protein migrated to and hydrolyzed the substrate, there is a halo with a whitened coloration that contrasts against the rest of the red-colored gel.

Method II – The citric pectin must be dissolved in gel buffer with the aid of a magnetic agitator, followed by the addition of acrylamide, bis-acrylamide and TEMED solutions. Crystals of ammonium persulfate are added immediately before the plate gel is overflown. After the run, incubate the gel for 1-2 hours with 100 mL of malic acid 0.1M, at 4°C, in order to cause a gradual change to pH 3.0. Such period allows the enzyme to interact with the pectin polymerized in the acrylamide gel in its suitable pH range. Wash with distilled water and color in Ruthenium red 0.02%, during 30 to 120 min. Wash with distilled water.

Result: against a redish gel, it is possible to notice the polygalacturonase activity due to the formation of clear, opaque or colorless areas.

3.3 Xylanase

In order to detect the xylanase activity in gel, the polyacrylamide must be polymerized with 0.5% xylan dissolved in the buffer of the electrophoresis to be performed (PAGE 4.5 or 8.9, depending on the isoelectric point of the enzyme under study). After the electrophoresis run the gel must be incubated in the temperature and in the reaction buffer which is mostly suitable for the xylanase under study for at least 1 hour. After this period, the gel is going to be stained with 1% Congo red ($C_{32}H_{22}N_6Na_2O_6S_2$) a sodium salt of benzidinediazo-bis-1-naphtylamine-4-sulfonic acid. Thus, in the region where the protein migrated to and hydrolyzed the substrate, there is a halo with a whitened coloration that contrasts against the rest of the red-colored gel. (Fig. 2C) (Sandrim et al., 2005; Damásio et al., 2011).

Xylanases may also be observed through zymograms (Fig. 2D). For so, the samples must be applied in SDS-PAGE without the addition of β-mercaptoethanol, any other reducing agent and without boiling. The advantage of the activity detection through zymograms is that effectively all active isoforms present there may be detected, once that regardless on the isoelectric point, all proteins will migrate in the gel, because of their molar mass.

In this case, the SDS gel must be performed with the running buffer commonly adopted in the protocols and, after the electrophoresis run, the proteins will have to be transferred to another gel composed by agarose and xylan, a substrate that is specific to the activity to be detected. Such transferring happens when there is a kind of "sandwich" with the polyacrylamide gel and the agarose + substrate gel. The transferring must happen overnight. After this period, the agarose + substrate gel, now also with the proteins to be analyzed, will have to be incubated in the buffer that is suitable for the isoforms under study, during one hour, following with the coloration suitable for the activity detection of the enzyme analyzed.

3.4 Proteases

The zymography may also be applied for proteases. It is a simple, quantitative and functional technique to analyze the activity of proteases (Leber & Balkwill, 1997). It consists basically of two stages, the separation through electrophoresis, followed by the activity detection of the enzyme in polyacrylamide gel, in non-reducing conditions (without treatment with DTT or β-mercaptoethanol) (Dong-Min et al., 2011). This technique has been used to evaluate the level of proteases in tissues or biological fluids, and it bears the advantage of distinguishing different kinds of enzymes due to the characteristic of mobility that each enzyme presents (Raser et al., 1995). The protease activity in zymography is observed as a clear band, indicating the substrate proteolysis after colored with Coomassie Brilliant Blue (Kim et al., 1998).

This methodology is widely employed for the detection of Matrix metalloproteinases (MMPs). However, it can also be employed for other types of proteases, with the need of adjustments in the methodology, such as the substitution of the substrate, generally gelatin for casein. Unfortunately, the zymography with casein is very little sensitive, when compared to the zymography with gelatin. Besides, casein migrates in the gel during the electrophoresis due to its relative low molar mass. That results in two clearly defined areas in the gel: the upper part, which still contains excess casein and the lower part, with less casein (Beurden and Von denHoff, 2005).

For the detection of proteases (Fig. 2E), the sample must be diluted in the sample buffer (5x) of the gel (0.4M Tris-HCl, pH 6.8; 5% SDS; 20% glycerol 0.03% Bromophenol blue). The samples cannot be boiled, because this process denatures the enzyme and it will no longer present activity (Kleiner & Stetler-Stevenson, 1994). The electrophoresis of the samples containing the protease must be performed according to Laemmli (1970). The gel concentration must be prepared according to the molar mass of its protease. The electrophoresis may be performed in constant 100V for 1-2 hours at 4°C.

For the development of the proteolytic activity, the gel must be incubated with 70 mL of buffer with the appropriate reaction pH, for 5 min., 4°C, 100 rpm. Following, the buffer must be removed and the gel must be incubated with 70 mL of Triton X-100 2.5% prepared in the reaction buffer. The gel must be kept at 100 rpm, for 30 min, 4°C. This step is for the removal of the SDS and the activation of the protease. Afterwards, the excess Triton X-100 must be removed. For so, add 70 mL of buffer with the appropriate reaction pH, incubate for 30 min., at 4°C, 100rpm.

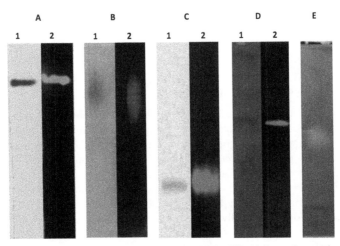

Fig. 2. Activity gels for different enzymes in SDS-PAGE 10%. (A1) α-glucosidase (a type of amylase) revealed with Coomassie Blue; (A2) α-glucosidase revealed with 10 mM Iodine solution and 14 mM potassium iodide; (B1) polygalacturonase revealed with silver solution; (B2) polygalacturonase activity revealed with 0.02% ruthenium red; (C1) xylanase revealed with silver solution; (C2) xylanase activity revealed with Congo red; (D1) Zymogram for xylanase revealed with silver solution and (D2) Congo red; (E) protease activity revealed with Coomassie Blue.

Following, remove the buffer and add the casein solution at 3%, prepared in a buffer with the enzyme reaction pH. The gel must be incubated for 30 min, at 4°C, for the diffusion of the casein to the gel. After that, the gel must be bathed at the enzyme reaction temperature for the period of 1-2 hours, so that the enzymatic reaction occurs (such period may be adjusted according to each enzyme and concentration).

The excess casein must be removed by bathing the gel for 5 times with distilled water at room temperature (García-Carreño et al., 1993 and Kleiner & Stetler-Steveson, 1994) with modifications.

For the coloration, the gel must be stained with a solution containing 40% ethanol, 10% acetic acid and 0.1% Coomassie brilliant blue R-250 (García-Carreño et al., 1993). In this stage, the gel shows a blue bottom and where the hydrolysis took place, there will be a white halo. The gel needs to have the excess Coomassie brilliant blue removed; hence, it will have to be discolored with a discoloring solution composed by 40% ethanol and 10% acetic acid (García-Carreño et al., 1993). The clear zones over the blue bottom indicate the protease activity.

Some enzymes need activators, such as Ca^{2+}, DTT and EDTA to show their activity. Those can be added together with the substrate. Other substrates, such as hemoglobin, bovine serum albumin, gelatin and collagen, may have their coloration improved through the use of other dyes, as for example, Amide black (García-Carreño et al., 1993).

4. Conclusion

The biological catalyzers present several advantages over their chemical similar, particularly, the regio and stereoselectivity that lead to the formation of products which are enantiomerically pure and in conformity with the norms established for the food, pharmaceutical and agriculture industry. The enzymes are efficient under the energetic point of view, operating in controlled pH, temperature and pressure. The development of the recombinant DNA technology enabling the expression of enzymes in different hosts has resulted in the production of more efficient biological catalyzers.

Several methods are attributed to enzymatic determinations. The most widely used are the colorimetric ones, where the reactions occur with specific substrates, generally leading to the formation of colored products, which can be easily quantified in spectrophotometers or through acrylamide gel electrophoresis, non-denaturing conditions. The detection of enzymatic activity through PAGE involves the migration potential of the enzyme in gel, which is influenced by the molar mass of the protein and its loads in specific pH. Thus, the visualization of the enzymatic activity in gels is seen as an advantaging condition, once that it is possible to consider that hardly did that enzyme migrate together with interferents or contaminating substances that could be in a crude extract and that could lead to errors in the enzymatic levels. PAGE for enzymatic activities can be considered as an elegant method that has been increasingly employed in researches.

The zymogram technique demands low enzyme concentrations, which can reach the order of nanograms. Several adaptations of substrates may be made in this technique, because the majority of substrates used are low-cost and yield good results. With this technique, we can infer how many types or isoforms of enzymes of a same class are present in the crude extract, for example, what the molar mass is and even in some cases, the quantification.

Hence, the proposal for the optimization of stages of enzymatic activity detection in electrophoresis must be conducted for each specific extract to be studied, yielding in this way high sensitivity and precision.

5. Acknowledgment

Dr. Maria de Lourdes T.M. Polizeli and Dr Hamilton Cabral are Research Fellows of CNPq. Dr Simone C. Peixoto-Nogueira, Dr Tony M. Silva and Dr Alexandre Maller are recipients of FAPESP Fellowship. This chapter concerns research data of the project National System for Research on Biodiversity (SISBIOTA-Brazil, CNPq 563260/2010-6/FAPESP number 2010/52322-3). We thank Dr. Abilio for the English technical assistance.

6. References

Aquino, A.C.M.M., Jorge, J.A., Terenzi, H.F. & Polizeli, M.L.T.M. (2001).Thermostable glucose-tolerant glucoamylase produced by the thermophilic fungus *Scytalidium thermophilum* 15.1. *Folia Microbiologica*, Vol. 45 (6), pp. 11-16.

Aquino, A.C.M.M., Jorge, J.A., Terenzi, H. F. & Polizeli, M.L.T.M. (2003). Studies on a thermostable alpha-amylase from the thermophilic fungus *Scytalidium thermophilum*. *Applied Microbiology and Biotechnology*, Vol. 61, pp. 323-328.

Betini, J.H.A., Michelin, M., Peixoto-Nogueira, S.C., Jorge, J.A., Terenzi, H.F. & Polizeli, M.L.T.M. (2009). Xylanases from *Aspergillus niger*, *Aspergillus niveus* and *Aspergillus ochraceus* produced under solid-state fermentation and their application in cellulose pulp bleaching. *Bioprocess and Biosystems Engineering*, Vol. 32, pp. 819-824.

Beurden Sanoek-van, P.A.M. & Von den Hoff, J.W. (2005). Zymographic techniques for the analysis of matrix metalloproteinases and their inhibitors. *BioTechniques*, Vol. 38, pp. 73-82.

Cabral, J.M.S., Aires-Barros, M.R. & Gama, M. (2003). *Engenharia Enzimática*, Lidel edições Técnicas, 972-757-272-3, Lousã.

Damásio, A.R.L., Silva, T.M., Almeida, F.B.R., Squina, F., Ribeiro, D.A. R., Leme, A.F.P., Segato, F., Prade, R.A., Jorge, J.A., Terenzi, H.F. & Polizeli, M.L.T.M. (2011). Heterologous expression of an *Aspergillus niveus* xylanase GH11 in *Aspergillus nidulans* and its characterization and application. *Process Biochemistry* Vol. 46, pp. 1236-1242.

Davis, B.J. (1964). Disc Electrophoresis. II. Method and application to human serum proteins. *Annals of the New York Academy of Sciences*. Vol.28, 121, pp. 404-427.

Dong-Min, C., Kim, K.E., Ahn, K.H., Park, C.S., Kim, D.H., Koh, H.B., Chun, H.K., Yoon, B.D., Kim, H.J., Kim, M.S. & Choi, N.S. (2011). Silver-stained fibrin zymography: separation of proteases and activity detection using a single substrate-containing gel. *Biotechnology Letters*, Vol. 33, pp. 1663-1666.

Facchini, F.D., Vici, A.C., Reis, V.R., Jorge, J.A., Terenzi, H.F., Reis, R.A. & Polizeli, M.L.T.M. (2011a). Production of fibrolytic enzymes by *Aspergillus japonicus* C03 using agro-industrial residues with potential application as additives in animal feed. *Bioprocess and Biosystems Engineering*, Vol. 34 (3), pp. 347-55.

Facchini, F.D.A., Vici, A.C., Benassi, V. M., Freitas, L.A.P., Reis, R.A., Jorge, J.A., Terenzi, H.F. & Polizeli, M.L.T.M. (2011b). Optimization of fibrolytic enzyme production by *Aspergillus japonicus* C03 with potential application in ruminant feed and their effects on tropical forages hydrolysis. *Bioprocess and Biosystems Engineering*, Vol. 34, pp. 1027–1038.

García-Carreño, F.L.; Dimes, L.E. & Haard, N.F. (1993). Substrate-gel electrophoresis for composition and molecular weight of proteinases or proteinaceous proteinase inhibitors. *Analytical Biochemistry* Vol. 214, pp. 65-69.

Godfrey, T & West, S. (1996). *Industrial Enzymology* (2^0 ed.), Stokton Press, 1561591637 0333594649, New York.

Gupta, R., Beg, Q.K. & Lorenz, P. (2002). Bacterial alkaline proteases: molecular approaches and industrial applications. *Applied Microbiology and Biotechnology*, Vol. 59, (April 2002), pp. 15–32, 0175-7598.

Gupta R, Gigras, P., Mohapatra, H., Goswam,i V.K. & Chauhan, B. (2003). Microbial α-amylases: a biotechnological perspective. *Process Biochemistry*, Vol 38, (june 2003), pp. (1599-616), 1359-5113.

Haefner, S., Knietsch, A., Scholten, E., Braun, J., Lohscheidt, M. & Zelder, O. (2005). Biotechnological production and applications of phytases. *Applied Microbiology and Biotechnology*. Vol. 68, (5), (September 2005), pp. 588-597, 1432-0614.

Hasan, F., Shah, A.A.A. & Hameed, A. (2006). Industrial applications of microbial lipases. *Enzyme and Microbial Technology*. Vol. 39, 2, (june 2006), pp. 235-251, 0141-0229.

Kim, SH., Choi, N.S. & Lee, W.Y. (1998). Fibrin zymography: a direct analysis of fibrinolytic enzymes on gels. *Analytical Biochemistry*, Vol. 263, pp. 115-116.

Kleiner, D.E. & Stetler-Stevenson, W.G. (1994). Quantitative zymography: detection of picogram quantities of gelatinases. *Analytical Biochemistry*, Vol. 218, pp. 325-329.

Laemmli, U.K. (1970). Cleavage of structural protein during the assembly of the head of bacteriophage T4. *Nature*, Vol. 227, pp. 680-685.

Leber, T.M. & Balkwill, F.R. (1997). Zymography: a single-step staining method for quantification of proteolytic activity on substrate gels. *Analytical Biochemistry*, Vol. 249, pp. 24-28.

Michelin, M., Peixoto-Nogueira, S.C., Betini, J.H,A., Silva, T.M., Jorge, J.A., Terenzi, H.F. & Polizeli, M.L.T.M. (2009). Production and properties of xylanases from *Aspergillus terricola* Marchal and *Aspergillus ochraceus* and their use in cellulose pulp bleaching. *Bioprocess and Biosystems Engineering*, Vol. 33, (7) (December 2009), pp. 813-821, 1615-7591.

Michelin, M., Silva, D. P., Ruzene, D. S., Vicente, A. A., Jorge, J. A., Terenzi, H. F., & Polizeli, M. L. T. M. (2011). Production of xylanolytic enzymes by *Aspergillus terricola* in stirred tank and airlift tower loop bioreactors. *Journal of Industrial Microbiology & Biotechnology*, DOI 10.1007/s10295-011-0987-7.

Polizeli, M.L.T.M., Rizzatti, A.C.S., Monti, R.; Terenzi, H.F., Jorge, J.A. & Amorim, D.S. Xylanases from fungi: properties and industrial applications. Review. (2005). *Applied Microbiology and Biotechnology*, Vol 67, pp. 577-591.

Polizeli, M.L.T.M. (2009). Properties and commercial applications of xylanases from fungi. *In: Mycotechnology: Current Trends and Future Prospects*, M.K.Rai (Editor). I.K. International Publisher, New Delhi., chapter 4, pp. 82-101.

Polizeli, M. L. T. M., Corrêa, E. C., Polizeli, A. M., & Jorge, J. A. (2011). Hydrolases from microorganisms used for degradation of plant cell wall and bioenergy production. *In: Routes to Cellulosic Ethanol*. Editors: Marcos S. Buckeridge & Gustavo H. Goldman. Springer Verlag, chapter 8, pp. 115-134.

Prakashan, N. (2008). *Biochemistry Basics and Applied* (3⁰ edition), Rachana Enterprises, 81-85790-52-3.

Raser, K.J., Posner, A. & Wang, K.K.W. (1995). Casein Zymography: A method to study μ-Calpain, m-Calpain, and their inhibitory agents. *Archives of Biochemistry and Biophysics*, n° 1, v. 319, pp. 211-216.

Reisfeld, R.A., Lewis, U.J. & Williams, D.E. (1962). Disk electrophoresis of basic proteins and peptides on polyacrylamide gels. *Nature* Jul. Vol. 21 (195), pp. 281-283.

Sandrim, V.C., Rizzatti, A.C.S., Terenzi, H.F., Jorge, J.A., Milagres, A.M.F. & Polizeli, M.L.T.M. (2005). Purification and biochemical characterization of two xylanases produced by *Aspergillus caespitosus* and their potential for the bleaching of kraft pulp. *Process Biochemistry* Vol. 40(5), pp. 1823-1828.

Silva, T.M., Maller, A., Damásio, A.R.L., Michelin, M., Ward, R.J., Hirata, I.Y., Jorge, J.A., Terenzi, H.F. & Polizeli, M.L.T.M. (2009). Properties of a purified thermostable glucoamylase from *Aspergillus niveus*. *Journal of Industrial Microbiology and Biotechnology*, Vol. 36, (august 2009a) pp. 1439-1446, 1367-5435.

Silva, T.M., Michelin, M., Damásio, A.R.L., Maller, A., Almeida, F.B.D.R., Ruller, R., Ward, R.J., Rosa, J.C., Jorge, J.A., Terenzi, H.F. & Polizeli, M.L.T.M. (2009b). Purification and biochemical characterization of a novel alpha-glucosidase from *Aspergillus niveus*. *Antonie van Leeuwenhoek*, Vol. 96, pp. 569-578.

Sterling, C. Crystal-structure of ruthenium red and stereochemistry of its pectic stain. (1970). *American Journal of Botany*, Vol. 57, pp. 172–175.

6

Molecular Electrophoretic Technique for Authentication of the Fish Genetic Diversity

Tsai-Hsin Chiu, Yi-Cheng Su, Hui-Chiu Lin and Chung-Kang Hsu

1Department of Food Science, National PengHu University of Science and Technology
2Seafood Research and Education Center, Oregon State University
3Penghu Marine Biology Research Center, Fisheries Research Institute, COA, EY
1,3Taiwan
2USA

1. Introduction

Cobia (*Rachycentron canadum*) is the sole representative of their family, the Rachycentridae They are distributed worldwide in tropical and subtropical seas, as the Atlantic and Pacific Oceans (Miao, et al., 2009). There are several species, including cobia, seabream, red progy, snappers, scads and groupers that are raised by cage culture in Taiwan. Among these cage-cultured fishes, cobia certainly takes a leading distribution in both annual total production (81.9%) and total value production (75.4%) as compared to the rest in Taiwan (Fisheries Agency 2006).

Giant grouper (*Epinephelus lanceolatus*) are also found in tropical and subtropical waters from the Indo-Western Pacific Ocean. It is one of the two largest species of groupers in the world. Due to its fast growth and high price, giant grouper currently is regarded as a favorite species for marine culture in Taiwan (Hseu, et al., 2004).

Red coral trout (*Plectropomus leopardus*) a reef-associated fish in Western Pacific, distributed from southern Japan to Australia and eastward to the Caroline Islands (Zhang, et al., 2010). Only few studies concerning population genetics of *Plectropomus leopardus* has been reported.

All of cobia, giant grouper, and red coral trout are high-valued fish market in Taiwan and neighboring countries, including China, Japan, and Vietnam. For the globalization of the seafood industry, seafood authentication and food safety are very important. We must know that the source of fish or accurately species of the fish. Traditional method to distinguish the fish species was observed the external traits. It can cause the error judgment. Today, DNA-based methods are also more frequently employed for food authentication (Lockley and Bardsley, 2000). It has proven to be reliable, sensitive and fast for many aspects of fish species and food authentication. Asensio et al. (2009) were suggesting that the species-specific PCR method could be potentially used by regulatory agencies as routine control assay for the commercial grouper fillets authentication. PCR-based methods commonly used for fish species identification include PCR-sequencing, random amplified polymorphic DNA (RAPD), inter simple sequence repeat (ISSR). Those methods are simplicity, specificity and sensitivity.

Recently, many researchers have reported for the assessment of genetic structure of aquaculture species such as red mullet (Mullus barbatus), Tropical abalone (Haliotis asinina), and suminoe oyster (Crassostrea ariakensis) using several kinds of molecular markers (Garoia et al., 2004; Tang et al., 2004; Zhang et al., 2005; Maltagliati, et al., 2006), including RAPD, ISSR, AFLP, and RFLP methods. Many molecular methods are available for studying various aspects of wild populations, captive broodstocks and interactions between wild cultured stocks of fish and other aquatic species (Okumus and Ciftci, 2003). Among those methods, RAPD and ISSR technology were cheaper, simple, and fast. And just only one primer could obtain the different profiles for genomic analysis (Welsh and McClelland, 1990). RAPD is simple, rapid and cheap, it have high polymorphism. RAPD analysis has been used to evaluate genetic diversity for species, subspecies and population identification in common carp (Bártfai, et al., 2003), Indian major carps (Barman, et al., 2003). The microsatellite method was already used to study of genetic diversity of other grouper (Antoro, et al., 2006; Ramirez, et al., 2006; Wang, et al., 2007). Zeng, et al. (2008) have report that genetic analysis of Malaysia and Taiwan wild populations of giant grouper by microsatellite method. Their results were shown polymorphic loci in those populations, but they didn't discriminate the wild and cultivated populations of giant grouper. Beside, genetic markers can be suitable for assessing the differences between culture stocks and wild population and monitoring the changes in the genetic variation (Okumuş, et al., 2003). Monitoring the genetic diversity of natural populations and fish raised in fish hatcheries is fundamentally important for species conservation. Molecular markers can be very useful in this context (Povh et al., 2008).

In our study, we try to identify the seafood products, including cobia, giant grouper, and red coral trout from cultivated and wild populations by molecular markers, and provide the fish population genetic diversity for seafood management and good monitoring for brood stock management.

2. Methods and material

2.1 Fish sampling and genomic DNA extraction

14 giant grouper (*Epinephelus lanceolatus*) and 14 cobia (*Rachycentron canadum*) were collected from Southern Taiwan, as Penghu Island, Kaohsiung, and Pengtung during 2007-2009. Those were selected in different cultured farms and local markets. 14 of red coral trout (*Plectropomus leopardus*) were collected from Penghu Island during 2009-2011. All samples were described in Table 1. All specimens were confirmed in the laboratory.

Approximately 1g of fish muscle tissue samples were cut into small pieces and pulverized in liquid nitrogen. The powdered fish samples were obtained and extracted genomic DNA using the QIAGEN® DNeasy Blood kit (QIAGEN Inc., Valencia, California) according to manufacturer's instructions. The extracted DNA concentration in 200 μl of sterile water and then the quality of DNA were assessed by a Qubit™ Fluorometer (invitrogen, USA). The DNAs were stored at -20ºC until PCR amplifications.

2.2 PCR-RAPD method

A total of 95 RAPD primers were used for PCR, which were shown in Table 2. Those sequences were obtained from University of British Columbia Biotechnology Laboratory,

No.	Species	Sampling sources	No.	Species	Sampling sources	No.	Species	Sampling sources
1	E. lanceolatus	Wild	R1	R. canadum	Wild	71411	P. leopardus	Wild
E2	E. lanceolatus	Cultivated	R2	R. canadum	Wild	71412	P. leopardus	Wild
E3	E. lanceolatus	Cultivated	R3	R. canadum	Cultivated	71413	P. leopardus	Wild
E4	E. lanceolatus	Cultivated	R4	R. canadum	Wild	82011	P. leopardus	Wild
E5	E. lanceolatus	Wild	R5	R. canadum	Cultivated	91321	P. leopardus	Wild
E6	E. lanceolatus	Cultivated	R6	R. canadum	Wild	91711	P. leopardus	Wild
E7	E. lanceolatus	Wild	R7	R. canadum	Cultivated	91712	P. leopardus	Wild
E8	E. lanceolatus	Wild	R8	R. canadum	Cultivated	71421	P. leopardus	Cultivated
E9	E. lanceolatus	Wild	R9	R. canadum	Wild	71422	P. leopardus	Cultivated
E10	E. lanceolatus	Cultivated	R10	R. canadum	Wild	71423	P. leopardus	Cultivated
E11	E. lanceolatus	Cultivated	R11	R. canadum	Wild	81621	P. leopardus	Cultivated
E12	E. lanceolatus	Cultivated	R12	R. canadum	Wild	80622	P. leopardus	Cultivated
E13	E. lanceolatus	Wild	R13	R. canadum	Wild	81623	P. leopardus	Cultivated
E14	E. lanceolatus	Wild	R14	R. canadum	Wild	91322	P. leopardus	Cultivated

Table 1. Specimens of *E. lanceolatus, R. canadum, and P. leopardus* fish analyzed and locality where they were collected

RAPD Analysis Kit (Amersham Pharmacia Biotech, Piscataway, NJ), and Operon primer kit (Operon, Advanced Biotechnologies). DNA amplification was performed in a final volume of 25 μl in the "Gene Amp PCR System 2720" thermal cycler (Applied Biosystems Inc., USA). The reaction mix contained 20 mM Tris-HCl, pH8.0, 50 mM KCl, 2 mM MgCl$_2$, 10 mM dNTPs each (dATP, dCTP, dGTP, dTTP), 20 μM of primer, 2.5 U *Taq*-polymerase (Promega, Co., Wisconsin, USA) and 1 μl of the 10 ng extracted DNA. The pre-amplification PCR procedure was: treatment at 94°C for 5 min, followed by 35 cycles of denaturation at 94°C for 30s, annealing at a primer-specific annealing temperature as table 2 for 30s and extension at 72°C for 30s, and final extension at 72°C for 10 min. The annealing temperature of Cobia was 36°C. A 10 μl of the PCR product were analyzed in a 2 % agarose gel in 0.5 X TBE. The electrophoresis was performed at a constant voltage of 150 V for 150 min and 250 V for 1 min. The gel was stained with ethidium bromide and visualized under UV light.

2.3 PCR-ISSR method

ISSR primers of this study were listed in the Table 3. A total 59 primers were screened. Pre-amplification PCR reaction was conducted in 25 μl reaction containing 12.5 μl PCR master mix (Promega, Co., Wisconsin, USA), 1μl each primer, 1 μl of the 10 ng extracted DNA, and 10.5 μl dH$_2$O. Then, the mixtures were subjected to 94°C for 5 min, followed by 35 cycles of denaturation at 94°C for 30s, 30s at a primer-specific annealing temperature as table 3,

Primer	Sequence of primer (5'-3')	Tm(℃)	references	Primer	Sequence of primer (5'-3')	Tm (℃)	references
RAPD16	GGTGGCGGGA	56	RAPD Primer Set#1, University of British Columbia	RAPD540	CGGACCGCGT	56	Set#6
RAPD17	CCTGGGCCTC	56		RPAD542	CCCATGGCCC	56	
RAPD22	CCCTTGGGGG	56		RAPD563	CGCCGCTCCT	56	
RAPD23	CCCGCCTTCC	56		RAPD571	GCGCGGCACT	56	
RAPD31	CCGGCCTTCC	56		RAPD584	GCGGGCAGGA	56	
RAPD34	CCGGCCCCAA	56		RAPD585	CCCGCGAGTC	56	
RAPD50	TTCCCCGCGC	56		RAPD592	GGGCGAGTGC	56	
RAPD56	TGCCCCGAGC	56		RAPD595	GTCACCGCGC	56	
RAPD63	TTCCCCGCCC	56		RAPD598	ACGGGCGCTC	56	
RAPD64	GAGGGCGGGA	56		RAPD601	CCGCCCACTG	56	Set#7
RAPD65	AGGGGCGGGA	56		RAPD603	ACCCACCGCG	56	
RAPD67	GAGGGCGAGC	56		RAPD606	CGGTCGGCCA	56	
RAPD70	GGGCACGCGA	56		RAPD615	CGTCGAGCGG	56	
RAPD71	GAGGGCGAGG	56		RAPD620	TTGCGCCCGG	56	
RAPD73	GGGCACGCGA	56		RAPD625	CCGCTGCAGC	56	
RAPD81	GAGCACGGGG	56		RAPD626	CCAAGCCCGG	56	
RAPD83	GGGCTCGTGG	56		RAPD640	CGTGGGGCCT	56	
RAPD84	GGGCGCGAGT	56		RAPD647	CCTGTGGGGG	56	
RAPD86	GGGGGGAAGG	56		RAPD769	GGGTGGTGGG	56	Set#8
RAPD87	GGGGGGAAGC	56		RAPD770	GGGAGGAGGG	56	
RAPD88	CGGGGGATGG	56		RAPD771	CCCTCCTCCC	56	
RAPD89	GGGGGCTTGG	56		RAPD772	CCCACCACCC	56	
RAPD94	GGGGGGAACC	56		primer1	GGTGCGGGAA	36	Ready-To-Go. RAPD Analysis Kit (Amersham Pharmacia Biotech, Piscataway, N J)
RAPD95	GGGGGGTTGG	56		primer 2	GTTTCGCTCC	36	
RAPD96	GGCGGCATGG	56		primer 3	GTAGACCCGT	36	
RAPD105	CTCGGGTGGG	56	RAPD Primer Set#2	primer 4	AAGAGCCCGT	36	
RAPD106	CGTCTGCCCG	56		primer 5	AACGCGCAAC	36	
RAPD115	TTCCGCGGGC	56		primer 6	CCCGTCAGCA	36	
RAPD149	AGCAGCGTGG	56		OPA1	CAGGCCCTTC	56	(Operon, Advanced Biotechnologies)
RAPD157	CGTGGGCAGG	56		OPA2	TGCCGAGCTG	56	
RAPD158	TAGCCGTGGC	56		OPA3	AGTCAGCCAC	56	
RAPD173	CAGGCGGCGT	56		OPA4	AATCGGGCTG	56	
RAPD174	AACGGGCAGC	56		OPA5	AGGGGTCTTG	56	
RAPD190	AGAATCCGCC	56		OPA6	GGTCCCTGAC	56	
RAPD196	CTCCTCCCCC	56		OPA7	GAAACGGGTG	56	
RAPD198	GCAGGACTGC	56		OPA8	GTGACGTAGG	56	
RAPD210	GCACCGAGAG	56	Set#3	OPA9	GGGTAACGCC	56	
RAPD211	GAAGCGCGAT	56		OPA10	GTGATCGCAG	56	
RAPD218	CTCAGCCCAG	56		OPA11	CAATCGCCGT	56	
RAPD241	GCCCGACGCG	56		OPA12	TCGGCGATAG	56	
RAPD245	CGCGTGCCAG	56		OPA13	CAGCACCCAC	56	
RAPD270	TGCGCGCGGG	56		OPA14	TCTGTGCTGG	56	
RAPD286	CGGAGCCGGC	56		S514	CAGGATTCCC	56	Portman International (China) Limited, Hong Kong
RAPD287	CGAACGGCGG	56		S1036	AAGGCACGAC	56	
RAPD319	GTGGCCGCGC	56	Set#4	S1040	CCTGTTCCCT	56	
RAPD480	GGAGGGGGGA	56	Set#5	S1042	TCGCACAGTC	56	
RAPD534	CACCCCCTGC	56	Set#6	S1201	CCATTCCGAG	56	
RAPD536	GCCCCTCGTC	56					

Table 2. RAPD primers of PCR amplification

extension at 72oC for 30s, and final extension at 72oC for 5 min before analysis by the electrophoresis as described previously.

Primer	Sequence of primer(5'-3')	Tm(°C)	References
ISSR1	$(GGAC)_3A$	48	Pazza et al. (2007)
ISSR2	$(GGAC)_3C$	48	
ISSR3	$(GGAC)_3T$	48	
ISSR4	$(TGTC)_4$	48	
ISSR5	$(GGAC)_4$	48	
ISSR6	$(GGAT)_4$	48	
ISSR7	$(TAGG)_4$	48	
ISSR8	$(GACA)_4$	48	
ISSR801	$(AT)_8T$	50	Liu et al. (2006)
ISSR817	$(CA)_8A$	48	
ISSR825	$(AC)_8T$	48	
ISSR842	$(CA)_8YG$	52	
ISSR848	$(CA)_8RG$	51	
ISSR850	$(GT)_8YC$	51	
ISSR855	$(AC)_8YT$	50	
ISSR856	$(AC)_8YA$	49	
ISSR858	$(TG)_8RT$	48	
ISSR859	$(TG)_8RC$	52	
ISSR860	$(TG)_8RA$	51	
ISSR888	$BDB(CA)_7$	52	
SAS1	$(GTG)_4GC$	55	Maltagliati et al.(2006)
SAS3	$(GAG)_4G$	55	
UBC809	$(AG)_8G$	55	
UBC811	$(GA)_8C$	55	
UBC827	$(AC)_8G$	55	
IT1	$(CA)_8GT$	55	
IT2	$(CA)_8AC$	55	
IT3	$(GAG)_4AG$	55	
PT1	$(GT)_8C$	55	
ISSR807	$(AG)_8T$	48	UBC Primer Set#9, University of British Columbia
ISSR819	$(GT)_8A$	48	
ISSR822	$(TC)_8A$	48	
ISSR831	$(AT)_8YA$	48	
ISSR834	$(AG)_8YT$	48	
ISSR843	$(CT)_8RA$	48	
ISSR852	$(TC)_8RA$	48	
ISSR861	$(ACC)_6$	48	
ISSR862	$(AGC)_6$	48	
ISSR868	$(GAA)_6$	48	
ISSR871	$(TA)_8RG$	48	
ISSR873	$(GACA)_4$	48	
ISSR877	$(TGCA)_4$	48	
ISSR9	$(GAG)_5RY$	55	Hou et al., 2006
ISSR10	$VBV(CA)_8$	54	
ISSR11	$VDV(GT)_8$	51	
ISSR12	$HVHT(GT)_7$	51	
ISSR13	$(CT)_8A$	49	
ISSR14	$(TG)_8GT$	48	
ISSR15	$(AG)_8TG$	54	
ISSR16	$(TC)_8C$	52	
ISSR17	$(TG)_8G$	53	
ISSR18	$(TG)_6R$	45	

ISSR19	(CA)$_6$RY	44
ISSR20	(GT)$_6$YR	44
ISSR21	(GT)$_6$AY	43
ISSR22	(ACTG)$_4$	52
ISSR23	(GACA)$_4$	48
ISSR24	(CAC)$_6$	57

Table 3. ISSR primers of PCR amplification

2.4 Genetic distances and phylogenetic analysis

Patterns from RAPD and ISSR methods were scored for the presence (1) or absence (0) of clear bands to analyze genetic similarities using the Dice coefficient of similarity. Similarity matrix cluster and phylogenetic analysis was used to reveal association among strains based on the unweighted pair group method with arithmetic averages (UPGMA) using the NTSYSpc software (Numerical taxonomy and multivariate analysis system, version 2.01b, State University of New York, Stony Brook, NY, USA) according to Rohlf (1997).

3. Results

3.1 RAPD method of giant group

A total of 14 giant grouper including cultivate and wild were obtained. Analysis on species of giant grouper in cultivate and wild. For RAPD method amplification, the species, For RAPD method, total 95 of RAPD oligonucletide primers were used to screening the genetic diversity of giant grouper. There are 21 RAPD primers (22.1%) have polymorphic bands. Total 279 bands were generated by those primers and 86 polymorphic bands (31%). The primer RAPD 115 (5'-TTCCGCGGGC-3') was got the more diversity than other primers, have 8 polymorphic bands (Fig 1). The RAPD 245 primer was generated the less bands, only

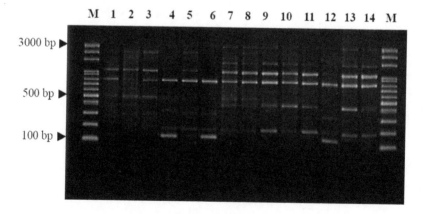

Fig. 1. RAPD profiles of the 14 *E. lanceolatus* fish obtained using the primer 115. Lanes M: Bio-100 bp DNA Ladder. Lanes 1-7: *E. lanceolatus* fish wild population (E1, E5, E7, E8, E9, E13, and E14); Lanes 8-14: *E. lanceolatus* fish cultivate population (E2, E3, E4, E6, E10 E11, and E12).

2 polymorphic bands. The sequence and PCR-RAPD condition were listed in Table 2. All primers were generated bands ranging in size from 100 to 3000 bp. The results shown that the ratio of polymorphic bands were between 13.3~66.7% by 21 RAPD primers. For dendrogram analysis, two groups were identified by RAPD 115 primer (Fig 2). E1, E5, E14, E7, E8, and E9 samples were clustered in group I, which were collected from wild population. Group II, which including E13, E10, E2, E3, E4, E6, E11, and E12 samples. For Group II, all samples were belonged to cultivated populations. For giant grouper, wild (seven samples) and cultivated (seven samples) populations of giant grouper can be discriminated by RAPD method.

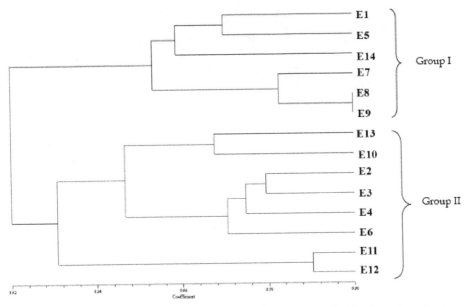

Fig. 2. UPGMA consensus dendrogram of dissimilarity among individuals analyzed using the primer RAPD 115.

3.2 ISSR method of giant group

Results of ISSR analysis, 59 primers were used in this study. According the results of ISSR method, 17 primers (29%) have polymorphic patterns. Total 166 bands were generated, 58 polymorphic bands (34.9%). The primer ISSR IT3 was got the more diversity than other primers, have 20 bands. The ISSR 15 primer was generated the less bands, only 3 bands. All the polymorphic patterns were ranged between 100~3000 bp. ISSR primer 868 (5'-(GAA)$_6$-3') was better distinguished than other primers. The result was shown in Fig 3. For giant grouper, the patterns of ISSR primer868 could discriminate giant grouper between wild and cultivated populations. For dendrogram analysis, four groups were clustered by ISSR primer868 primer (Fig 4). Among those groups, Group I, Group III, and Group IV were collected from wild population. Samples were clustered in Group II were from cultivated populations. We also found that the results of ISSR method have the same tend to RAPD method. ISSR method was more discriminate ability than RAPD method.

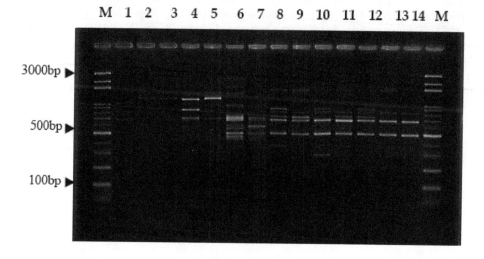

Fig. 3. ISSR profiles of the 14 giant grouper fish obtained using the primer ISSR 868. Lanes M: Bio-100bp DNA Ladder. Lanes 1-7: giant grouper fish wild population (E1, E5, E7, E8, E9, E13, and E14); Lanes 8-14: giant grouper fish cultivated population (E2, E3, E4, E6, E10, E11, and E12).

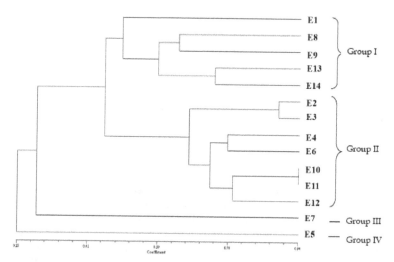

Fig. 4. UPGMA consensus dendrogram of dissimilarity among individuals analysed using the primer ISSR 868.

3.3 RAPD and ISSR methods of cobia

Ninety-five RAPD primers and 59 ISSR primers were used for PCR amplification. The results were shown that all the cobia samples were the same patterns and no polymorphic bands. The primer ISSR UBC809 and RAPD31 were generated 12 to 17 bands and in size from 100 to 2000 bp. The results were also shown in Fig 5 and Fig 6.

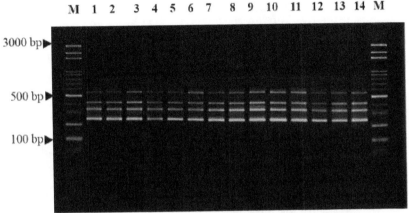

Fig. 5. ISSR profiles of the 14 R. *canadum* fish obtained using the primer UBC 809. Lanes M: Bio-100 bp DNA Ladder. Lanes: 1, 2, 4, 6, 9, 10, 11, 12, 13, and 14 (R1, R2, R4, R6, R9, R10, R11, R12, R13, and R14) R. *canadum* fish wild population; Lanes: 3, 5, 7, and 8 (R3, R5, R7, and R8) R. *canadum* fish cultivate population.

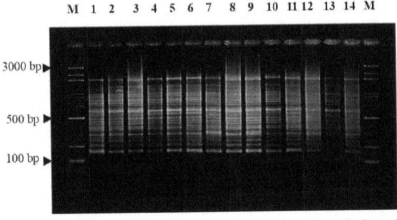

Fig. 6. RAPD profiles of the 14 R. *canadum* fish obtained using the primer 31. Lanes M: Bio-100 bp DNA Ladder. Lanes : 1, 2, 4, 6, 9, 10, 11, 12, 13, and 14 (R1, R2, R4, R6, R9, R10, R11, R12, R13, and R14) R. *canadum* fish wild population; Lanes: 3, 5, 7, and 8 (R3, R5, R7, and R8) R. *canadum* fish cultivate population.

Sequence variability of mitochondrial DNA regions was low between the six cobia (Garnet, et al., 2002). These results were also similar in our study. Both RAPD and ISSR methods have no different patterns. Hence, this could provide more useful information of molecular genetic data in population and stock enhancement studies.

3.4 ISSR methods of red coral trout

For screening ISSR primers, the ISSR primer15 (ISSR15: 5'-(AG)$_8$TG -3') was better distinguished than other primers. The result was shown in Fig 7. The primer ISSR 15 was generated 10 to 16 bands and in size from 200 to 2000 bp. For dendrogram analysis, three groups were identified. Group I, the 71412, 71413, 71422, and 82011 were clustered in the group. Group II, including 81621, 91321, 71411, and 91712 samples; group III including 71423, 80622, 91322, 71421, and 81623 samples. All the nodes of the dendrograms ranged from 90 to 100%. The result was shown in Fig 8.

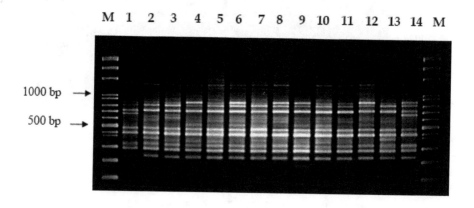

Fig. 7. ISSR profiles of the 14 red coral trout obtained using the primer ISSR 15. Lanes M: Bio-100 bp DNA Ladder. Lanes: 4, 5, 6, 8, 9, 10, and 12 (71411, 71412, 71413, 82011, 91321, 91711, and 91712) red coral trout fish wild population; Lanes: 1, 2, 3, 7, 11, 13, and 14 (71421, 71422, 71423, 81621, 80622, 81623, and 91322) red coral trout fish cultivate population.

4. Discussions

RAPD and ISSR methods were generally used for genetic diversity and populations study; those methods also could be used to analyze the breeding relationship. For species identification and genetic resource/diversity analysis, RAPD and microsatellitsa method were recommended (Liu & Cordes, 2004). The RAPD techniques has been used for discrimination of populations of species of the genus Barbus, grouper, Nile perch and wreck fish, salmonids, among others (Partis & Wells, 1996; Callejas & Ochando, 2001; Asensio et al., 2002; Jin, Cho, Seong, Park, Kong & Hong, 2006). Genetic analysis with RAPD markers is relatively easy, fast, and efficient. RAPD analysis, however, may not be practical for identifying interbreed species (Martinez, Elvevoll & Haug, 1997). SSRs are inherited in a co-dominant fashion. This allows one to discriminate between homo- and heterozygous state, and increases the efficiency of genetic mapping and population genetic studies. ISSR markers have recently been used successfully for genetic analysis in hatchery and wild *Paralichthys olivaceus* strains. It indicates that molecular marker systems contribute greater levels of capability for the detection of polymorphism, and provide a better solution for the assessment of genetic variations (Shikano, 2005; Liu, Chen, Li & Li, 2006).

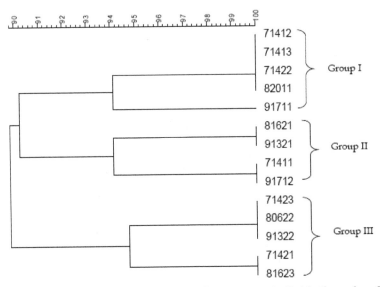

Fig. 8. UPGMA consensus dendrogram of dissimilarity among individuals analysed using the primer ISSR 15.

According to our results, RAPD and ISSR method can be effectively discriminate giant grouper from different sources, such as cultivate and wild, even if fish are from different cultivate farms. But cobia and red coral trout were less discriminate ability. The study found that ISSR and RAPD methods were positively high correlations. Giant grouper species have highly genetic diversity. A comparison of RAPD and ISSR patterns in 14 giant grouper samples, ISSR primers have higher polymorphism and fewer bands than those of RAPD primers. It could provide simple and convenient method to discriminate genetic variation of giant grouper samples. In this study, ISSR method could distinguish genetic variation within specie and different populations. Some reports also have suggested that ISSR may reveal a much higher numbers of polymorphic fragments per primer than those of RAPD (Esselman, et al., 1999). Among these markers, microsatellite DNAs have revolutionized the use of molecular genetic markers in the applications mentioned before, and the markers are destined to dominate this type of studies in the coming years (Asensio, 2007). It also has been revealed as important tools in studies regarding the genetic structure of populations, phylogeographic relations and phylogenetic reconstruction in fish (Antunes, et al., 2010).

5. Conclusion

We developed DNA molecular marker techniques which could be used to generate information for fish genetic diversity, species identification, trace genetic variation between different individuals in aquaculture, authenticate fish, fishery products and provide good reference resources for species sources and relationships.

6. Acknowledgment

This research project was supported by the Council of Agriculture (COA) (97AS-4.1.2-AI-I2).

7. References

Antoro, S.; Na-Nakom, U. & Koedprang, W. (2006). Study of Genetic Diversity of Orange-Spotted, *Epinephelus coioides*, from Thailand and Indonesia Using Microsatellite Markers. *Marine Biotechnology*, Vol.8, pp. 17-26, ISSN 1432-1793

Antunes, R. S. P.; Gomes, V. N.; Prioli, S. M. A. P.; Prioli, R. A.; Júlio Jr.; H. F. ; Prioli, L. M.; Agostinho, C. S. & Prioli, A. J. (2010). Molecular Characterization and Phylogenetic Relationships Among Species of the Genus *Brycon* (Characiformes: Characidae) from Four Hydrographic Basins in Brazil. *Genetics and Molecular Research*, Vol.9, pp. 674-684, ISSN 1676-5680

Asensio, L.; González, I.; Fernández, A.; Rodríguez, M. A.; Lobo, E.; Hernández, P. E.; García, T. & Martín, R. (2002). Application of Random Amplified Polymorphic DNA (RAPD) Analysis for Identification of Grouper (*Epinephelus guaza*), Wreck fish (*Polyprion americanus*), and Nile perch (*Lates niloticus*) Fillets. *Journal of Food Protection*, Vol.65, pp,432-435, ISSN 0362-028x

Asensio, L. (2007). PCR-Based Methods for Fish and Fishery Products Authentication. *Trends in Food Science and Technology*, Vol.18, pp. 558-566, ISSN 0924-2244

Asensio, L.; González, I.; Rojas, M.; García, T. & Martín, R. (2009). PCR-Based Methodology for the Authentication of Grouper (*Epinephelus marginatus*) in Commercial Fish Fillets. *Food control*, Vol.20, pp. 618-622, ISSN 0956-7135

Barman, H. K.; Barat, A.; Yadav, B. M.; Banerjee, S.; Meher P. K.; Reddy, P. V. G. K. & Jana, R. K. (2003). Genetic Variation Between Four Species of Indian Major Carps as Revealed by Random Amplified Polymorphic DNA Assay. *Aquaculture*, Vol. 217, pp. 115-123, ISSN 0044-8486

Bártfai, R.; Egedi, S.; Yue, G. H.; Kovács, B.; Urbányi, B.; Tamás, G.; Horváth, L. & Orbán, L. (2003). Genetic Analysis of Two Common Carp Broodstocks by RAPD and Microsatellite Markers. *Aquaculture*, Vol. 219, pp. 157-167, ISSN 0044-8486

Callejas, C. & Ochando, M. D. (2001). Molecular Identification (RAPD) of the Eight Species of the Genus Barbus (*Cyprinidae*) in the Iberian Peninsula. *Journal of Fish Biology*, Vol. 59, pp. 1589-1599, ISSN 1095-8649

Esselman, E. J.; Jianqiang, L.; Crawford, D. J.; Windus, J. L. & Wolfe, A. D. (1999). Clonal Diversity in the Rare *Calamagrostis porteri* ssp. *insperata* (Poaceae): Comparative Results for Allozymes and Random Amplified Polymorphic DNA and Intersimple Sequence Repeat Markers. *Molecular Ecology*, Vol. 8, pp. 443-453, ISSN 1365-294X

Garber, A. F.; Grater, W. D.; Stuck, K. C. & Franks, J. S. (2002). Characterization of the Mitochondrial DNA Control Region of Cobia, *Rachycentron vanadium*, from Mississippi Coastal Waters. *Gulf and Caribbean Fisheries Institute*, Vol. 53, pp. 570-578, ISSN 0072-9019

Garoia, F.; Guarniero, I.; Piccinetti, C. & Tinti, F. (2004). First Microsatellite Loci of Red Mullet (Mullus barbatus) and Their Application to Genetic Structure Analysis of Adriatic Chared Stock. *Marine biotechnology*, Vol. 6, pp. 446-452, ISSN 1436-2236

Hou, L.; Lü, H.; Zou, X.; Bi, X.; Yan, D. & He, C. (2006). Genetic Characterizations of Mactra Veneriformis (Bivalve) Along the Chinese Coast Using ISSR-PCR Marker. *Aquaculture*, Vol. 261, pp. 865-871, ISSN 0044-8486

Hseu, J.-R.; Hwang, P.-P. & Ting, Y.-Y. (2004). Morphometric Model and Laboratory Analysis of Intracohort Cannibalism in Giant Grouper *Epinephelus lanceolatus* Fry. *Fisheries Science*, Vol. 70, pp. 482-486, ISSN 1444-2906

Jin, L. G.; Cho, J. G.; Seong, K. B.; Park, J. Y.; Kong, I. S. & Hong, Y. K. (2006). 18 rRNA Gene Sequences and Random Amplified Polymorphic DNA Used in Discriminating Manchurian Trout from Other Freshwater Salmonids. *Fisheries Science*, Vol. 72, pp. 903-905, ISSN 1444-2906

Liu, Y.-G.; Chen, S.-L.; Li, J. & Li, B.-F. (2006). Genetic Diversity in Three Japanese Flounder (*Paralichthys olivaceus*) Populations Revealed by ISSR Markers. *Aquaculture*, Vol. 255, pp. 565-572, ISSN 0044-8486

Liu, Z. J. & Cordes, J. F. (2004). DNA Marker Technologies and Their Applications in Aquaculture Genetics. *Aquaculture*, Vol. 238, pp. 1-37, ISSN 0044-8486

Lockley, A. & Bardsley, R. (2000). DNA-Based Methods for Food Authentication. *Trends in Food Science & Technology*, Vol. 11, pp. 67-77, ISSN 0924-2244

Maltagliati, F.; Lai, T.; Casu, M.; Valdesalici, S. & Castelli, A. (2006). Identification of Endangered Mediterranean Cyprinodontiform Fish by Means of DNA Inter-Simple Sequence repeats (ISSRs). *Biochemical systematics and Ecololgy*, Vol. 34, pp. 626-634, ISSN 0305-1978

Martinez, I.; Elevoll, E. O. & Haug, T. (1997). RAPD Typing of North-East Atlantic Minke Whale (Balaenoptera acutorostrata). *ICES Journal of Marine Science*, Vol. 54, pp. 478-484, ISSN 1095-9289

Miao, S.; Jen, C. C.; Huang, C. T. & Hu S.-H. (2009). Ecological and Economic Analysis for Cobia *Rachycentron canadum* Commercial Cage Culture in Taiwan. *Aquaculture International*, Vol. 17, pp. 125–141, ISSN 0967-6120

Okumus, Í. & Çiftci, Y. (2003). Fish Population Genetics and Molecular Marker: II - Molecular Markers and Their Applications in Fisheries Aquaculture. *Turkish Journal of Fisheries and Aquatic Sciences*, Vol. 3, pp. 51-79, ISSN 1303-2712

Partis, L. & Wells, R. J. (1996). Identification of Fish Species Using Rrandom Amplified Polymorphic DNA (RAPD). *Molecular and Cellular Probes*, Vol. 10, pp. 435-441, ISSN 0890-8508

Pazza, R.; Kavalco K. F.; Prioli So^nia M. A. P.; Prioli, A. J. & Bertollo, L. A. C. (2007). Chromosome Polymorphism in Astyanax Fasciatus (Teleostei, Characidae), Part 3: Analysis of the RAPD and ISSR Molecular Markers. *Biochemical Systematics and Ecololgy*, Vol. 35, pp. 843-851, ISSN 0305-1978

Povh, J. A.; Lopera-Barrero, N. M.; Ribeiro, R. P.; Lupchinski Jr. E.; Gomes, P. C. & Lopes, T. S. (2008). Genetic Monitoring of Fish Repopulation Pprograms Using Molecular Markers. *Ciencia e Investigación Agraria*, Vol. 35, pp. 1-10, ISSN 0718-1620

Ramirez, M. A.; Pareicia-Acevedo, J. & Planas, S. (2006). New Microsatellite Resources for Groupers (Serranidae). *Molecular Ecology Notes*, Vol. 6, pp. 813-817, ISSN 1755-0998

Rohlf, F. J. (1997). "NTSYS-pc v. 2.01b Numerical Taxonomy and Multivariate Analysis System." Applied Biostatistics, Inc., Setauket, NY.

Shikano, T. (2005). Marker-Based Estimation of Heritability for Body Color Variation in Japanese Flounder Paralichthys Olivaceus. *Aquaculture*, Vol. 249, pp. 95-105, ISSN 0044-8486

Tang, S.; Tassanakajon, A.; Klinbunga, S.; Jarayabhand, P. & Menasveta, P. (2004). Population Structure of Tropical Abalone (Haliotis asinine) in Coastal Waters of Thailand Determined Using Microsatellite Markers. *Marine Biotechnology*, Vol. 6, pp. 604-611, ISSN 1432-1793

Wang, S. F.; Du, J. Y. & Wang, J. (2007). Identification of *Epinephelus malabaticus* and *Epinephelus coioides* Using DNA Markers. *Acta Oceanology Sinica*, Vol. 26, pp. 122-129, ISSN 1869-1099

Welsh, J. & McClelland, M. (1990). Fingerprinting Genomes Using PCR Arbitrary Primers. *Nucleic Acids Research*, Vol. 18, pp. 7213-7218, ISSN 1362-4962

Zeng, H.-S.; Ding, S.-X.; Wang, J. & Su, Y.-Q. (2008). Characterization of Eight Polymorphic Microsatellite Loci for the Giant Grouper (*Epinephelus lanceolatus* Bloch). *Molecular Ecology Resources*, Vol. 8, pp. 805-807, ISSN 1755-0998

Zhang, Q.; Allen Jr. S. K. & Reece, K. S. (2005). Genetic Variation in Wild and Hatchery Stocks of Suminoe Oyster (Crassostrea ariakensis) Assessed by PCR-RFLP and Microsatellite Markers. *Marine Biotechnology*, Vol. 7, pp. 588-599, ISSN 1432-1793

Zhang, J.; Liu, H. & Song, Y. (2010). Development and Characterization of Polymorphic Microsatellite Loci for a Threatened Reef Fish *Plectropomus leopardus*. Conservation Genetics Resources, Vol. 2 (Supplement 1), pp. 101-103, ISSN 1877-7260

Part 2

Electrophoresis Application in Bacteriology, Parasitology, Mycology and Public Health

7

Application of Molecular Typing Methods to the Study of Medically Relevant Gram-Positive Cocci

Laura Bonofiglio, Noella Gardella and Marta Mollerach
Department of Microbiology, Immunology and Biotechnology, University of Buenos Aires
Argentina

1. Introduction

The development of molecular genotyping methods has been a landmark in the possibility of classifying microorganisms below the species level. The ability to differentiate efficiently related bacterial isolates is essential for the control of infectious diseases and has become a necessary technology for clinical microbiology laboratories.

Strain typing is an integral part of epidemiological investigations of bacterial infections. Typing methods fall into two broad categories: phenotypic and genotypic methods. Phenotypic methods are those that characterize the products of gene expression in order to differentiate strains. Properties such as biochemical profiles, antimicrobial susceptibility profiles, bacteriophage types, and antigens present on the cell surface are examples of phenotypic methods that can be used for typing isolates. Since they involve gene expressions, these properties have a tendency to vary, based on changes in growth conditions and growth phase, being often difficult to detect.

Methods for distinguishing among bacterial strains have profoundly changed over the last years mainly due to the introduction of molecular technology. Genotypic strain typing methods are based on the analysis of differences in the chromosomal and extrachromosomal nucleic acid sequences between strains. Molecular epidemiology of infectious diseases integrates practices and principles of molecular biology with those of epidemiology (Tenover et al. 1997).

Investigations of presumed outbreaks of bacterial infections in hospitals often require strain typing data to identify outbreak-related strains and to distinguish epidemic from endemic or sporadic isolates.

All typing systems can be characterized in terms of typeability, reproducibility, discriminatory power, ease of performance, and ease of interpretation. For each isolate, the system should provide an interpretable result, preferably based on objective criteria. Ideally, results should be reproducible from day to day and from laboratory to laboratory and should allow differentiation of unrelated strains. Additionally, the method should be standardized and if possible should be technically simple, cost-effective, and rapid (van Belkum, et al. 2007).

The results of bacterial strain typing have many different applications including outbreak investigation and surveillance in clinical care settings and public health investigations and also within other contexts such as food and pharmaceutical industries and environmental analysis.

The aim of this chapter is to provide an overview of the methods available for analyzing bacterial isolates, focusing on those methods employed for typing *Streptococcus pneumoniae* and *Staphylococcus aureus*. Different molecular approaches have been used to better understand the epidemiology of these medically relevant gram-positive cocci (Willems. et al. 2011).

2. Genotypic methods

The application of molecular biology tools to infectious disease epidemiology is perhaps just as revolutionary in advancing knowledge and concepts in epidemiology. Genotypic typing methods assess genome variation in bacterial isolates.

The advantages of nucleic acid-based typing systems lie in that they are less likely to be affected by growth conditions or the laboratory manipulations to which organisms are subjected. Undoubtedly, genetic materials undergo changes due to natural or artificial selective pressures, but this mechanism is exactly the basis for their typeability.

Compared to the classical phenotypic typing techniques, genotypic typing techniques have several advantages such as general applicability and a high discriminatory power.

A molecular technique must take into consideration the relative accumulation of variation (short or long term) of a targeted set of genes in a pathogen. Nearly all the typing systems can be grouped into variants of just three basic analytical procedures: (i) PCR, (ii) the use of restriction enzymes, and (iii) nucleic acid sequencing. These procedures allow for the use of common equipment and standard reagents to analyze many different types of infectious agents. In addition, genotypic characterization of pathogens facilitates standardization of information storage and data analyses, interpretation, and communication, which are all amenable to computer-assisted manipulations.

2.1 PCR-based typing methods

In the last years, a number of PCR-based strategies have been developed for use as typing tools. PCR can be readily performed with commercially available supplies and there is little variation in the reagents and equipment needed to perform PCR assays from different microorganisms. The major advantages of PCR-based techniques are speed and simplicity.

2.1.1 Repetitive element sequence-based PCR (rep-PCR)

A variety of repetitive DNA sequence elements have been identified in bacterial pathogens, which have been exploited to develop strategies for bacterial typing. Rep-PCR is a simple PCR-based technique that targets multiple copies of repetitive elements in the bacterial genome to generate DNA fingerprints (Versalovic et al. 1991). Primers designed to anneal in the outward direction, near the end of these repetitive elements bind to multiple non-coding, repetitive sequences interspersed throughout the bacterial genome. Multiple DNA fragments between those sites (interrepeat fragments) are amplified. Since the number and location of the repetitive elements are variable, the sizes and number of effectively amplified fragments vary depending on the strain (Figure 1).

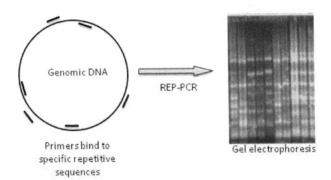

Fig. 1. Schematic representation of REP-PCR assay. On the right BOX-PCR patterns of *S. pneumoniae* isolates using BOXAR1 primer.

Two different Rep-PCR have been used for typing enteric bacteria: a 38-bp repetitive extragenic palindromic element (REP) and a 126-bp enterobacterial repetitive intergenic consensus (ERIC) sequence (Versalovic et al. 1991), whose function has not yet been elucidated.

A BOX repetitive element is a highly conserved repeated DNA element that has been identified in the *Streptococcus pneumoniae* (pneumococcus) chromosome. Although the function of this element has not yet been completely understood, it has been demonstrated that the presence of a BOX element is associated with variation in colony opacity of the pneumococcus (Saluja & Weiser 1995). BOX-PCR has been effectively used for typing *S. pneumoniae* as well as other bacterial species (van Belkum et al. 1996).

Several genetic elements have been used for developing Rep-PCR to type *Staphylococcus aureus*. The element IS256 occurs in the genome either independently or as part of the composite transposon Tn4001, IS256 insertion position is strain-specific and spaced close enough to allow amplification of polymorphic inter-IS256 element sequences (Deplano et al. 1997). Another element used for this methodology is RW3A, a repetitive sequence initially found in *Mycoplasma pneumoniae*, which also generates strain-specific DNA fragments when *S. aureus* DNA is used as template (van der Zee et al. 1999).

2.1.2 Randomly Amplified Polymorphic DNA-PCR (RAPD-PCR) or Arbitrarily-Primed PCR (AP-PCR)

Randomly Amplified Polymorphic DNA-PCR (RAPD-PCR), also referred to as Arbitrarily-Primed PCR (AP-PCR), is a variation of the PCR technique employing a single, generally short primer, that is not targeted to amplify a specific bacterial DNA sequence. Low annealing temperatures are used during amplification, allowing imperfect hybridization at multiple chromosomal locations. When the primer binds in two sites on opposite strands, at the proper orientation and with sufficient affinity to allow the initiation of polymerization, the amplification of the fragment between those sites will occur. The amplified products will be various different-sized fragments that can be resolved by conventional agarose gel electrophoresis (Figure 2).

Although the method is much faster than many other typing methods, it is much more susceptible to technical variation. Slight variations in the reaction conditions or the reagents can lead to difficulty in result reproducibility and in the band patterns generated. Therefore, trying to make comparisons among potential outbreak strains can be very problematic (van Belkum et al. 1995). When RAPD-PCR is tightly controlled, it can provide a high level of discrimination, especially when multiple amplifications with different primers are performed.

PCR-based typing methods are simplest and rapid genotyping methods, but is remarkable for its susceptibility to minor variations in experimental conditions.

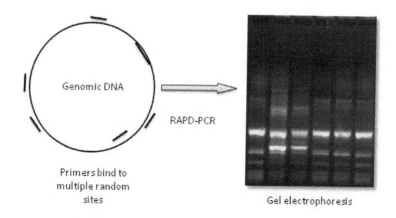

Fig. 2. Schematic representation of RAPD-PCR assay. On the right RAPD-PCR patterns of *Enterococcus faecalis* isolates using D8635 primer.

2.2 Based on enzymatic restriction of chromosomal DNA

2.2.1 Restriction endonuclease analysis of chromosomal DNA by hybridization with a nucleic acid probe (Southern blotting)

Following digestion with high frequency restriction endonucleases, chromosomal DNA is separated into different-sized fragments by conventional agarose gel electrophoresis, but this type of polymorphism is difficult to interpret due to the high number of fragments generated. However, interpretation of these polymorphisms can be facilitated by a Southern blot hybridization technique. By this methodology, fragments are separated by electrophoresis and transferred to a nitrocellulose or nylon membrane and hybridized using specific chemically or radioactively-labeled probes (Figure 3). DNA probes are designed for specific sequences that are found in multiple copies and in different positions of the chromosome. One of the most frequently used probes is ribosomal RNA (16s rRNA) because most species have more than one chromosomal rRNA operon distributed around the chromosome. This particular technique is denominated ribotyping. In recent years, the use of this technique has declined, mainly due to its limited discriminatory power compared to other techniques. Several DNA probes have been employed for the study of methicillin-resistant *S. aureus* (MRSA) outbreaks, including various insertion sequences, such as *IS431*, *IS256* and *mecA* gene.

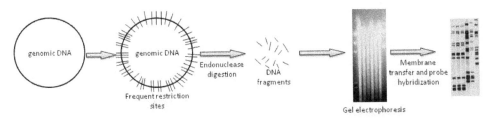

Fig. 3. Schematic representation of restriction endonuclease analysis by Southern-blotting assay. On the right *IS6110*-restriction fragment length of *Mycobacterium tuberculosis* isolates.

2.2.2 Pulsed-field gel electrophoresis (PFGE) of chromosomal DNA

Pulsed-field gel electrophoresis is based on the digestion of bacterial DNA with restriction endonucleases that recognize few sites along the chromosome, generating large DNA fragments (30-800 Kb) that cannot be effectively separated by conventional electrophoresis. The basis for PFGE separation is the size-dependent time-associated reorientation of DNA migration achieved by periodic switching of the electric field in different directions. The DNA fragments will form a distinctive pattern of bands in the gel, which can be analyzed visually and electronically (Figure 4 A). Bacterial isolates with identical or very similar band patterns are more likely to be related genetically than bacterial isolates with more divergent band patterns.

This technique is laborious and includes several steps, requires good standardization and takes at least two days for obtention of results. Procedures will differ to some extent depending on the organism that is being analyzed.

Regarding DNA preparation, PFGE requires intact DNA for restriction endonuclease treatment. The risk of mechanical breakage to DNA molecules during the extraction procedure is avoided by embedding intact organisms into agarose plugs where cells are enzymatically lysed and cellular proteins digested. After endonuclease treatment, the agarose plugs containing the digested DNA are then submitted to PFGE (Figure 4B). The choice of the restriction enzyme for DNA digestion and pulse-time switching parameters for PFGE are critical variables for the obtention of restriction profiles to show well- resolved fragments.

Recent protocols can be completed in as little as two days through shortcuts such as the direct addition of lytic enzymes to the agarose mixture before the blocks are cast and also high temperature short-term washes which facilitate the extraction of unwanted compounds (Goering 2010; Halpin et al. 2010).

Isolates with identical PFGE patterns were considered to represent the same epidemiological type. Isolates differing by one genetic event were considered epidemiologically-related subtypes, expecting that a single genetic event could occur in the chromosome of an organism as it moved from patient to patient. Isolates differing by two genetic events were also deemed to be potentially related, while three or more chromosomal differences were thought to represent an epidemiologically-significant difference (unrelated isolates). Van Belkum suggested a more conservative approach where only nosocomial isolates differing by a single genetic event (up to four differences in the PFGE restriction fragment pattern) were considered related subtypes. The terminology within both proposed formats was left intentionally vague, understanding that molecular typing is only one component of epidemiological evaluation which must include other available clinical data for accurate analysis (Tenover et al. 1995; van Belkum et al. 2007; Goering 2010).

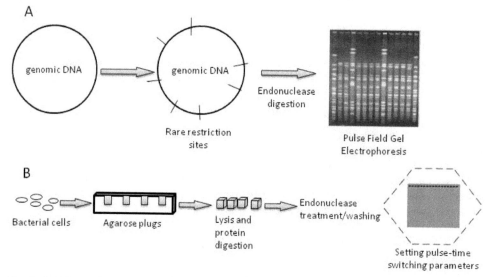

Fig. 4. A. Schematic representation of pulse field gel electrophoresis. On the right PFGE of *SmaI*-digested genomic DNA of *S. aureus* isolates. B. Sequence of steps involved in PFGE.

Furthermore, isolates with more uniform PFGE profiles require more conservative interpretation. The fact that two strains share the same pattern does not prove that they are epidemiologically related. The establishment of an epidemiologic relationship depends on the frequency with which the "indistinguishable" pattern is seen among epidemiologically-unrelated isolates and correlation with clinical and epidemiological information. If common contact between two patients with strains having the same pulsed-field gel electrophoresis (PFGE) type can be established, the chances are greater that an epidemiologic link could be ascribed. Thus, the greatest power of PFGE typing lies in showing strain dissimilarity rather than in proving similarity or relatedness. These considerations must be taken into account for banding pattern analysis from other molecular typing methods.

In some instances, initial unsatisfactory PFGE results may be aided by the use of an alternative restriction enzyme (Kam et al. 2008; Bosch et al.) or, in more difficult situations, the use of one or more additional typing methods (van Belkum et al. 2007).

The intra- and interlaboratory reproducibility of this method depends on understanding and controlling variables (Cookson et al. 1996; van Belkum et al. 1998; te Witt et al. 2010).

This success is due to an emphasis on standardized quality control especially in major areas of potential PFGE variability such as DNA sample preparation, choice of restriction enzyme, and electrophoresis conditions.

PFGE has been applied to a wide range of microorganisms and has remarkable discriminatory power and reproducibility. It is currently considered the strain typing method of choice for many commonly encountered pathogens. However, one of the main notable limitations is the need for specialized and relatively expensive equipment.

2.3 DNA sequencing-based methods

Genotyping methods based on DNA sequencing discriminate among bacterial strains directly from polymorphisms in their DNA considering the original sequence of nucleotides.

2.3.1 Single-locus sequence typing

Sequencing of a single genetic locus has been used for epidemiological studies of many bacterial species, yielding valuable typing results. In this approach, it is essential to select highly variable gene sequences. Valuable typing results have been obtained for *S. pyogenes* by DNA sequencing of 150 nucleotides coding for the N-terminal end of M protein (*emm* typing) (Beall et al. 2000). Another example is spa typing for *S.aureus* that consists in sequencing of the X region of the protein A gene (*spa typing*). This technique is widely used for subtyping methicillin-resistant *S. aureus* (MRSA) strains (Shopsin, 1999, 2000; Shopsin & Kreiswirth 2001; Harmsen et al. 2003), (Figure 5).

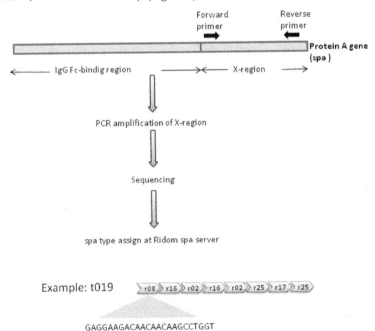

Fig. 5. Sequence of steps involved in *spa* typing.

2.3.2 Multi-locus sequence typing (MLST)

MLST is a genotyping method based on the measurement of DNA sequence variation in a set of housekeeping genes (usually seven genes) whose sequences are constrained because of the essential function of the proteins they encode. This method was proposed in 1998 as a general approach to provide accurate, portable data that were appropriate for the epidemiological investigation of bacterial pathogens and which also reflected their evolutionary and population biology (Maiden et al. 1998).

MLST schemes have been developed for several species and databases containing the allelic profiles of a great number of strain types with corresponding clinical information that can be readily consulted over the Internet (http://www.mlst.net/ and http://pubmlst. org/), (Aanensen & Spratt 2005). Additional information such as date, place of isolation and antibiotype is included in the database when a strain is deposited so this database is continuously expanding as new STs are identified and additional nucleotide sequence data are deposited.

Internal fragments of the seven housekeeping genes are amplified by PCR from chromosomal DNA using the primer pairs described in the web site. The amplified fragments are directly sequenced in each direction. The sequences at each of the seven *loci* are then compared with all the known alleles at that *locus,* and a number representing a previously described allele (or a new one) is assigned to the *locus.* For a given isolate, alleles present at each gene position are combined into an allelic profile and assigned a sequence type (ST) designation (Maiden et al. 1998). Relationships among isolates are assessed by comparisons of allelic profiles: closely related isolates have identical STs, or STs that differ at a few *loci,* whereas unrelated isolates have unrelated STs (Figure 6).

A number of clustering algorithms have been employed to analyze the data in the MLST scheme, including UPGMA (unweighted pair group method with arithmetic mean) and eBURST analyses (Feil et al. 2004).

The original conception of MLST used the allele number as the primary unit of analysis (Enright & Spratt 1998; Maiden et al. 1998) which was appropriate for organisms where horizontal genetic exchange is common. However, MLST data can also be interpreted by tree-building approaches that use nucleotide substitutions rather than allelic changes as the unit of analysis; this is more pertinent to bacteria where mutational change predominates over genetic exchange in the evolution of variants.

An important advantage of MLST is that results are unambiguous and easily and unequivocally exchangeable, much more so than images of agarose gel electrophoresis patterns. MLST drawbacks are practical, including limited accessibility and high cost. It is a relatively expensive technique available for the characterization of bacterial isolates, mainly in reference or research laboratories. However, MLST is increasingly applied as an informative typing tool that enables international comparison of isolates. It has been applied to problems as diverse as the emergence of antibiotic-resistant variants (Crisostomo et al. 2001; Enright et al. 2002), the association of particular genotypes with virulence (Brueggemann et al. 2003) or antigenic characteristics (Meats et al. 2003) and also the global spread of disease caused by novel variants (Albarracin Orio et al. 2008). In addition to these medically-motivated epidemiological analyses, MLST data have been exploited in evolutionary and population analyses (Jolley et al. 2000) that estimate recombination and mutation rates (Feil et al. 2001) and in investigation of the evolutionary relationships among bacteria that are classified as belonging to the same genus (Godoy et al. 2003).

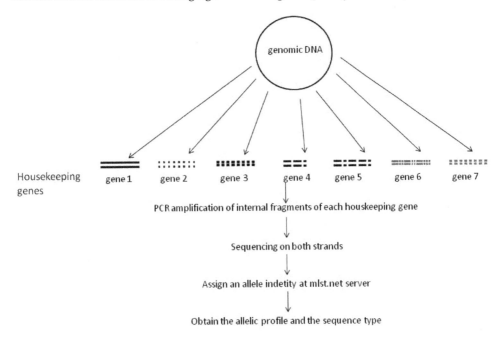

Fig. 6. Sequence of steps involved in MLST scheme. Adapted from Vazquez et al. 2004.

2.4 Analysis of results obtained by molecular epidemiology

Comparison and interpretation of raw data generated by molecular typing methods, such as gel electrophoresis band patterns, sequence alignments, or hybridization matrix patterns could be performed by visual analysis when there are few strains. However, if the analysis includes many strains, the comparison turns out to be very difficult. Therefore computer programs have become indispensable in molecular epidemiological investigations.

Computer programs that compare band sequences or patterns employ clustering algorithms that can generate dendrograms or trees illustrating the arrangement of the clusters produced. For pattern recognition, such as electrophoretic banding patterns or hybridization matrices, additional programs are needed to capture, digitize, and normalize the patterns.

There are different commercially available platforms for databasing and gel analysis that have been developed for computer-assisted analysis such as BioNumerics, GelCompar (Applied Maths, Sint-Martens-Latem, Belgium), Diversity Database Fingerprinting Software (Bio-Rad Laboratories, Hercules, Ca). Treecon (Van de Peer and De Wachter 1994).

3. Epidemiologic applications of bacterial typing techniques

A more comprehensive knowledge of the evolution and the epidemiology of bacterial pathogens had been obtained by combination of genetic, phenotypic, spatial and temporal data.

Multiple techniques have been developed to assess genomic differences among different isolates or clones of the same species, such as PCR-based methods and PFGE. These methods present portability problems and limited comprehension of the processes by which variation occurs. However, DNA sequence-based techniques generate portable differentiation in bacterial populations that can be used to understand their phylogenetic history. Extensive genomic and phenotypic diversity exists within populations of microbial pathogens of the same species. This diversity reflects the evolutionary divergence arising from mutations and gene flux. These distinctive characters are scored by typing systems which are designed to optimize discrimination between epidemiologically related and unrelated isolates of the pathogen of interest (Maslow and Mulligan 1996; Struelens 1996).

Epidemiologic typing systems can be used for outbreak investigations to confirm and delineate the transmission patterns of one or more epidemic clone(s), to test hypotheses about the sources and transmission vehicles of these clones and to monitor the reservoirs of epidemic organisms. Typing also contributes to epidemiologic surveillance and evaluation of control measures by documenting the prevalence over time and the circulation of epidemic clones in infected populations. Clearly, different requirements will be needed for these distinct applications (Maslow and Mulligan 1996; Struelens 1996).

Typing can be undertaken at two different levels, depending on the situation: i) short term or local epidemiology, when organisms are recovered in a defined setting over a short period of time, which is used to study nosocomial outbreaks, local transmission and carriage, and the relationship between isolates associated with carriage and infection in a given geographic area, ii) long term or global epidemiology, when strains are recovered from one geographic area related to those isolated worldwide or strains recovered at different times.

Local epidemiology is applied to study outbreaks with the aim to characterize that the increase in incidence of infection is caused by enhanced transmission of a specific strain. In this framework, typing methods are applied to investigate the sources of contamination and the route of transmission. Accurate application of bacterial typing will support appropriate control measures designed to contain or interrupt the outbreak and prevent further spread of disease. Typing may also be used for isolates cultured from the same patient over time to help define whether a second episode of infection is due to relapse or reinfection. PCR fingerprinting is the simplest and most rapid genotypic method for local application; however, PCR typing is very susceptible to minor variations in experimental conditions and reagents. Therefore, the method is more appropriate for the comparison of a limited number of samples processed simultaneously and run on one gel. By contrast, PFGE has good reproducibility and is highly discriminatory. Therefore, PFGE is considered the current gold standard for outbreak and local epidemiology studies.

At a different level, collaborative studies have been performed to define major internationally disseminated bacterial clones of important human pathogens. Currently, MLST in combination with PFGE is the most appropriate strategy for long term epidemiology and have reached useful conclusions from infectious disease surveillance data. The evaluation of global population genetic structure, genetic evolution, genetic diversity and pathogenicity has been successfully developed within this framework.

For eukaryotes, clones are genetically identical organisms. However, in bacterial epidemiology, the clone concept is of an even more pragmatic nature, denoting isolates obtained during real outbreaks with common features (e.g. multiple antibiotic-resistant isolates) from different geographic locations, the so-called epidemic clones.

The threshold of marker similarity used for definition of a clone need to be adjusted to the species studied, the typing system used, the environmental selective pressure and the time and space scale of the study (Tibayrenc 1995; Struelens 1996). Mutation rate and gene flux vary between species, pathovars and environments. In vivo micro-evolution of most pathogens remains poorly understood. Subclonal evolution and emergence of variants that occur in individual hosts or during prolonged transmission can be recognized by several high resolution molecular typing systems, like, for instance, macrorestriction analysis by pulsed-field gel electrophoresis (Struelens 1996).

4. Strategies applied for surveillance and typing of relevant gram-positive pathogens

4.1 Methicillin-resistant *Staphylococcus aureus*

Staphylococcus aureus is recognized as one of the most important human pathogens. *It* has shown great ability to acquire resistance to different antimicrobial agents. The first isolation of methicillin-resistant *S. aureus* (MRSA) was reported in 1960 and since then, the prevalence of this pathogen has increased.

Methicillin resistance is conferred by the *mecA* gene which codes for an additional penicillin-binding protein named PBP 2a; this protein has reduced affinity to β-lactam agents. This gene is located in a mobile genetic element of variable size known as staphylococcal cassette chromosome *mec* (SCC*mec*). So far, eight types and several subtypes of SCC*mec* have been characterized (Deurenberg & Stobberingh 2008; Chambers & Deleo 2009).

The incidence of MRSA varies geographically throughout the world. MRSA has emerged as an important pathogen among hospitalized patients. Most hospital-acquired infections caused by methicillin-resistant *Staphylococcus aureus* (HA-MRSA) are associated with a relatively small number of epidemic clones that spread over different continents. According to the Sistema Informático de Resistencia (Asociación Argentina de Microbiología, Buenos Aires, Argentina), MRSA strains are among the most prevalent nosocomial pathogens (http://www.aam.org.ar) in Argentina, whereas the Brazilian clone, the pediatric clone and the Cordobés clone have been found to be the main clones associated with HA-MRSA infections (Corso et al. 1998; Sola et al. 2002; Gardella et al. 2005).

However, since 1990, MRSA has been recognized as a cause of infections in people without established risk factors for HA-MRSA, such as recent hospitalization, surgery, residence in a long-term care facility, receipt of dialysis, or presence of invasive medical devices (Fridkin et al. 2005; Chambers & Deleo 2009). These infections are thought to be acquired in the community and are referred to as community-associated MRSA infections (CA-MRSA). This term has also been used to refer to MRSA strains with bacteriological characteristics considered typical of isolates recovered from patients with CA-MRSA infections (Salgado et al. 2003). HA-MRSA strains are generally resistant to antibiotics other than β-lactams, whereas typical CA-MRSA strains are only resistant to methicillin. HA-MRSA isolates frequently harbor SSC*mec* types-I, II and III whereas CA-MRSA strains carry types IV and V (Ma et al. 2002; Naimi et al. 2003).

The Panton-Valentine leukocidin (PVL) toxin has been described as a genetic marker of CA-MRSA isolates, rarely identified in HA-MRSA isolates (Ma et al. 2002; Naimi et al. 2003). Several studies have demonstrated that the presence of PVL genes is associated with *S. aureus* recovered from patients suffering from primary skin infections (Lina et al. 1999), severe necrotizing pneumonia, and increased complications of hematogenous osteomyelitis; however, the role of PVL in the pathogenesis of *S. aureus* infections has not yet been fully elucidated.

The spectrum of disease caused by CA-MRSA appears to be similar to that of methicillin-susceptible *Staphylococcus aureus* (MSSA) in the community. Skin and soft tissue infections (SSTIs), specifically furuncles (abscessed hair follicles or "boils"), carbuncles (coalesced masses of furuncles), and abscesses, are the most frequently reported clinical manifestations (Fergie & Purcell 2001; Baggett et al. 2003; Fridkin et al. 2005). Less commonly, MRSA has been associated with severe and invasive staphylococcal infections in the community, including necrotizing pneumonia, bacteremia, osteomyelitis, toxic shock syndrome, and meningitis (Deurenberg & Stobberingh 2008).The rapid emergence of these infections has been one of the most unexpected events in bacterial infectious diseases in the recent years.

Distinct genetic lineages associated with CA-MRSA infections have been determined by typing and their geographic dissemination evaluated in different countries. In Latin America, CA-MRSA has been described several times (Ma et al. 2005; Ribeiro et al. 2005; Alvarez et al. 2006; Gardella et al. 2008).

Typing of MRSA strains is necessary for proper epidemiological investigations of sources and modes of transmission of these strains in hospitals, and the design of appropriate control measures and the application of different typing methods have contributed to understanding the emergence of MRSA in the community. Phenotyping methods generally have limited discriminatory power and poor typeability; therefore, a number of molecular techniques have been developed for *S. aureus* typing, namely restriction fragment length polymorphism (RFLP) analysis techniques, including ribotyping and Southern blot analysis with probes for mobile elements present in multiple copies in the staphylococcal genome, like insertion sequences (IS256,IS257, IS431 and IS1181) and transposons (Tn554 and Tn4001) (Wei et al. 1992; Tenover et al. 1994; Kreiswirth et al. 1995).

Among PCR methods, rep-PCR and RAPD-PCR analysis were found to be epidemiologically useful, but interlaboratory studies showed that reproducibility is an important drawback for these techniques (Saulnier et al. 1993; van Belkum et al. 1995; Deplano et al. 1997; van der Zee et al. 1999). Pulsed-field gel electrophoresis (PFGE) analysis is an accurate and discriminating method which is now used as the reference method for *S. aureus* typing in some reference centers (Bannerman et al. 1995). However, PFGE analysis is costly and technically demanding and still requires interlaboratory standardization (Cookson et al. 1996). PFGE proved to be a highly discriminatory and sensitive technique in microepidemiological (local or short term) and macroepidemiological (national, continental, or long term) surveys (Struelens et al. 1993; McDougal et al. 2003). Nevertheless, some authors have argued that the stabilities of PFGE markers may be insufficient for the reliable application of PFGE to long-term or macroepidemiological studies (Blanc et al. 2002).

Sequence-based methods such as multilocus sequence typing (MLST) has proved to be adequate for long-term global epidemiology and the study of the recent evolution of *S. aureus*

(Enright, 2000, 2002). Another useful technique is the Staphylococcal cassette chromosome *mec* (SCC*mec*) typing, based on the molecular characterization by multiplex PCR of the mobile genetic element carrying the methicillin-resistant gene (*mecA*) (Oliveira and de Lencastre 2002). The combination of the MLST type and the SCC*mec* type, defined as the "clonal type," is now used for the international nomenclature of MRSA clones (Enright et al. 2002).

Moreover, single-locus DNA sequencing of repeat regions of the *coa* (coagulase) gene and the *spa* gene (protein A), respectively, could be used for reliable and accurate MRSA typing (Shopsin, 1999 2000; Tang et al. 2000; Shopsin & Kreiswirth 2001; Harmsen et al. 2003). Spa typing is especially interesting for rapid typing of MRSA in a hospital setting since it offers higher resolution than coa typing (Shopsin et al. 2000). The repeat region of the *spa* gene is subject to spontaneous mutations, as well as to loss and gain of repeats. Repeats are assigned an alpha-numerical code, and the spa type is deduced from the order of specific repeats. There is a good correlation between clonal groupings determined by MLST and the respective spa types (Harmsen et al. 2003).

We performed different studies to characterize MRSA clones within diverse scenarios of Argentina. In 2005, we demonstrated the replacement of the multiresistant MRSA "Brazilian" clone (SCC*mec* III, ST239) by the "Cordobes" clone (SCC*mec* I, ST5), a MRSA clone susceptible to rifampin, minocycline and trimethoprim/sulfamethoxazole in two university hospitals. Isolates were characterized by using RAPD-PCR and PFGE and SCCmec typing (Gardella et al. 2005).

Later, we analyzed community-associated methicillin-resistant *Staphylococcus aureus* (CA-MRSA) isolates recovered from patients suffering from different types of infections. All CA-MRSA isolates carried SCC*mec* type IV. Four major clones were detected in Argentina by PFGE. The largest cluster was named CAA clone: Pulsotype A, spa type 311, ST 5, LPV (+) (Gardella et al. 2008) and two isolates of this clone were recovered from two cases of acute bacterial meningitis (von Specht et al. 2006), (Figure 7).

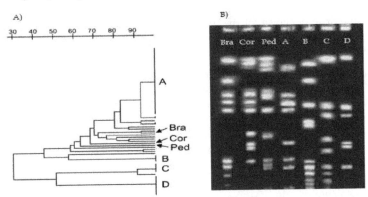

Fig. 7. (A) Dendrogram of pulsed-field electrophoresis banding pattern of CA-MRSA isolates and the 3 clonal types most prevalent in Argentinean hospitals: Brazilian clone (Bra), Cordobes clone (Cor), Pediatric clone (Ped). Similarity coefficient was calculated by using Dice coefficient, and cluster analysis was performed by the unweighted pair-group method. Four major pulsotypes were coded from A, B, C and D and representative HA-MRSA strains of prevalent clones in Argentina.

We also characterized CA-MRSA strains isolated from skin and soft-tissue infections in isolates recovered from Uruguay in 2005. In that study, we identified three major groups of CA-MRSA strains (1, 2, and 4) that were defined according to phenotypic and genotypic characteristics. The most frequent group, G1, showed a PFGE pattern identical to that of CA-MRSA strains previously isolated in Uruguay and Brazil; these strains are still producing SSTI, illustrating the stability of this emergent pathogen over time, as well as its excellent adaptation to the community environment (Pardo et al. 2009).

During the 2008 school- year period we conducted the first epidemiological study of *S. aureus* carriage in Argentina. Carriage was investigated in all children attending the last year of kindergarten in a city of Buenos Aires province, Argentina. Of 316 healthy children, 31.0% carried *S. aureus*, including 14 MRSA carriers (4.4%). All MRSA isolates carried the SCCmec type IV cassette. Eight of those 14 carriers were closely related to the CAA clone, which was responsible for the most severe community-acquired MRSA infections caused in our country (PFGE A, SCCmec IV, *spa* t311, ST5), (Gardella et al.).

Our results should serve as a warning for the health system since the main clone circulating in the community presents epidemic characteristics and also possesses a genetic background (ST5) of demonstrated plasticity and efficiency to be established as prevalent in the hospital environment.

4.2 Molecular epidemiology of *Streptococcus pneumoniae*

Streptococcus pneumoniae is a human pathogen of increasing clinical relevance causing important diseases such as meningitis, pneumonia, bacteremia and otitis media. Surveillance has become progressively more important because of the worldwide distribution of penicillin-resistant and multidrug-resistant pneumococci clones in the last 15 years.

S. pneumoniae with resistance to one or more antibiotics has been isolated since 1990. Penicillin is the drug of choice for the treatment of pneumococcal infections. The resistance mechanism includes the modification of penicillin-binding proteins (PBP), which in general is associated to cephalosporin resistance.

The clinical relevance of multiple antibiotic-resistant pneumococcal strains has led to the creation of a network. The Pneumococcal Molecular Epidemiology Network (PMEN) was established in 1997 with the aim of global surveillance of antibiotic-resistant *S.pneumoniae* and the standardization of nomenclature and classification of resistant clones (http://www.sph.emory.edu/PMEN/index.html). The PMEN also includes major invasive antibiotic-susceptible clones that have a wide geographic spread. Up to date, there are currently 43 clones described by the PMEN. Of those, three penicillin- resistant and two penicillin- susceptible PMEN clones have been detected in Argentina http://www.sph.emory.edu/PMEN/(Zemlickova et al. 2005). Clones to be included in the Network have to be subjected to PFGE, MLST and PBP fingerprinting to confirm that they differ from previously accepted ones. PBP fingerprinting is a typing technique that includes PCR amplification of the *pbp1a*, *pbp2b* and *pbp2x* genes with previously described primers (Gherardi et al. 2000). Amplified genes are digested with *HaeIII+DdeI* (*pbp1a*) and *HaeIII+RsaI* (*pbp2b* and *pbp2x*) restriction enzymes and electrophoresed on 3% gels. This technique has been used in many reports over the last 20 years to study the molecular epidemiology of *S. pneumoniae* resistance to β-lactams (Zhang et al. 1990; Munoz et al. 1991). For macrolide-resistant strains, tests for *erm* and *mef* genes have to be performed (http://www.sph.emory.edu/PMEN). The PMEN network has included BOX-PCR typing in the guidelines for the recognition of pneumococcal clones (McGee et al. 2001).

In addition, the use of modern typing methods, mainly MLST, has greatly helped track the geographical spread of specific *S. pneumoniae* strains and follow the dynamics of microbial populations over time. The application of all these techniques have shown that the spread of multiresistant international clones defined by the PMEN is the main cause of increase in pneumococcal resistance to β-lactams and other drugs (Munoz et al. 1991; Klugman 2002; Smith et al. 2006; Sadowy et al. 2007; Soriano et al. 2008). In pneumococcus, each serotype may typically be made up of a number of clones, which are not closely related and are not equivalent in terms of antibiotic resistance.

Furthermore, molecular methods showed that the evolution of penicillin-resistance and multiresistance is a phenomenon in which the acquisition and/or alteration of molecular targets is mainly a consequence of intergenic change and that *S. pneumoniae* diversity has largely been driven by recombination (Hermans et al. 1997; Enright & Spratt 1998; Enright et al. 1999; Descheemaeker et al. 2000; McGee et al. 2001).

As multiresistance has increased the difficulties of treating this serious bacterial infection, prevention through vaccination has become even more important. There are at least 91 known pneumococcus capsular types, with 23 capsular types included in the current pneumococcal polysaccharide (adult) vaccine and 13 types included in the current conjugate (child) vaccine. To overcome serotype specificity of actual vaccines, upcoming pneumococcal vaccines should offer a different approach to the prevention of pneumococcal disease and the decrease in carriage. Several proteins have been identified as possible candidates to develop more appropriate vaccines. One of them , the pneumococcal surface protein A (PspA), is a surface virulence factor, antigenically variable yet cross-reactive that interferes with complement-mediated clearance of pneumococci (McDaniel et al. 1991; Tu et al. 1999).Since 1993, six Latin-American countries have been participating in an epidemiological surveillance study conducted by the Pan American Health Organization (PAHO) in order to determine the relative prevalence of capsular types and antimicrobial resistant patterns of *S. pneumoniae* causing invasive infections in children <5 years of age. One of these studies showed that, the prevalence of penicillin resistant *S. pneumoniae* (PRSP) in Argentina was 24.4%, which was significantly associated with the expansion of serotype 14 clone that had been previously described in Europe expressing serotype 9V (Rossi et al. 1998). A similar situation was encountered in Uruguay in the same period (Camou et al. 1998).

Ongoing surveillance programs for invasive pneumococcal disease also monitor the appropriateness of existing vaccine formulations and provide valuable information on which to base the formulation and application of new vaccines that are currently under development. In this framework, from 1993 to 2000, with the participation of the Argentinean *Streptococcus pneumoniae* Working Group, 1293 invasive isolates were studied to determine capsular type distribution and antimicrobial susceptibility. We selected a sample of 149 strains, having the same serotype distribution as in the total collection, in order to characterize the distribution of PspA variants among Argentinean invasive isolates recovered from children less than 6 years of age. The genetic relatedness among the isolates of the major serotypes was also evaluated by BOX-PCR because it is a quick molecular method that is suitable for investigation of genetic relatedness of pneumococcal strains and provides results whose interpretation is relatively unambiguous (Hermans et al. 1995; van Belkum et al. 1996). This study provided epidemiological information about the PspA family distribution and the genetic diversity of Argentinean *S. pneumoniae* isolates and informed of the potential coverage of a PspA- based vaccine. It was the first insight into diversity of PspA within strains circulating in Argentina (Mollerach et al. 2004). Family 1 PspA was detected in 54.4% of the isolates, 41.6% of which were family 2 and 4.0% expressed both family 1 and family 2 PspAs. This observation indicates that a PspA vaccine containing only family 1 and family 2 PspAs should be able to cover the bulk of the strains in this region. Box typing revealed the Argentinian strains were from at least 10 clonally related groups.

In some cases, a strong association between one PspA type and a certain capsular type was found. For example, serotype 1 and 5 and the majority of isolates of penicillin-susceptible serotype 14 isolates exhibited PspA family 1. On the other hand, serotypes 7 F, serotype14 PRSP and the majority of type 9V isolates were assigned to PspA family 2. BOX-PCR analysis revealed genetic homogeneity of serotype 14 PRSP and serotype 5 isolates. Antibiotype suggests correlation with the Spain[9V]-3 clone and Colombia[5]-19 clone, respectively. These clones had been previously described in the region (Gamboa et al. 2002; Brandileone et al. 2004; Zemlickova et al. 2005).

Nowadays, the sequencing of DNA allows to compare the results between laboratories and to obtain a global look for the situation of the circulating multidrug-resistant clones. This effort includes a database that contains information concerning the clones that are currently widespread in different parts of the world (http://spneumoniae.mlst.net).

In the framework of PAHO in Latin American countries, surveillance data revealed that penicillin-nonsusceptible *S. pneumoniae* (PNSP) type 6B increased from 15.8 % in the period between 1993-1997, to 67.3 %   in 1998-2002 (p<0.001). Serotype 6 ranks fourth among capsular types causing invasive diseases in Argentinean patients under 6 years of age, and it has been included in the heptavalent conjugate polysaccharide vaccine licensed in Argentina in 2001 and also in the 13-valent introduced in 2010 (Organización Panamericana de la Salud 2007; Ruvinsky et al. 2008). This serotype is a frequent cause of invasive diseases (Riedel et al. 2007; Gabastou et al. 2008; Darabi et al.). We characterized the population of penicillin non-susceptible *S. pneumoniae* type 6B strains isolated from pediatric patients in Argentina between 1993-2002 with the use of molecular typing methods including BOX-PCR, PFGE and MLST (Bonofiglio et al. 2011) (Figure 8). The results of the study showed that the increase in penicillin resistance in serotype 6B may be partly explained by the entrance of the Poland[6B]-20 clone, which is a PMEN clone not previously described in Argentina. Our findings showed that the Poland[6B]-20 clone established in 1999; and the use of BOX-PCR and PFGE subtypes suggested that horizontal transfer or differentiation events had occurred after the common lineage became established. Dissemination of this clone could be traced through demographic data, as isolates representative of the clone had been recovered in different regions of Argentina and its expansion is also responsible for the emergence of erythromycin-resistance in *S. pneumoniae* serotype 6B. The pneumococcal MLST database currently contains information of 81 strains of the Poland[6B]-20 clone.

Other similar studies were carried out by Sadowy et al, who analyzed isolates recovered in Poland in the period 2003-2005 using serotyping, MLST and sequencing of *murM* and *pspA* alleles. They demonstrated that the vast majority of the isolates (90.7%) belong to international multiresistant clones whereas, the Spain 9V -ST156 clonal complex being the most prevalent. Moreover, this clone has evolved rapidly, as demonstrated by the observed number of STs , the use of another approach of MLST (multiple locus variable-number-tandem repeat analysis) and the polymorphism of *pbp* and *pspA* genes (coding for penicillin-binding proteins and the pneumococcal surface protein A, respectively) (Sadowy et al.).

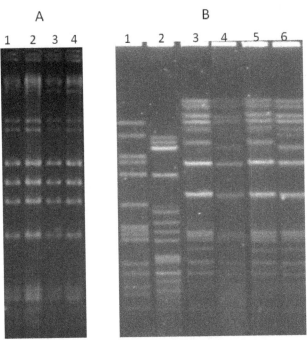

Fig. 8. A) BOX-PCR patterns of 4 Argentinean isolates belonging to Poland[6B]-20 clone. B). Pulse field gel electrophoresis of *Sma*I digested DNA of *S. pneumoniae*. Lane1: *S.pneumoniae* R6, lane2: *S.pneumoniae* 6B (1994), lanes3-6: *S.pneumoniae* isolates belonging to Poland[6B]-20 clone recovered in Argentina (2000).

The relationship between PspA and ST was also explored (Qian et al.). This author analyzed 171 invasive *S. pneumoniae* isolates from Chinese children in 11 hospitals between 2006 and 2008. He found that Family 1 and family 2 PspAs were prevalent and that strains with the same ST always presented the same PspA family.

5. Conclusion

This chapter has reviewed some of the most popular molecular methods for the epidemiological typing of two medically relevant gram-positive cocci, discussing their principles, strengths and weaknesses. We have described several examples of our recent work showing the application of molecular typing techniques to the study of two relevant pathogens. The examples we have considered herein include a relative clonal species such as *S.aureus*, and on the other hand, a pathogen showing a high recombination rate, such as *S. pneumoniae*.

Molecular epidemiology has enormous potential in understanding the evolution of bacterial populations and can help establish appropriate control measures and interventions, including the use of vaccines, therapeutics, public health actions and ongoing pathogen surveillance.

6. Acknowledgments

This work was supported in part by grants from the University of Buenos Aires, Buenos Aires, Argentina, and Agencia Nacional de Promoción Científica y Tecnológica (Buenos Aires, Argentina; PICT 1634) to MM. MM and LB are members of "Carrera del Investigador", CONICET, Argentina.

The authors thank clinical microbiologists for providing isolates and clinical data. In particular, Martha von Spetch and Mabel Regueira; as well as Alejandra Corso for providing the pediatric clone isolate.

7. References

Aanensen, D. M. & B. G. Spratt (2005).The multilocus sequence typing network: mlst.net. *Nucleic. Acids Res.* 33(Web Server issue): W728-33.1362-4962 (Electronic)

Albarracin Orio, A. G., P. R. Cortes, M. Tregnaghi, G. E. Pinas & J. R. Echenique (2008).A new serotype 14 variant of the pneumococcal Spain^{9V}-3 international clone detected in the central region of Argentina. *J. Med. Microbiol.* 57(Pt 8): 992-9.0022-2615 (Print)

Alvarez, C. A., O. J. Barrientes, A. L. Leal, G. A. Contreras, L. Barrero, S. Rincon, L. Diaz, N. Vanegas & C. A. Arias (2006).Community-associated methicillin-resistant *Staphylococcus aureus*, Colombia. *Emerg. Infect. Dis.* 12(12): 2000-1.1080-6040 (Print)

Baggett, H. C., T. W. Hennessy, R. Leman, C. Hamlin, D. Bruden, A. Reasonover, P. Martinez & J. C. Butler (2003).An outbreak of community-onset methicillin-resistant *Staphylococcus aureus* skin infections in southwestern Alaska. *Infect. Control Hosp. Epidemiol.* 24(6): 397-402.0899-823X (Print)

Bannerman, T. L., G. A. Hancock, F. C. Tenover & J. M. Miller (1995).Pulsed-field gel electrophoresis as a replacement for bacteriophage typing of *Staphylococcus aureus*. *J. Clin. Microbiol.* 33(3): 551-5.0095-1137 (Print)

Beall, B., G. Gherardi, M. Lovgren, R. R. Facklam, B. A. Forwick & G. J. Tyrrell (2000).*emm* and *sof* gene sequence variation in relation to serological typing of opacity-factor-positive group A streptococci. *Microbiology* 146 (Pt 5): 1195-209.1350-0872 (Print)

Blanc, D. S., P. Francioli & P. M. Hauser (2002).Poor value of pulsed-field gel electrophoresis to investigate long-term scale epidemiology of methicillin-resistant *Staphylococcus aureus*. *Infect. Genet. Evol.* 2(2): 145-8.1567-1348 (Print)

Bonofiglio, L., M. Regueira, J. Pace, A. Corso, E. Garcia & M. Mollerach (2011).Dissemination of an erythromycin-resistant penicillin-nonsusceptible *Streptococcus pneumoniae* Poland(6B)-20 clone in Argentina. *Microb. Drug Resist.* 17(1): 75-81.1931-8448 (Electronic)

Bosch, T., A. J. de Neeling, L. M. Schouls, K. W. van der Zwaluw, J. A. Kluytmans, H. Grundmann & X. W. Huijsdens (2010).PFGE diversity within the methicillin-resistant *Staphylococcus aureus* clonal lineage ST398. *BMC Microbiol* 10: 40.1471-2180 (Electronic)

Brandileone, M. C., A. L. Andrade, E. M. Teles, R. C. Zanella, T. I. Yara, J. L. Di Fabio & S. K. Hollingshead (2004).Typing of pneumococcal surface protein A (PspA) in *Streptococcus pneumoniae* isolated during epidemiological surveillance in Brazil: towards novel pneumococcal protein vaccines. *Vaccine* 22(29-30): 3890-6.0264-410X (Print)

Brueggemann, A. B., D. T. Griffiths, E. Meats, T. Peto, D. W. Crook & B. G. Spratt (2003).Clonal relationships between invasive and carriage *Streptococcus pneumoniae* and serotype- and clone-specific differences in invasive disease potential. *J. Infect. Dis.* 187(9): 1424-32.0022-1899 (Print)

Camou, T., M. Hortal & A. Tomasz (1998).The apparent importation of penicillin-resistant capsular type 14 Spanish/French clone of *Streptococcus pneumoniae* into Uruguay in the early 1990s. *Microb. Drug Resist.* 4(3): 219-24.1076-6294 (Print)

Cookson, B. D., P. Aparicio, A. Deplano, M. Struelens, R. Goering & R. Marples (1996).Inter-centre comparison of pulsed-field gel electrophoresis for the typing of methicillin-resistant *Staphylococcus aureus*. *J. Med. Microbiol.* 44(3): 179-84.0022-2615 (Print)

Corso, A., I. Santos Sanches, M. Aires de Sousa, A. Rossi & H. de Lencastre (1998).Spread of a methicillin-resistant and multiresistant epidemic clone of *Staphylococcus aureus* in Argentina. *Microb Drug Resist* 4(4): 277-88.1076-6294 (Print)

Crisostomo, M. I., H. Westh, A. Tomasz, M. Chung, D. C. Oliveira & H. de Lencastre (2001).The evolution of methicillin resistance in *Staphylococcus aureus*: similarity of genetic backgrounds in historically early methicillin-susceptible and -resistant isolates and contemporary epidemic clones. *Proc. Natl. Acad. Sci. U. S. A.* 98(17): 9865-70.0027-8424 (Print)

Chambers, H. F. & F. R. Deleo (2009).Waves of resistance: *Staphylococcus aureus* in the antibiotic era. *Nat. Rev. Microbiol.* 7(9): 629-41.1740-1534 (Electronic)

Darabi, A., D. Hocquet & M. J. Dowzicky (2010).Antimicrobial activity against *Streptococcus pneumoniae* and *Haemophilus influenzae* collected globally between 2004 and 2008 as part of the Tigecycline Evaluation and Surveillance Trial. *Diagn. Microbiol. Infect. Dis.* 67(1): 78-86.1879-0070 (Electronic)

Deplano, A., M. Vaneechoutte, G. Verschraegen & M. J. Struelens (1997).Typing of *Staphylococcus aureus* and *Staphylococcus epidermidis* strains by PCR analysis of inter-IS256 spacer length polymorphisms. *J Clin Microbiol* 35(10): 2580-7.0095-1137 (Print)

Descheemaeker, P., S. Chapelle, C. Lammens, M. Hauchecorne, M. Wijdooghe, P. Vandamme, M. Ieven & H. Goossens (2000).Macrolide resistance and erythromycin resistance determinants among Belgian *Streptococcus pyogenes* and *Streptococcus pneumoniae* isolates. *J. Antimicrob. Chemother.* 45(2): 167-73.0305-7453 (Print)

Deurenberg, R. H. & E. E. Stobberingh (2008).The evolution of *Staphylococcus aureus. Infect Genet Evol* 8(6): 747-63.1567-1348 (Print)

Enright, M. C., N. P. Day, C. E. Davies, S. J. Peacock & B. G. Spratt (2000).Multilocus sequence typing for characterization of methicillin-resistant and methicillin-susceptible clones of *Staphylococcus aureus*. *J. Clin. Microbiol.* 38(3): 1008-15.0095-1137 (Print)

Enright, M. C., A. Fenoll, D. Griffiths & B. G. Spratt (1999).The three major Spanish clones of penicillin-resistant *Streptococcus pneumoniae* are the most common clones recovered in recent cases of meningitis in Spain. *J. Clin. Microbiol.* 37(10): 3210-6.0095-1137 (Print)

Enright, M. C., D. A. Robinson, G. Randle, E. J. Feil, H. Grundmann & B. G. Spratt (2002).The evolutionary history of methicillin-resistant *Staphylococcus aureus* (MRSA). *Proc. Natl. Acad. Sci. U. S. A.* 99(11): 7687-92.0027-8424 (Print)

Enright, M. C. & B. G. Spratt (1998).A multilocus sequence typing scheme for *Streptococcus pneumoniae*: identification of clones associated with serious invasive disease. *Microbiology* 144 (Pt 11): 3049-60.1350-0872 (Print)

Feil, E. J., E. C. Holmes, D. E. Bessen, M. S. Chan, N. P. Day, M. C. Enright, R. Goldstein, D. W. Hood, A. Kalia, C. E. Moore, J. Zhou & B. G. Spratt (2001).Recombination within natural populations of pathogenic bacteria: short-term empirical estimates and long-term phylogenetic consequences. *Proc. Natl. Acad. Sci. U. S. A.* 98(1): 182-7.0027-8424 (Print)

Feil, E. J., B. C. Li, D. M. Aanensen, W. P. Hanage & B. G. Spratt (2004).eBURST: inferring patterns of evolutionary descent among clusters of related bacterial genotypes from multilocus sequence typing data. *J. Bacteriol.* 186(5): 1518-30.0021-9193 (Print)

Fergie, J. E. & K. Purcell (2001).Community-acquired methicillin-resistant *Staphylococcus aureus* infections in south Texas children. *Pediatr. Infect. Dis. J.* 20(9): 860-3.0891-3668 (Print)

Fridkin, S. K., J. C. Hageman, M. Morrison, L. T. Sanza, K. Como-Sabetti, J. A. Jernigan, K. Harriman, L. H. Harrison, R. Lynfield & M. M. Farley (2005).Methicillin-resistant *Staphylococcus aureus* disease in three communities. *N. Engl. J. Med.* 352(14): 1436-44.1533-4406 (Electronic)

Gabastou, J. M., C. I. Agudelo, M. C. Brandileone, E. Castaneda, A. P. de Lemos & J. L. Di Fabio (2008).Characterization of invasive isolates of *S. pneumoniae, H. influenzae,* and *N. meningitidis* in Latin America and the Caribbean: SIREVA II, 2000-2005. *Rev. Panam. Salud Publica* 24(1): 1-15.1020-4989 (Print)

Gamboa, L., T. Camou, M. Hortal & E. Castaneda (2002).Dissemination of *Streptococcus pneumoniae* clone Colombia[(5)]-19 in Latin America. *J. Clin. Microbiol.* 40(11): 3942-50.0095-1137 (Print)

Gardella, N., S. Murzicato, S. Di Gregorio, A. Cuirolo, J. Desse, F. Crudo, G. Gutkind & M. Mollerach (2011).Prevalence and characterization of methicillin-resistant *Staphylococcus aureus* among healthy children in a city of Argentina. *Infect. Genet. Evol.* 11(5): 1066-71.1567-7257 (Electronic)

Gardella, N., R. Picasso, S. C. Predari, M. Lasala, M. Foccoli, G. Benchetrit, A. Famiglietti, M. Catalano, M. Mollerach & G. Gutkind (2005).Methicillin-resistant *Staphylococcus aureus* strains in Buenos Aires teaching hospitals: replacement of the multidrug resistant South American clone by another susceptible to rifampin, minocycline and trimethoprim-sulfamethoxazole. *Rev. Argent. Microbiol.* 37(3): 156-60.0325-7541 (Print)

Gardella, N., M. von Specht, A. Cuirolo, A. Rosato, G. Gutkind & M. Mollerach (2008).Community-associated methicillin-resistant *Staphylococcus aureus,* eastern Argentina. *Diagn. Microbiol. Infect. Dis.* 62(3): 343-7.0732-8893 (Print)

Gherardi, G., C. G. Whitney, R. R. Facklam & B. Beall (2000).Major related sets of antibiotic-resistant Pneumococci in the United States as determined by pulsed-field gel electrophoresis and pbp1a-pbp2b-pbp2x-dhf restriction profiles. *J. Infect. Dis.* 181(1): 216-29.0022-1899 (Print)

Godoy, D., G. Randle, A. J. Simpson, D. M. Aanensen, T. L. Pitt, R. Kinoshita & B. G. Spratt (2003).Multilocus sequence typing and evolutionary relationships among the causative agents of melioidosis and glanders, *Burkholderia pseudomallei* and *Burkholderia mallei. J. Clin. Microbiol.* 41(5): 2068-79.0095-1137 (Print)

Goering, R. V. (2010).Pulsed field gel electrophoresis: a review of application and interpretation in the molecular epidemiology of infectious disease. *Infect. Genet. Evol.* 10(7): 866-75.1567-7257 (Electronic)

Halpin, J. L., N. M. Garrett, E. M. Ribot, L. M. Graves & K. L. Cooper (2010).Re-evaluation, optimization, and multilaboratory validation of the PulseNet-standardized pulsed-field gel electrophoresis protocol for *Listeria monocytogenes. Foodborne Pathog. Dis.* 7(3): 293-8.1556-7125 (Electronic)

Harmsen, D., H. Claus, W. Witte, J. Rothganger, D. Turnwald & U. Vogel (2003).Typing of methicillin-resistant *Staphylococcus aureus* in a university hospital setting by using novel software for spa repeat determination and database management. *J. Clin. Microbiol.* 41(12): 5442-8.0095-1137 (Print)

Hermans, P., M. Sluijter, T. Hoogenboezem, H. Heersma, A. v. Belkum & R. d. Groot (1995).Comparative Study of Five Different DNA Fingerprint Techniques for Molecular Typing of *Streptococcus pneumoniae* Strains. *J. Clin. Microbiol.* 33: 1606-1612

Hermans, P. W., M. Sluijter, S. Dejsirilert, N. Lemmens, K. Elzenaar, A. van Veen, W. H. Goessens and R. de Groot (1997).Molecular epidemiology of drug-resistant pneumococci: toward an international approach. *Microb. Drug Resist.* 3(3): 243-51.1076-6294 (Print)

Jolley, K. A., J. Kalmusova, E. J. Feil, S. Gupta, M. Musilek, P. Kriz & M. C. Maiden (2000).Carried meningococci in the Czech Republic: a diverse recombining population. *J. Clin. Microbiol.* 38(12): 4492-8.0095-1137 (Print)

Kam, K. M., C. K. Luey, M. B. Parsons, K. L. Cooper, G. B. Nair, M. Alam, M. A. Islam, D. T. Cheung, Y. W. Chu, T. Ramamurthy, G. P. Pazhani, S. K. Bhattacharya, H. Watanabe, J. Terajima, E. Arakawa, O. A. Ratchtrachenchai, S. Huttayananont, E. M. Ribot, P. Gerner-Smidt & B. Swaminathan (2008).Evaluation and validation of a PulseNet standardized pulsed-field gel electrophoresis protocol for subtyping *Vibrio parahaemolyticus*: an international multicenter collaborative study. *J. Clin. Microbiol.* 46(8): 2766-73.1098-660X (Electronic)

Klugman, K. P. (2002).The successful clone: the vector of dissemination of resistance in *Streptococcus pneumoniae. J. Antimicrob. Chemother.* 50 Suppl S2: 1-5.0305-7453 (Print)

Kreiswirth, B. N., S. M. Lutwick, E. K. Chapnick, J. D. Gradon, L. I. Lutwick, D. V. Sepkowitz, W. Eisner & M. H. Levi (1995).Tracing the spread of methicillin-resistant *Staphylococcus aureus* by Southern blot hybridization using gene-specific probes of *mec* and Tn554. *Microb Drug Resist* 1(4): 307-13.1076-6294 (Print)

Lina, G., Y. Piemont, F. Godail-Gamot, M. Bes, M. O. Peter, V. Gauduchon, F. Vandenesch & J. Etienne (1999).Involvement of Panton-Valentine leukocidin-producing *Staphylococcus aureus* in primary skin infections and pneumonia. *Clin. Infect. Dis.* 29(5): 1128-32.1058-4838 (Print)

Ma, X. X., A. Galiana, W. Pedreira, M. Mowszowicz, I. Christophersen, S. Machiavello, L. Lope, S. Benaderet, F. Buela, W. Vincentino, M. Albini, O. Bertaux, I. Constenla, H. Bagnulo, L. Llosa, T. Ito & K. Hiramatsu (2005).Community-acquired methicillin-resistant *Staphylococcus aureus*, Uruguay. *Emerg. Infect. Dis.* 11(6): 973-6.1080-6040 (Print)

Ma, X. X., T. Ito, C. Tiensasitorn, M. Jamklang, P. Chongtrakool, S. Boyle-Vavra, R. S. Daum & K. Hiramatsu (2002).Novel type of staphylococcal cassette chromosome *mec* identified in community-acquired methicillin-resistant *Staphylococcus aureus* strains. *Antimicrob. Agents Chemother.* 46(4): 1147-52.0066-4804 (Print)

Maiden, M. C., J. A. Bygraves, E. Feil, G. Morelli, J. E. Russell, R. Urwin, Q. Zhang, J. Zhou, K. Zurth, D. A. Caugant, I. M. Feavers, M. Achtman & B. G. Spratt (1998).Multilocus sequence typing: a portable approach to the identification of clones within populations of pathogenic microorganisms. *Proc. Natl. Acad. Sci. U. S. A.* 95(6): 3140-5.0027-8424 (Print)

Maslow, J. & M. E. Mulligan (1996).Epidemiologic typing systems. *Infect. Control Hosp. Epidemiol.* 17(9): 595-604.0899-823X (Print)

McDaniel, L. S., J. S. Sheffield, P. Delucchi & D. E. Briles (1991).PspA, a surface protein of *Streptococcus pneumoniae*, is capable of eliciting protection against pneumococci of more than one capsular type. *Infect. Immun.* 59(1): 222-8.0019-9567 (Print)

McDougal, L. K., C. D. Steward, G. E. Killgore, J. M. Chaitram, S. K. McAllister & F. C. Tenover (2003).Pulsed-field gel electrophoresis typing of oxacillin-resistant *Staphylococcus aureus* isolates from the United States: establishing a national database. *J. Clin. Microbiol.* 41(11): 5113-20.0095-1137 (Print)

McGee, L., L. McDougal, J. Zhou, B. G. Spratt, F. C. Tenover, R. George, R. Hakenbeck, W. Hryniewicz, J. C. Lefevre, A. Tomasz & K. P. Klugman (2001).Nomenclature of major antimicrobial-resistant clones of *Streptococcus pneumoniae* defined by the pneumococcal molecular epidemiology network. *J. Clin. Microbiol.* 39(7): 2565-71.0095-1137 (Print)

Meats, E., A. B. Brueggemann, M. C. Enright, K. Sleeman, D. T. Griffiths, D. W. Crook & B. G. Spratt (2003).Stability of serotypes during nasopharyngeal carriage of *Streptococcus pneumoniae*. *J. Clin. Microbiol.* 41(1): 386-92.0095-1137 (Print)

Mollerach, M., M. Regueira, L. Bonofiglio, R. Callejo, J. Pace, J. L. Di Fabio, S. Hollingshead & D. Briles (2004).Invasive *Streptococcus pneumoniae* isolates from Argentinian children: serotypes, families of pneumococcal surface protein A (PspA) and genetic diversity. *Epidemiol. Infect.* 132(2): 177-84.0950-2688 (Print)

Munoz, R., T. J. Coffey, M. Daniels, C. G. Dowson, G. Laible, J. Casal, R. Hakenbeck, M. Jacobs, J. M. Musser, B. G. Spratt and Tomasz, A. (1991) Intercontinental spread of a multiresistant clone of serotype 23F *Streptococcus pneumoniae*. *J. Infect. Dis.* 164(2): 302-6.0022-1899 (Print)

Naimi, T. S., K. H. LeDell, K. Como-Sabetti, S. M. Borchardt, D. J. Boxrud, J. Etienne, S. K. Johnson, F. Vandenesch, S. Fridkin, C. O'Boyle, R. N. Danila & R. Lynfield (2003).Comparison of community- and health care-associated methicillin-resistant *Staphylococcus aureus* infection. *J. A. M. A.* 290(22): 2976-84.1538-3598 (Electronic)

Oliveira, D. C. & H. de Lencastre (2002).Multiplex PCR strategy for rapid identification of structural types and variants of the mec element in methicillin-resistant *Staphylococcus aureus*. *Antimicrob. Agents Chemother.* 46(7): 2155-61.0066-4804 (Print)

Organización Panamericana de la Salud (2007). Informe regional de SIREVA II: datos por país y por grupos de edad sobre las características de los aislamientos de

Streptococcus pneumoniae,Haemophilus influenzae y *Neisseria meningitidis*, en procesos invasivos, 2000–2005. Washington, D.C.

Pardo, L., V. Machado, M. Mollerach, M. I. Mota, L. P. Tuchscherr, P. Gadea, N. Gardella, D. O. Sordelli, M. Vola, F. Schelotto & G. Varela (2009).Characteristics of Community-Associated Methicillin-Resistant *Staphylococcus aureus* (CA-MRSA) Strains Isolated from Skin and Soft-Tissue Infections in Uruguay. *Int. J. Microbiol.* 2009: 472126.1687-9198 (Electronic)

Qian, J., K. Yao, L. Xue, G. Xie, Y. Zheng, C. Wang, Y. Shang, H. Wang, L. Wan, L. Liu, C. Li, W. Ji, Y. Wang, P. Xu, S. Yu, Y. W. Tang & Y. Yang (2011).Diversity of pneumococcal surface protein A (PspA) and relation to sequence typing in *Streptococcus pneumoniae* causing invasive disease in Chinese children. *Eur. J. Clin. Microbiol. Infect. Dis.*1435-4373 (Electronic)

Ribeiro, A., C. Dias, M. C. Silva-Carvalho, L. Berquo, F. A. Ferreira, R. N. Santos, B. T. Ferreira-Carvalho & A. M. Figueiredo (2005).First report of infection with community-acquired methicillin-resistant *Staphylococcus aureus* in South America. *J. Clin. Microbiol.* 43(4): 1985-8.0095-1137 (Print)

Riedel, S., S. E. Beekmann, K. P. Heilmann, S. S. Richter, J. Garcia-de-Lomas, M. Ferech, H. Goosens & G. V. Doern (2007).Antimicrobial use in Europe and antimicrobial resistance in *Streptococcus pneumoniae. Eur. J. Clin. Microbiol. Infect. Dis.* 26(7): 485-90.0934-9723 (Print)

Rossi, A., A. Corso, J. Pace, M. Regueira & A. Tomasz (1998).Penicillin-resistant *Streptococcus pneumoniae* in Argentina: frequent occurrence of an internationally spread serotype 14 clone. *Microb. Drug Resist.* 4(3): 225-31.1076-6294 (Print)

Ruvinsky, R., M. Regueirea, A. Corso, S. Fossari, M. Moscoloni, S. García & L. Di-Fabio (2008).*Streptococcus pneumoniae* Invasive Diseases:Surveillance of serotypes in children in Argentina. *Abstract presented at the 46th Annual Interscience Conference on Antimicrobial Agents and Chemotherapy (ICAAC).* Washington DC, USA. October 25-28. Abstract no. G1-2087.

Sadowy, E., R. Izdebski, A. Skoczynska, P. Grzesiowski, M. Gniadkowski & W. Hryniewicz (2007).Phenotypic and molecular analysis of penicillin-nonsusceptible *Streptococcus pneumoniae* isolates in Poland. *Antimicrob. Agents Chemother.* 51(1): 40-7.0066-4804 (Print)

Sadowy, E., A. Kuch, M. Gniadkowski & W. Hryniewicz (2010).Expansion and evolution of the *Streptococcus pneumoniae* Spain[9V]-ST156 clonal complex in Poland. *Antimicrob. Agents Chemother.* 54(5): 1720-7.1098-6596 (Electronic)

Salgado, C. D., B. M. Farr & D. P. Calfee (2003).Community-acquired methicillin-resistant *Staphylococcus aureus*: a meta-analysis of prevalence and risk factors. *Clin. Infect. Dis.* 36(2): 131-9.1537-6591 (Electronic)

Saluja, S. K. & J. N. Weiser (1995).The genetic basis of colony opacity in *Streptococcus pneumoniae*: evidence for the effect of box elements on the frequency of phenotypic variation. *Mol. Microbiol.* 16(2): 215-27.0950-382X (Print)

Saulnier, P., C. Bourneix, G. Prevost & A. Andremont (1993).Random amplified polymorphic DNA assay is less discriminant than pulsed-field gel electrophoresis for typing strains of methicillin-resistant *Staphylococcus aureus*. *J. Clin. Microbiol.* 31(4): 982-5.0095-1137 (Print)

Shopsin, B., M. Gomez, S. O. Montgomery, D. H. Smith, M. Waddington, D. E. Dodge, D. A. Bost, M. Riehman, S. Naidich & B. N. Kreiswirth (1999).Evaluation of protein A gene polymorphic region DNA sequencing for typing of Staphylococcus aureus strains. J. Clin. Microbiol. 37(11): 3556-63.0095-1137 (Print)

Shopsin, B., M. Gomez, M. Waddington, M. Riehman & B. N. Kreiswirth (2000).Use of coagulase gene (coa) repeat region nucleotide sequences for typing of methicillin-resistant Staphylococcus aureus strains. J. Clin. Microbiol. 38(9): 3453-6.0095-1137 (Print)

Shopsin, B. & B. N. Kreiswirth (2001).Molecular epidemiology of methicillin-resistant Staphylococcus aureus. Emerg. Infect. Dis. 7(2): 323-6.1080-6040 (Print)

Smith, A. J., J. Jefferies, S. C. Clarke, C. Dowson, G. F. Edwards & T. J. Mitchell (2006).Distribution of epidemic antibiotic-resistant pneumococcal clones in Scottish pneumococcal isolates analysed by multilocus sequence typing. Microbiology 152(Pt 2): 361-5.1350-0872 (Print)

Sola, C., G. Gribaudo, C. M. S. Groups, A. Vindel, L. Patrito & J. L. Bocco (2002).Identification of a novel methicillin-resistant Staphylococcus aureus epidemic clone in Córdoba, Argentina, involved in nosocomial infections. Journal of Antimicrobial Microbiology 40(4): 1427-1435

Soriano, F., F. Cafini, L. Aguilar, D. Tarrago, L. Alou, M. J. Gimenez, M. Gracia, M. C. Ponte, D. Leu, M. Pana, I. Letowska & A. Fenoll (2008).Breakthrough in penicillin resistance? Streptococcus pneumoniae isolates with penicillin/cefotaxime MICs of 16 mg/L and their genotypic and geographical relatedness. J. Antimicrob. Chemother. 62(6): 1234-40.1460-2091 (Electronic)

Struelens, M. J. (1996).Consensus guidelines for appropriate use and evaluation of microbial epidemiologic typing systems. Clin. Microbiol. Infect. 2(1): 2-11.1469-0691 (Electronic)

Struelens, M. J., R. Bax, A. Deplano, W. G. Quint & A. Van Belkum (1993).Concordant clonal delineation of methicillin-resistant Staphylococcus aureus by macrorestriction analysis and polymerase chain reaction genome fingerprinting. J. Clin. Microbiol. 31(8): 1964-70.0095-1137 (Print)

Tang, Y. W., M. G. Waddington, D. H. Smith, J. M. Manahan, P. C. Kohner, L. M. Highsmith, H. Li, F. R. Cockerill, 3rd, R. L. Thompson, S. O. Montgomery & D. H. Persing (2000).Comparison of protein A gene sequencing with pulsed-field gel electrophoresis and epidemiologic data for molecular typing of methicillin-resistant Staphylococcus aureus. J. Clin. Microbiol. 38(4): 1347-51.0095-1137 (Print)

te Witt, R., A. van Belkum, W. G. MacKay, P. S. Wallace & W. B. van Leeuwen (2010).External quality assessment of the molecular diagnostics and genotyping of meticillin-resistant Staphylococcus aureus. Eur. J. Clin. Microbiol. Infect. Dis. 29(3): 295-300.1435-4373 (Electronic)

Tenover, F. C., R. Arbeit, G. Archer, J. Biddle, S. Byrne, R. Goering, G. Hancock, G. A. Hebert, B. Hill, R. Hollis, W. R. Javis, B. Kreiswirth, W. Eisner, J. Maslow, L. K. McDougal, J. M. Miller, M. Mulligan & M. A. Pfaller.(1994).Comparison of traditional and molecular methods of typing isolates of Staphylococcus aureus. J. Clin. Microbiol. 32(2): 407-15.0095-1137 (Print)

Tenover, F. C., R. D. Arbeit, R. V. Goering, P. A. Mickelsen, B. E. Murray, D. H. Persing & B. Swaminathan (1995).Interpreting chromosomal DNA restriction patterns produced by pulsed-field gel electrophoresis: criteria for bacterial strain typing. *J. Clin. Microbiol.* 33(9): 2233-9.0095-1137 (Print).

Tenover, F.C., Arbeit, R.D., Goering, R. V. (1997).How to select and interpret molecular strain typing methods for epidemiological studies of bacterial infections: a review for healthcare epidemiologists. Molecular Typing Working Group of the Society for Healthcare Epidemiology of America. Infect. Control Hosp. Epidemiol.18(6):426-39.0899-823X (print).

Tibayrenc, M. (1995).Population genetics and strain typing of microorganisms: how to detect departures from panmixia without individualizing alleles and loci. *C. R. Acad. Sci. III.* 318(1): 135-9.0764-4469 (Print)

Tu, A. H., R. L. Fulgham, M. A. McCrory, D. E. Briles & A. J. Szalai (1999).Pneumococcal surface protein A inhibits complement activation by *Streptococcus pneumoniae. Infect. Immun.* 67(9): 4720-4.0019-9567 (Print)

van Belkum, A., J. Kluytmans, W. van Leeuwen, R. Bax, W. Quint, E. Peters, A. Fluit, C. Vandenbroucke-Grauls, A. van den Brule, H. Koeleman, W. Melchers, J. Meis, A.Elaichouni, M. Vaneechoutte, F. Moonens, N. Maes, M. Struelens, F. Tenover & H. Verbrugh (1995).Multicenter evaluation of arbitrarily primed PCR for typing of *Staphylococcus aureus* strains. *J. Clin. Microbiol.* 33(6): 1537-47.0095-1137 (Print)

van Belkum, A., M. Sluijuter, R. de Groot, H. Verbrugh & P. W. Hermans (1996).Novel BOX repeat PCR assay for high-resolution typing of *Streptococcus pneumoniae* strains. *J. Clin. Microbiol.* 34(5): 1176-9.0095-1137 (Print)

van Belkum, A., P. T. Tassios, L. Dijkshoorn, S. Haeggman, B. Cookson, N. K. Fry, V. Fussing, J. Green, E. Feil, P. Gerner-Smidt, S. Brisse & M. Struelens (2007).Guidelines for the validation and application of typing methods for use in bacterial epidemiology. *Clin. Microbiol. Infect.* 13 Suppl 3: 1-46.1198-743X (Print)

van Belkum, A., W. van Leeuwen, M. E. Kaufmann, B. Cookson, F. Forey, J. Etienne, R. Goering, F. Tenover, C. Steward, F. O'Brien, W. Grubb, P. Tassios, N. Legakis, A. Morvan, N. El Solh, R. de Ryck, M. Struelens, S. Salmenlinna, J. Vuopio-Varkila, M. Kooistra, A. Talens, W. Witte & H. Verbrugh (1998).Assessment of resolution and intercenter reproducibility of results of genotyping *Staphylococcus aureus* by pulsed-field gel electrophoresis of SmaI macrorestriction fragments: a multicenter study. *J Clin Microbiol* 36(6): 1653-9.0095-1137 (Print)

van der Zee, A., H. Verbakel, J. C. van Zon, I. Frenay, A. van Belkum, M. Peeters, A. Buiting & A. Bergmans (1999).Molecular genotyping of *Staphylococcus aureus* strains: comparison of repetitive element sequence-based PCR with various typing methods and isolation of a novel epidemicity marker. *J. Clin. Microbiol.* 37(2): 342-9.0095-1137 (Print)

Vázquez,J. A., Berrón, S.(2004).Multilocus sequence typing: el marcador molecular de la era de Internet.Enfermedades Infecciosas y Microbiología Clínica (2004). Enfermedades infecciosas y microbiología clínica.22 (2): 113-120.0213005X

Versalovic, J., T. Koeuth & J. R. Lupski (1991).Distribution of repetitive DNA sequences in eubacteria and application to fingerprinting of bacterial genomes. *Nucleic Acids. Res.* 19(24): 6823-31.0305-1048 (Print)

von Specht, M., N. Gardella, P. Tagliaferri, G. Gutkind & M. Mollerach (2006).Methicillin-resistant *Staphylococcus aureus* in community-acquired meningitis. *Eur. J. Clin. Microbiol. Infect. Dis.* 25(4): 267-9.0934-9723 (Print)

Wei, M. Q., E. E. Udo & W. B. Grubb (1992).Typing of methicillin-resistant *Staphylococcus aureus* with IS256. *FEMS Microbiol. Lett.* 78(2-3): 175-80.0378-1097 (Print).

Willems, R. J., Hanage, W.P., Bessen, D. E., Feil, E. J. (2011).Population biology of Gram-positive pathogens: high-risk clones for dissemination of antibiotic resistance.FEMS Microbiol. Rev.35(5):872-900.0168-6445 (print).

Zemlickova, H., M. I. Crisostomo, M. C. Brandileone, T. Camou, E. Castaneda, A. Corso, G. Echaniz-Aviles, M. Pasztor & A. Tomasz (2005).Serotypes and clonal types of penicillin-susceptible *Streptococcus pneumoniae* causing invasive disease in children in five Latin American countries. *Microb. Drug Resist.* 11(3): 195-204.1076-6294 (Print)

Zhang, Q. Y., D. M. Jones, J. A. Saez Nieto, E. Perez Trallero & B. G. Spratt (1990).Genetic diversity of penicillin-binding protein 2 genes of penicillin-resistant strains of *Neisseria meningitidis* revealed by fingerprinting of amplified DNA. *Antimicrob Agents Chemother* 34(8): 1523-8.0066-4804 (Print).

8

Molecular and Proteolytic Profiles of *Trypanosoma cruzi* Sylvatic Isolates from Rio de Janeiro-Brazil

Suzete A. O. Gomes[1,2] et al.*
*¹Laboratório de Biologia de Insetos, GBG
Universidade Federal Fluminense-UFF, Rio de Janeiro, RJ
²Laboratório de Transmissores de Leishmanioses
Setor de Entomologia Médica e Forense IOC-FIOCRUZ-Rio de Janeiro, RJ
Brazil*

1. Introduction

Chagas disease, also known as American trypanosomiasis, has its epidemiology conditioned to the (i) triatominae vectors, (ii) etiologic agent, *Trypanosoma cruzi*, and (iii) sylvatic and sinantropic reservoirs, the mammals. Social factors associated with economic factors, such as industry development, population growth and rural area colonization, which lead directly to ecological imbalance, provide favorable conditions for the disease establishment (Barretto, 1967; Ávila-Pires, 1976).

In 1909, Carlos Chagas releases his discovery on a new human disease, the American trypanosomiasis, subsequently known as Chagas disease. Carlos Chagas described the etiologic agent, the protozoan belonging to the Trypanosomatidae family *Trypanosoma cruzi*, and its insect vector belonging to the Hemiptera order, Triatominae subfamily, the so-called kissing bug (Chagas, 1909; Lent & Wygodzinsky, 1979).

The natural history of the Chagas disease probably started millions of years ago probably as a sylvatic enzooty, and it is still present in different areas from Brazilian territory. The arrival of men in these areas, as well as comprehensive deforestation caused by extensive farming during the past 300 years has caused triatomine insects, formerly sylvatic animal blood-sucking bugs, to meet men (Ferreira et al., 1996; Coura, 2007). Hence, the disease was characterized as a zoonosis, when men invaded the sylvatic habitat, deforesting and changing the ecological balance, and making triatomine bugs access to the residences.

* Danielle Misael[2], Cristina S. Silva[2], Denise Feder[1], Alice H. Ricardo-Silva[2], André L. S. Santos[3], Jacenir R. Santos-Mallet[2] and Teresa Cristina M. Gonçalves[2]
¹Laboratório de Biologia de Insetos, GBG, Universidade Federal Fluminense-UFF, Rio de Janeiro, RJ, Brasil
²Laboratório de Transmissores de Leishmanioses, Setor de Entomologia Médica e Forense, IOC-FIOCRUZ-Rio de Janeiro, RJ, Brasil
³Laboratório de Estudos Integrados em Bioquímica Microbiana, Instituto de Microbiologia Paulo de Góes (IMPG), Bloco E-subsolo, Universidade Federal do Rio de Janeiro (UFRJ), Rio de Janeiro, RJ, Brasil

Therefore, the transmission cycle of *T. cruzi* is comprised by a sylvatic cycle, in which the parasite circulates among mammals and sylvatic vectors, and a domiciliary cycle, in which the infection is ensued by the contact of mammals, sylvatic vectors and sinantropic animals with domestic and domiciled animals, including men (Barretto, 1979).

Human Chagas disease, an antropozoonosis that evolved from a zoonosis, is strongly related with men's social class, type of work and habitation (Dias, 2000). During the 70's, the disease endemic area achieved at least 2,450 Brazilian cities, 771 of which were detected to have *Triatoma infestans*, the main disease vector in Brazil. At that time, there were over five million people affected by the disease in the country, with an incidence of approximately one hundred thousand new cases yearly and mortality above ten thousand deaths yearly. Less than five percent of blood banks used to control donors and over seven hundred cities had their homes infected by *T. infestans*. This situation led scientists to press the government to prioritize a national program against the disease. Homes from endemic areas were sprinkled with the appropriate insecticide and, in accordance with law; mandatory screening of blood donors was implemented throughout the country (Dias et al., 2002). The control program of the main vector in Brazil was recognized in 2006, with a certificate from the World Health Organization (WHO) for virtual elimination of *T. infestans* in Brazil (Dias, 2006). As the main vector was eliminated, currently there is a concern that other Triatominae species, formerly deemed secondary in the disease transmission, such as *Triatoma brasiliensis*, *Triatoma pseudomaculata* and *Panstrongylus megistus*, take the place of *T. infestans* in some locations, therefore becoming potential disease vectors in Brazil (Coura, 2009).

Despite the great progress in controlling vector and transfusion transmission in the countries from the Southern Cone, transmission is ongoing in other parts of the continent, and the issue of already infected people, most of whom are in the chronic phase of the disease, is still a challenge to public health (Urbina, 1999). Currently Chagas disease affects between twelve and fourteen million people in Latin America, and at least 60 million people live in areas with transmission risk (WHO, 2002). In Brazil, the disease notification became compulsory as per Ordinance V of Health Surveillance Secretary of Ministry of Health dated February 21, 2006.

2. Triatomines

The first report of triatomine existence was recorded by the Spanish Francisco López de Gomara, in 1514, when mentioning Darién region he said: "Hay muchas garrapatas y chinches com alas", apparently referring to *Rhodnius prolixus* (Stål, 1859) (León, 1962). *Cimex rubrofasciatus* (*Triatoma rubrofasciata*), was described in 1773 by De Geer, and later assigned by Laporte as the type species of *Triatoma* genus (Lent & Wygodzinsky, 1979). In Brazil, the first report of triatomine in domicile was possibly *Panstrongylus megistus* (Burmeister, 1835) (Gardner, 1942). However, the identification of *Trypanosoma cruzi* sylvatic isolates is contemporary to the discovery of this parasite and Chagas disease by Carlos Chagas in 1909. When they went to Lassance, Minas Gerais, Brazil, for malaria epidemics study, he identified flagellated forms in the intestine of triatomine of *Conorhinus megistus* (*Panstrongylus megistus*) in humans and cats, referring to them as *Schizotrypanum cruzi* (Chagas, 1909). Later Chagas (1912) isolated the parasite in armadillos (*Tatusia novemcincta*, now called *Daysipus novemcinctus*), identifying the *T. cruzi* sylvatic reservoirs, and in the same ecotope he found infected *Triatoma geniculata* (*Panstrongylus geniculatus*) specimens, establishing the disease sylvatic cycle (Coura & Dias, 2009).

Between 1913 and 1924 it became evident that the disease was not restricted to Brazil, being diagnosed in other countries in Central and South Americas, such as El Salvador, Venezuela, Peru and Argentina (Talice et al., 1940; Zeledón, 1981). In subsequent studies, Coura & Dias, 2009 mentions that Chagas (1924) demonstrated *T. cruzi* transmission cycle in the Amazon region with the identification of this parasite in monkeys of *Saimiri scirius* species.

In Rio de Janeiro state, the first Triatominae occurrence dated 1859, when Stal described *Conorhinus vitticeps* species, now called *Triatoma vitticeps*. At that time, Rio de Janeiro was assigned as type location, without defining whether it referred to the city or state.

Following this finding, Neiva (1914) recorded the occurrence of *T. vitticeps* in Conceição de Macabu, formerly Macaé city district, presently Conceição de Macabu city. Due to information accuracy, Lent (1942) suggested it would be considered as the type location of *T. vitticeps*.

Subsequently, Pinto (1931, as cited in Lent, 1942) pointed out its presence in Magé, and Lent (1942) in Nova Friburgo, at Secretario location in Petrópolis city and at Federal District, which was Rio de Janeiro at that time. In Minas Gerais state, it was observed by the first time by Martins et al (1940), and in Espírito Santo state, as mentioned by Lent (1942).

In Rio de Janeiro state other species were also found. Guimarães and Jansen (1943) collected *Panstrongylus megistus* specimens in a building by the hill, and identified *Trypanosoma cruzi* sylvatic reservoir (skunk), but did not find the sylvatic focus. Dias (1943) listed Chagas disease transmitters in Rio de Janeiro as being *Panstrongylus megistus*, *Panstrongylus geniculatus* (Latreille, 1811), *Triatoma vitticeps* (Stal, 1859), *Triatoma oswaldoi* (Neiva & Pinto, 1923), *Triatoma infestans* (Klug) and *Triatoma rubrofasciata* (De Geer, 1773), first recording the occurrence of *Schizotrypanum sp*-infected *P. megistus* in two districts in the capital of Republic (Santa Tereza and Botafogo). In 1953, in a survey performed at Araruama and Magé, Dias stated it was a relevant issue for the State, while Bustamante & Gusmão 1953 pointed out the presence of *T. infestans* at Resende and Itaverá cities. New findings have been identified, such as that of Coura et al. (1966), who found *P. megistus*, *Triatoma tibiamaculata* and *T. rubrofasciata* in three districts at Rio de Janeiro city, and that of Aragão & Souza (1971), who signalized the presence of *T. infestans* colonizing domiciles at two cities in Baixada Fluminense. In the same year, Coura et al. (1966) described some autochthonous instances of *T. infestans*-transmitted Chagas disease at Baixada Fluminense, and Becerra-Fuentes et al. (1971) recorded *T. rubrofasciata* occurrence at Morro do Telégrafo in the former Guanabara state. Silveira et al. (1982) performed an entomologic inquiry at Duque de Caxias and Nova Iguaçu cities (RJ), and only found *T. infestans* species. Ferreira et al. (1986) verified the occurrence of *T. vitticeps*, and positivity for *T. cruzi*-like forms, in 12 cities, of which the one with the highest incidence for both observations was Triunfo location at Santa Maria Madalena city. In 1989, a *P. geniculatus* specimen was found in a domicile at São Sebastião do Alto city (RJ) (personal communication with Teresa Cristina M. Gonçalves). The occurrence of *Rhodnius prolixus* (Stål, 1859) in Teresópolis was pointed out by Pinho et al. (1998), which caused questioning, once this species was restricted to the northern region of the country. Nowadays it is known this species does not occur in Brazil (Monteiro et al., 2000, 2003). *T. vitticeps* was found in Poço das Antas, Silva Jardim city, by Lisbôa et al. (1996), and in Santa Maria Madalena by Gonçalves et al. (1998). In both locations, biological and morphological characterization of *T. cruzi* isolates, obtained for both triatomine bugs and vertebrate hosts, confirmed the maintenance of enzootic disease form. In the period from 2008 to 2010 *T. vitticeps* was pointed out at Cantagalo, Tanguá, Trajano de Morais, and São Fidélis cities (Oliveira et al., 2010).

In Espírito Santo, where *T. vitticeps* incidence was also signalized, the rates of infection by *T. cruzi*-like forms were assessed in specimens collected in the domicile: 4% by Santos et al. (1969) at Alfredo Chaves (ES); 25.2% by Silveira et al. (1983) at Cachoeiro do Itapemirim and Guarapari (ES); 35.2% by Ferreira et al. (1986) in 12 cities from Rio de Janeiro state; 64.70% by Sessa & Carias (1986) in 19 cities from Espírito Santo state; and 70.2% and 51.8%, respectively, for females and males, by Dias et al. (1989).

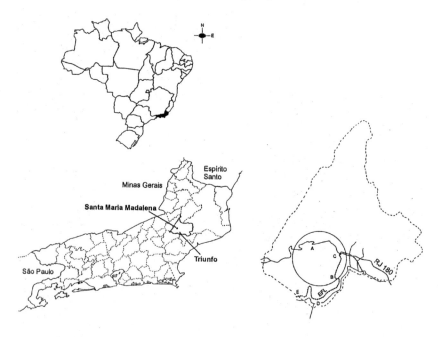

Fig. 1. Studied area and sites of capture of *Triatoma vitticeps* in Triunfo, Santa Maria Madalena, Municipal district, State of Rio de Janeiro, Brazil.

Data from National Health Foundation ("FUNASA") signalized *T. vitticeps* presence in the northern region of Rio de Janeiro state, and the number of notifications on adult form occurrence was increasing (Lopes et al., 2009; Dias et al., 2010). Although studies regarding *T. vitticeps* biology have suggested that this species would not represent a major concern from epidemiologic point of view (Dias, 1956; Heitzmann-Fontenelle, 1980; Silva, 1985; Diotaiuti et al., 1987; Gonçalves et al., 1988, 1989), reports of this species frequently invading the domicile with high *T. cruzi* infection rates (Gonçalves et al., 1998, Gonçalves, 2000) indicated its study was required. With sylvatic habit and unknown habitat, this species ecobiology was studied in further details at Triunfo district, Santa Maria Madalena city (RJ), in three areas (A, B and C) (Figure 1). Of the triatomine bugs collected, 68 *T. cruzi* samples were isolated, which showed heterogeneity in which refers to biology, histopathogenesis and differential expression of surface enzymes.

2.1 *Trypanosoma cruzi*

Trypanosoma cruzi (Figure 2) is a flagellated protozoan belonging to Trypanosomatidae family (Kent, 1880), Kinetoplastida order, *Trypanosoma* genus (Chagas, 1909a; Coura, 2006). Kinetoplastida order was established as a function of the presence of a single cytoplasmic structure, the kinetoplast (Wallace, 1966), where mitochondrial DNA or k-DNA is concentrated. Its form, size, and position are important for characterizing the different evolution forms of the parasite (Vickerman, 1985).

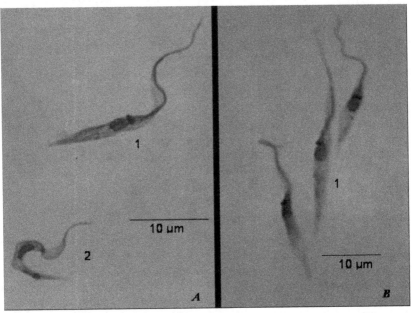

Fig. 2. Epimastigote (1) and tripomastigote (2) forms of *Trypanosoma cruzi* sylvatic isolates from Trinfo, Santa Maria Madalena municipal district, State of Rio de Janeiro – Brazil.

It is a euryxene and digenetic trypanosomatid, since part of its life cycle occurs inside a vertebrate or invertebrate host (Hoare, 1964). Vertebrate and invertebrate hosts are represented, respectively, by domiciled or domestic mammals and sylvatic triatomines.

The parasite cycle can be summarized as follows: the triatomine vector usually defecates during or at the end of blood sucking, eliminating metacyclic trypomastigote forms of *T. cruzi* on the vertebrate hosts. These forms found in dejections can penetrate the host through a continuity skin solution or skin mucosa. Inside the host cell, trypomastigotes transform into amastigotes and, approximately 35 hours later, the binary division begins. After five days, amastigotes transform into trypomastigotes, and as soon as they have long flagella, the cell disrupts releasing these forms into the bloodstream, so that they infect other cells or achieve different organs (Sousa, 2000). In triatomines, the blood-sucking trypomastigote forms ingested during hematophagy differentiate into epimastigotes in the digestive tract. Another differentiation occurs in the digestive tract, more specifically in its final portion and in rectus, when epimastigotes transform into metacyclic trypomastigotes, which is infectious for the vertebrate host and eliminated with the feces (Zeledón et al., 1977; Garcia & Azambuja, 2000).

T. cruzi is found as a parasite in a considerable number of mammals and in a wide range of tissues and niches in these hosts (Deane et al., 1984). Such eclecticism has characterized *T. cruzi* as one of the most successful microorganism in presenting parasitary life (Jansen et al., 1999). Therefore, this protozoan comprises a wide set of heterogeneous populations that circulate through very diverse vertebrate and invertebrate hosts, with a variation of different genotype predominance. The parasite has several morphological, physiological and ecological variations, and also in which refers to its infectivity and pathogenicity (Miles et al., 1978, 1980, 2009), which can warrant the various clinical manifestation forms of Chagas disease observed in different geographic regions (Miles et al., 1981a). Many studies have been performed seeking molecular markers that could correlate the parasite genotype with varying types of this infirmity clinical manifestation. Several works tried to clarify the multiple factors related with population epidemiology and genetics.

T. cruzi has a great phenotypic and genotypic variability in its strains, and therefore this protozoan has the ability to perform genetic exchanges through an unusual mechanism of nuclear fusion, forming a polyploidy progeny, which can suffer recombination among alleles, and after losing its chromosome, can return to diploid status. Some studies provided strong evidence that sexual reproduction is absent in *T. cruzi*, and that its population structure is clonal (Gaunt et al., 2003; Lewis et al., 2009).

3. Molecular profile of *T. cruzi* populations

Early investigations on the genetic of *T. cruzi* populations are based on electrophoretic profiling of isoenzymes (zimodeme analysis), a technique used to explore the genetic diversity of microorganisms. Enzymatic electrophoresis uses soluble raw-materials and extracts from an organism to assess the activity of a protein, and its product is revealed by means of a colorimetric reaction. Under controlled conditions, differences in isoenzymatic mobility imply genetic differences (Miles, 1985; Miles & Cibulkis, 1986). Toye (1974) was the first to use isoenzymes to classify trypanosomas from the New World, reporting differences among *T. cruzi* samples. By the end of the 70's and beginning of the 80's, several studies on isoenzymatic variability among *T. cruzi* populations were performed in Brazilian Northeast, and later in different regions within the country, by employing six enzymes: ALT (alanine aminotransferase), AST (aspartate aminotransferase), glucose phosphate isomerase (GPI), glucose-6-dehydrogenase phosphate (G6PDH), malic enzyme (ME) and phosphoglucomutase (PGM), characterizing three enzymatic profiles belonging to parasite groups called zymodemes I (Z1), II (Z2) and III (Z3). Z1 and Z3 are related with the sylvatic transmission cycle and Z2 with the domestic transmission cycle of the parasite (Miles et al., 1977, 1978, 1980, 1981a, b). As the number of analyzed isoenzymes has been amplified and sub-populations circulating among domestic and sylvatic vertebrates and invertebrates have been studied, an elevated degree of *T. cruzi* heterogeneity was verified (Miles et al., 1980; Bogliolo et al., 1986; Tibayrenc et al., 1986; Tibayrenc & Ayala, 1988; Barnabe et al., 2000).

With technologic advancement and the discovery of new molecular biology tools, it was possible to study the diversity of *T. cruzi* by means of DNA analysis, allowing for molecular characterization of this parasite strains (Devera et al., 2003). Therefore, the genetic diversity was corroborated by randomly amplified polymorphic DNA (RAPD) and restriction fragment length polymorphism (RFLP) analyses, DNA fingerprinting, microsatellites and molecular karyotyping (reviewed by Zingales et al., 1999). Analyses of gene sequences with lowest evaluative rates, such as ribosomal RNA genes, classic evolution markers and mini-exon genes, indicated dimorphism in *T. cruzi* isolates, rating them into two groups (Souto et al., 1996). Mini-exon gene that is present in Kinetoplastid nuclear genome at approximately 200 copies in a tandem type array is composed by three different regions: exon, intron and intergenic regions. Exon is a highly preserved sequence between de order compounds, added to nuclear messenger RNA post-transcription (Devera et al., 2003). Intron is moderately preserved between species of the same genus or sub-genus, and the intergenic region is particularly different among species. In *T. cruzi*, the amplification of mini-exon intergenic region by Polimerase Chain Reaction (PCR) allowed us to classify the different isolates into two main taxonomic groups: *T. cruzi* I and *T. cruzi* II (Fernandes, 1996; Souto et al., 1996; Fernandes et al., 1998). Thereafter, PCR amplification assay were standardized, allowing for rapid molecular typing, which started to be broadly used. Thereby the use of multiplex PCR based on intergenic region allowed us to classify the isolates as *T. cruzi* I, *T. cruzi* II, *T. cruzi* Z3 or *T. rangeli* with 200, 250, 150 pb and 100 pb, respectively (Fernandes et al., 2001a).

Aiming at standardizing double lines and hybrid isolates, a committee settled the lines were referred to as *T. cruzi I* and *T. cruzi II* "groups" (Zingales et al., 1999). Such denomination was not attributed to hybrid isolates, and additional studies are recommended to better characterize them (Zingales, 2011). From hybrid isolate gene sequence analysis, it has been shown that events of genetic exchanges with these parasites originated four distinct isolate groups (Sturm & Campbell, 2009). Thus, by using multilocus enzyme electrophoresis (MLEE) and RAPD markers, it was suggested that the group *T. cruzi* II was divided into five subgroups, including the four hybrid groups (Freitas et al., 2006; Brisse et al., 2000). *T. cruzi III*, a third ancestral group, was proposed from the analysis of microsatellites and mitochondrial DNA.

In 2009, the scientific community felt the need to standardize once again *T. cruzi* groups' nomenclature, aiming at clarifying questions on biology, eco-epidemiology and pathogenicity (Zingales et al., 2009). In this respect, it was recommended that *T. cruzi* was divided into six groups (*T. cruzi I–VI*), and that each group was called Discreet Taxonomic Units (DTUs) I, IIa, IIb, IIc, IId, IIe (Figure 3), defined as groups of isolates that are genetically similar and can be identified through molecular or immune markers (Tibayrenc, 1998), with DTU I corresponding to *T. cruzi* line I and DTU IIb corresponding to *T. cruzi* line II, and sub-lines IIa and IIc-e associated with hybrid strains and those belonging to zymodeme 3 (Brisse et al., 2000). The distribution of haplotypes from five nuclear genes and one satellite DNA was analyzed in isolates that were representative of the six DTUs by net genealogy and Bayesian phylogeny. Such data indicated that DTUs *T. cruzi I* and *T. cruzi II* are monophyletic and the other DTUs have different combinations of *T. cruzi I* and *T. cruzi II* haplotypes and DTU-specific haplotypes (Tomazi et al., 2009; Ienne et al., 2010). One of the possible interpretations for this observation is that *T. cruzi I* and *T. cruzi II* are two different species and that DTUs II-IV are hybrid resulting from independent hybridization/genomic combination events (Zingales, 2011).

In this setting, the characterization of these parasites extracted from different hosts aim at helping clarify the biological meaning and repercussion of this variability for clinics and for Chagas disease epidemiology (Lainson et al., 1979). However, the great majority of studies performed are related to parasite populations belonging to TCI and TCII groups, with scarce works performed with Z3 group.

- *T. cruzi* I
- **T. cruzi IIa/Z3**
- *T. cruzi* IIb
- *T. cruzi* IIc
- *T. cruzi* IId
- *T. cruzi* IIe

Fig. 3. General pattern of distribution of *T. cruzi* lineages and sublineages; the sylvatic isolates from Rio de Janeiro (extended map showing in green Triunfo, Santa Maria Madalena municipal district) were typed as *T. cruzi* IIa/Z3. (Adaptated map by Noireau F. Vet. Res. (2009)).

3.1 *T. cruzi* isolates from Rio de Janeiro

Therefore, this work was performed from *T. cruzi* samples isolated from *Triatoma vitticeps* (Figure 1) by Gonçalves in 2000, at Triunfo location, 2nd district of Santa Maria Madalena city, Rio de Janeiro state (Figure 2). Four hundred sixty five (465) *Triatoma vitticeps* specimens were collected: 294 females, 156 males, and 15 nymphs from five different areas:

area A, located at 250-meter altitude and 3.5 km distant from the district headquarters, very modified by deforestation for banana farming; area B, located at 130-meter altitude and 4 km distant from the headquarters, placed in a valley with preserved vegetation (secondary forest). These areas are 2-km distant to each other, separated by a mountain (Figure 3). Area C, the district headquarters, at 40-meter distance, was totally modified by pasture formation, and areas D and E were totally preserved and placed at 10 and 12-km distances from the headquarters, respectively. *T. cruzi* isolates used in this study were extracted from triatomines captured from areas A, B and F (Table 1). Area F was located in Vista Alegre, a city neighboring Conceição de Macabu, at Northern region of Rio de Janeiro State (Gonçalves, 2000).

Isolates (Samples)	Area	Host	Geographical origin
SMM10	A	Tv	Triunfo
SMM53	A	Tv	Triunfo
SMM88	A	Tv	Triunfo
SMM98	A	Tv	Triunfo
SMM36	B	Tv	Triunfo
SMM82	B	Tv	Triunfo
SMM1	F	HCD	Conceição de Macabu

SMM (Santa Maria Madalena)
Tv – *Triatoma vitticeps*; HCD (Haemoculture of the swiss mouse) – the parasites were inoculated in mice and was done haemoculture.

Table 1. *Trypanosoma cruzi* samples isolated from *Triatoma vitticeps* captured on the State of Rio de Janeiro, Brazil

Those *T. cruzi* samples isolated from *Triatoma vitticeps*, collected in Rio de Janeiro State, were classified by our group as Z3 based on mini-exon gene (Santos-Mallet et al., 2008) and showed great heterogeneity regarding growth curve and mouse virulence patterns (Silva, 2006), susceptibility to benznidazole (Sousa, 2009), total protein pattern and proteolytic activity profile (Gomes et al., 2006; Gomes et al., 2009). This heterogeneity observed in samples collected from the same region leads to questionings on how this diversity could influence the parasite-host cell interaction.

3.2 Molecular profile of *T. cruzi* isolates from Rio de Janeiro

The results obtained by means of molecular analysis revealed that the isolates have similar profiles, except for sample SMM1 (area F). Samples SMM10, SMM53, SMM88, SMM98 (area A), SMM36 and SMM82 (area B) revealed the presence of 150 bp, indicating that they belong to the zymodeme III group (Z3; Figure 4). Likewise, sample SMM1 from area F showed similarity to Z3 (150 bp), but also presented another band that may be related to the TcII profile (250 bp) and was very similar to the reference strain CL Brener (Figure 4). The phylogenetic position of Z3 has been much debated. According to some authors, the numerical taxonomy based on 24 isoenzymatic Z3 profiles is more closely associated with Z1 (TcII) than with Z2 (TcI) (Ready & Miles, 1980). However, other works place Z3 in an intermediate position between Z1 and Z2 (Stothard et al., 1998). Our study revealed one isolate (SMM1) with a hybrid profile associated with Z3 and TcII. This result may corroborate the hypothesis that this isolate is the product of a mixture of parasite populations, since the vector in wild environments may feed on several vertebrate hosts. This complexity was demonstrated in the State of Rio de Janeiro by Fernandes et al. (1999), who showed a preferential association of the two lineages of *T. cruzi* with different hosts. They suggest that the vector *T. vitticeps* is involved in the transmission cycle among mammals infected by lineage 2 in the municipality of Teresópolis, and in the transmission cycle of primates in municipality of Silva Jardim. The hybrid profile found in these samples may indicate a possibility that the vector *T. vitticeps* does not only participate in the wild cycle of the disease.

The main purpose of typing of isolates of *T. cruzi* is to identify strains with different epidemiological and/or clinical characteristics of Chagas disease. Our results corroborate other descriptions in the literature, and contribute to the knowledge and records of the profile of some additional wild isolates of *T. cruzi* in regions not yet affected by the disease. Added to the complexity observed between the isolates is the finding that the Z3 profile is divided into two groups, called Z3a and Z3b (Mendonça et al., 2002). Our laboratory is interested in investigating whether such a dichotomy occurs among the Z3 isolates obtained from *T. vitticeps* in this area of study.

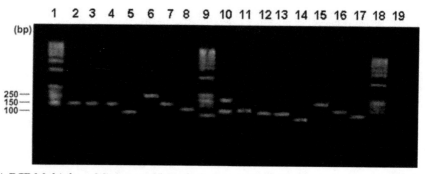

Fig. 4. PCR Multiplex – Mini-exon. The gel of agarose for electrophoresis was amplified using isolates of *Trypanosoma cruzi* of reference that possess approach bands of TCI, compared to TCII, Z3 and *Trypanosoma rangeli* and with *T. cruzi* sylvatics isolates from Rio de Janeiro. The isolates was performed using 25 ng of genomic DNA extracted using the phenol–chloroform method. Five primers were used: for Tc1 (5'-TTG CTC GCA CAC TCG GCT GCAT-3'), for Tc2 (5'-ACA CTT TCT GTG GCG CTG ATC G-3'), for Z3 (CCG CGW ACA ACC CCT MAT AAA AAT G-3'), for Tr (CCT ATT GTG ATC CCC ATC CCC ATC TTC G-3'), and for the mini-exon (5' TAC CAA TAT AGT ACAGAA ACT G-3'). Lane 1. Molecular weight marker (100bp DNA ladder), 2. SMM98, 3. SMM36, 4. SMM82, 5. *T. rangeli*, 6. CL Brener, 7. DM28c, 8. JJ, 9. Molecular weight marker (100bp DNA ladder), 10.SMM1, 11. SMM10, 12. SMM53, 13. SMM88, 14. *T. rangeli*, 15. CL Brener, 16. DM28c, 17. JJ, 18. Molecular weight marker (100bp DNA ladder), 19. negative control (no DNA added). bp = base pairs.

3.3 Proteolytic enzymes

Despite the existing knowledge of this flagellate genome and its main families of proteins, little is known about these parasites isolated from triatomines captured in the field, as well *T. cruzi* in mammals of wild origin. Proteolytic enzymes are reported to play an important role in determining the virulence of these microorganisms.

Proteases are essential for all life forms. They are involved in a multitude of physiological reactions, ranging from simple digestion of proteins for nutritional purposes, to highly-regulated metabolic cascades (e.g. proliferation and growth, differentiation, signaling and death pathways), and are essential for homeostatic control in both prokaryote and eukaryote cells (Rao et al., 1998). Proteases are also essential molecules in viruses, bacteria, fungi and protozoa, for their colonization, invasion, dissemination and evasion of host immune responses, mediating and sustaining the infectious disease process. Collectively, proteases participate in different steps of the multifaceted interaction events between microorganism and host structures, being considered as virulent attributes. Consequently, the biochemical characterization of these proteolytic enzymes is of interest not only for understanding proteases in general, but also for understanding their roles in microbial infections, and thus, their use as targets for rational chemotherapy of microbial diseases (Santos, 2010) (dos Santos, 2011).

Proteases are subdivided into two major groups, depending on their site of action: exopeptidases and endopeptidases. Exopeptidases cleave the peptide bond proximal to the amino (NH_2) or carboxyl (COOH) termini of the proteinaceous substrate, whereas endopeptidases cleave peptide bonds within a polypeptide chain. Based on their site of action at the NH_2 terminal, the exopeptidases are classified as aminopeptidases, dipeptidyl peptidases or tripeptidyl peptidases that act at a free NH_2 terminus of the polypeptide chain and liberate a single amino acid residue, a dipeptide or a tripeptide, respectively. Carboxypeptidases or peptidyl peptidases act at the COOH terminal of the polypeptide chain and liberate a single amino acid or a dipeptide (which can be hydrolyzed by the action of a dipeptidase). Carboxypeptidases can be further divided into three major groups: serine, metallo and cysteine carboxypeptidases, based on the functional group present at the active site of the enzymes. Similarly, endopeptidases are classified according to essential catalytic residues at their active sites in: serine, metallo, glutamic, threonine, cysteine and aspartic endopeptidases. Conversely, there are a few miscellaneous proteases that do not precisely fit into the standard classification (dos Santos, 2010, 2011).

Cysteine peptidases from parasitic protozoa have been characterized as factors of virulence and pathogenicity in several human and veterinary diseases. T. cruzi contains a major cysteine peptidase named cruzipain (also known as cruzain or GP57/51), which is present in different developmental forms of the parasite, although at variable levels (Dos Reis et al., 2006). Cruzipain is a papain-like peptidase that shares biochemical characteristics with both cathepsin L and cathepsin B (Cazzulo et al., 1990b). Cysteine peptidases have already been detected in many species of Trypanosomatidae, and are regarded as essential for the survival of several parasitic protozoa. The enzyme has been shown to be lysossomal, and is located in an epimastigote-specific pre-lysossomal organelle called the 'reservossome', which contains proteins that are digested during differentiation to metacyclic trypomastigotes (Soares et al., 1992). Some authors have suggested a second location of enzyme isoforms in the plasma membrane, associated with a glycosylphosphatidylinositol (GPI) anchor (Elias et al., 2008). These isoforms were present in epimastigotes, amastigotes and trypomastigotes, and reacted with polyclonal anti-cruzipain sera, thereby becoming an immunodominant antigen that is recognized by the sera of human patients with chronic Chagas disease (Martínez et al., 1991). Recently, the peptidase expression analysis of fresh field sylvatic isolated strains of T. cruzi showed a heterogeneous profile of cysteine proteolytic activities in the main phylogenetic groups TCI and TCII (Fampa et al., 2008).

Gomes et al (2009) investigated the production of peptidases, especially cruzipain, as well as the protein surface distribution in four newly sylvatic isolates of *T. cruzi* belonging to the Z3 genotype.

3.4 Proteolytic profile of *T. cruzi* isolates from Rio de Janeiro

The differences in peptidase expression between TCI and TCII phylogenetic groups have recently been investigated. Since *T. cruzi* isolates from sylvatic triatomines were included in the third phylogenetic group, named Z3, our investigation contributes to investigate the expression of surface polypeptides and the major cysteine peptidase from the Z3 parasite population, thereby furthering understanding on the genetic variability in the pathogenesis of Chagas disease. In this context, we carried out an identification of the protein profile and peptidase from epimastigotes (replicative forms of this parasite) of sylvatic isolates of *T. cruzi* (classified as Z3) from triatomines captured in Santa Maria Madalena (SMM) in the State of Rio de Janeiro. The separation of soluble whole proteins revealed a different protein profile, with approximately 35 polypeptides presenting apparent molecular masses from 118 to 25 kDa in all the samples. The proteolytic activity was determined by zymograms analysis of all the samples, using SDS-polyacrylamide gel electrophoresis containing gelatin as substrate. Our main results demonstrate a major band of 45 kDa sensible to E-64, a powerful cysteine peptidase inhibitor, in all the samples. In order to confirm this data, western blotting was performed using the anti-cruzipain polyclonal antibody. These findings showed a strong polypeptide band with an apparent molecular mass between 40 and 50 kDa in all the sylvatic isolates: SMM10; SMM53; SMM88 and SMM98 respectively and also Dm28c (Figure 5).

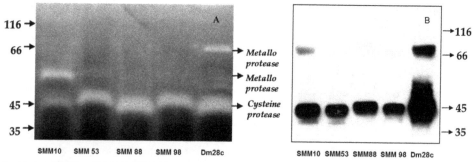

Fig. 5. A – Gelatin-SDS-PAGE showing the proteolytic activity profiles of *T. cruzi* sylvatic isolates. Parasites (SMM10, SMM53, SMM88, SMM98, and Dm28c) grown for 7 days were harvested and lysed by SDS. The gel was incubated in 50 mM sodium phosphate buffer, pH 5.5, supplemented with 2 mM DTT for 40 h at 37°C; B- Western blotting showing the reactivity of cellular polypeptides of *T. cruzi* sylvatic isolates with the anti-cruzipain polyclonal antibody. Numbers on the left indicate the relative molecular mass markers, expressed in kilodaltons.

These results show the presence of a main cysteine peptidase, cruzipain, in the sylvatic isolates of *T. cruzi* from Santa Maria Madalena, in the State of Rio de Janeiro (Gomes et al., 2009). We also observed another gelatinolyti activity of 66 kDa that was recognized by the anti-cruzipain antibody, probably a cruzipain isoform; since cruzipain is a high mannose-type glycoprotein containing about 10% carbohydrate, its molecular mass can be estimated from the sequence, considering two high-mannose oligosaccharide chains, as about 40 kDa. However, this enzyme can present anomalous behavior in SDS-PAGE, yielding apparent molecular mass values of 35 to 60 kDa depending on the experimental conditions. The cysteine peptidases from parasites, including *T. cruzi*, have proven to be valuable targets for chemotherapy. Due to the biological importance of cruzipain in the life cycle of *T. cruzi*, many studies have sought to build specific inhibitors against the active core of this enzyme, in order to obtain a new drug capable of providing protection against human infection by *T. cruzi*.

4. Conclusion

Trypanosoma cruzi shows considerable heterogeneity among populations isolated from sylvatic and domestic cycles. Despite of knowledge concerning the genome of these flagellated organisms and their main protein families, very little is known about these parasites isolated from triatomine bugs captured from field, as well as *T. cruzi* extracted from sylvatic mammals. In this context, we do hereby highlight the importance of molecular studies on *T. cruzi* sylvatic isolates collected by blood culture from vertebrate hosts and/or from triatomine vectors, *Triatoma vitticeps*, in Triunfo location, 2nd district of Santa Maria Madalena city, Northern region of Rio de Janeiro State, Brazil. The results of our investigations with *T. cruzi* samples isolated from sylvatic triatomine insects revealed that these parasites belong to a phylogenetic group called ZIII, and proteolytic analyzes evidenced the presence of a key peptidase cysteine, cruzipain, in all samples of sylvatic *T. cruzi* isolates from Santa Maria Madalena - Rio de Janeiro (Brazil), which was confirmed by anti-cruzipain antibody recognition. Taken together, our results can corroborate in understanding the role of proteolytic enzymes in determining the virulence of these microorganisms, as well as genetic variability of Z3 population in Chagas disease pathogenesis.

5. Acknowledgment

The authors would like to thank all the members of Setor de Entomologia Forense from Laboratório de Transmissores de Leishmanioses at Instituto Oswaldo Cruz- FIOCRUZ for the encouragement and help, especially to Prof. Catarina Macedo Lopes, who helped and made some figures of this chapter. The financial support CAPES, CNPq, FAPERJ and FIOCRUZ.

6. References

Aragão, M.B. & Souza, S.A.(1967). Triatoma infestans colonizado em domicílios da baixada fluminense, Estado do Rio de Janeiro, Brasil. *Revista da Sociedade Brasileira de Medicina Tropical*, Vol.5, No.1, (August 1971), ISSN 0037-8682.
Ávila-Pires, F.D. (1976). Ecology of small mammals in relation to sylvan and domestic transmission cycles. In new approaches in American tripanosomiasis research. *Pan American Health organization Scientific Publication*, Vol. 318, (March 1976), pp.301-306, ISSN 1020-4989.

Barnabé, C.; Brisse, S. & Tibayrenc, M. (1908). Population structure and genetic typing of *Trypanosoma cruzi*, the agent of Chagas disease: a multilocus enzyme electrophoresis approach. *Parasitology*, Vol.120, No.5, (May 2000), pp.513-526, ISSN 0031-1820.

Barretto, M.P. (1967). Estudos sobre reservatórios e vectores silvestres do *Trypanosoma cruzi*. XXII. Modificações dos focos naturais da tripanossomose americana e suas conseqüências. *Revista da Sociedade Brasileira de Medicina Tropical*, Vol.1, (November 1967), pp.167-173, ISSN 0037-8682.

Barretto, M.P. (1979). Epidemiologia. In: *Trypanosoma cruzi e doença de Chagas*, Brener, Z. and Andrade, Z.A. pp. 89-15, Guanabara Koogan ISBN 85-277-0563-X, Rio de Janeiro, Brasil.

Becerra-Fuentes, F.; Coura, J.R. & Ferreira, L.F. (1967). Observações sobre o *Triatoma rubrofasciata* (De Geer, 1773) no Estado da Guanabara. *Revista da Sociedade Brasileira de Medicina Tropical*, Vol. 5, (may 1971), pp. 47-54, ISSN 0037-8682.

Bogliolo, A.R.; Chiari, E.; Silva-Pereira, R.O. & Silva-Pereira, A.A. (1981). A comparative study of *Trypanosoma cruzi* enzyme polymorphism in South America. *Brazilian Journal of Medical and Biological Research*, Vol. 19, No.6, (jan 1986), pp.673-683, ISSN 1678-4510.

Brisse, S.; Barnabe, C. & Tibayrenc, M. (1971). Identification of six *Trypanosoma cruzi* phylogenetic lineages by random amplified polymorphic DNA and multilocus enzyme electrophoresis. *International Journal for Parasitology*, Vol.30, No.1, (January 2000), pp.35–44, ISSN 0020-7519.

Cazzulo, J.J.; Hellman, U., Couso, R., Parodi, A.J. (1980). Amino acid andcarbohydrate composition of a lysosomal cysteine proteinase from *Trypanosoma cruzi*. Absence of phosphorylated mannose residues. *Molecular and Biochemical Parasitology*, Vol.38, No.1, (January 1990b), pp.41–48, ISSN 0166-6851.

Chagas, C. (1909). Nova tripanossomíase humana: Estudos sobre a morfologia e o ciclo evolutivo do *Schizotrypanum cruzi* n. gen., n. sp., agente etiológico de nova entidade mórbida do homem. *Memórias do Instituto Oswaldo Cruz*, Vol.1, No.2, (August 1909), pp.159-218, ISSN 0074-0276.

Chagas, C. (1912). Sobre um *Trypanosoma* do tatu, *Tatusia novemcincta*, transmitido pelo *Triatoma geniculata* Latr (1811). Possibilidade de ser o tatu um depositário do *Trypanosoma cruzi* no mundo exterior. Nota prévia. *Braz Med*, Vol.26, (August 1912), pp.305-306.

Chagas, C. (1924). Infection naturelle dês singes du Pará (*Chrysotrix sciureus*) par *Trypanosoma cruzi*. *Comp Rend Séanc Soc Biol Ses Fin*, Vol.90, (February 1924), pp.873-876.

Coura J.R. (1967).Transmission of chagasic infection by oral route in the natural history of Chagas disease. *Revista da Sociedade Brasileira de Medicina Tropical*, Vol.39, No.Suppl 3, (November/December 2006), pp. 113-117, ISSN 0037-8682.

Coura, J.R. & Dias, J.C.P. (2009). Epidemiology, control and surveillance of Chagas disease - 100 years after its discovery. *Memórias do Instituto Oswaldo Cruz*, Vol.104, Suppl. 1, (July 2009), pp.31-40, ISSN 0074-0276.

Coura, J.R. (1909). Chagas disease: what is known and what is needed-a background article. *Memórias do Instituto Oswaldo Cruz*, Vol.102, Suppl. I, (October 2007), pp.113-122, ISSN 0074-0276.

Coura, J.R.; Ferreira, L.F.; Silva, F.R. (1959). Triatomíneos no Estado da Guanabara e suas relações com o domicílio humano. *Revista do Instituto de Medicina Tropical de São Paulo*, Vol.8, No.4, (July/August 1966), pp.162-166, ISSN 0036-4665.

Deane, M.P.; Lenzi, H.L. & Jansen, A.M. (1909). *Trypanosoma cruzi*: vertebrate and invertebrate cycles in the same mammals host, the opossum *Didelphis marsupialis*. *Memórias do Instituto Oswaldo Cruz*, Vol.79, No.4, (October/December 1984), pp.513-515, ISSN 0074-0276.

Devera, R.; Fernandes, O. & Coura, J.R. (1909). Should *Trypanosoma cruzi* be called 'cruzi' complex? A review of the parasite diversity and potential of selecting population after in vitro culturing and mice infection. *Memórias do Instituto Oswaldo Cruz*, Vol.98, No.1, (January 2003), pp.1-12, ISSN 0074-0276.

Dias, C.M.G.; Bezerra, I.O.; Moza, P.G.; Braga, J O.; Silva, J.L.; Santos, H.R.; Souza, M.S.B.; Fonseca Filho, M.; Pacheco, S R R.; Gonçalves, T.C.M.; Santos-Mallet, J.R.; Lopes, C.M. (2010). Investigação do risco de Doença de Chagas no município Trajano de Morais – Região Serrana do Estado do Rio de Janeiro. 26ª Reunião de Pesquisa Aplicada em Doença de Chagas/ 14ª Reunião de Pesquisa Aplicada em Leishmanioses. pp.17.

Dias, E. (1909). Observações sobre eliminação de dejeções e tempo de sucção em alguns triatomíneos sul-americanos. *Memórias do Instituto Oswaldo Cruz*, Vol. 54, No.1, (June 1956) pp.115-124, ISSN 0074-0276.

Dias, E. (1909). Presença do *Panstrongylus megistus* infectado por *Schizotrypanum* no Rio de Janeiro, DF. *Memórias do Instituto Oswaldo Cruz*, Vol.38, No.2, (April 1943), pp.177-180, ISSN 0074-0276.

Dias, J.C. (1967). Notes about of *Trypanosoma cruzi* and yours bio-ecology characteristics with agents of the transmission by meals. *Revista da Sociedade Brasileira de Medicina Tropical*, Vol.39, No.4, (Jul-Aug 2006), 370-375, ISSN 0037-8682.

Dias, J.C.P. (2000). Chagas disease control and the natural history of human Chagas disease: a possible interaction? *Memórias do Instituto Oswaldo Cruz*, Vol.95, Suppl. II, pp.14-20, ISSN 0074-0276.

Dias, J.C.P.; Feitosa, V.R.; Ferraz Filho, N.A.; Rodrigues, V.L.C.; Alencar, A.S. & Sessa, P.A. (1909). Fonte alimentar e potencial vetorial de *Triatoma vitticeps* (Stal, 1859), com relação à doença de Chagas humana no Estado do Espírito Santo, Brasil (Hemiptera, Reduviidae). *Memórias do Instituto Oswaldo Cruz*, Vol.84, Suppl IV, (November 1989) pp.165-173, ISSN 0074-0276.

Dias, J.C.P.; Silvera, A.C. & Schofield, C.J. (1909). The impact of Chagas disease control in Latin América. A Review. *Memórias do Instituto Oswaldo Cruz*, Vol.97, No.5, (July 2002), pp.603-612, ISSN 0074-0276.

Diotaiuti, L.; Bronfen, E.; Perilo, M.M.; Machado, G.B.N. & Loiola, C.F. (1967). Aspectos do comportamento do *Triatoma vitticeps* na transmissão da doença de Chagas. *Revista da Sociedade Brasileira de Medicina Tropical*, Vol.20, Suppl., (February 1987), pp.87, ISSN 0037-8682.

Dos Reis, F.C.G.; Júdice, W.A.F.; Juliano, M.A.; Juliano, J.; Scharfstein, J.; Lima, A.P.C.A. (1977). The substrate specificity of cruzipain 2, a cysteine protease isoform from *Trypanosoma cruzi*. *FEMS Microbiology Letters*, Vol.259, No.2, (June 2006), pp.215-220, ISSN 0378-1097.

Dos Santos A.L. (2009) HIV aspartyl protease inhibitors as promising compounds against Candida albicans. *World Journal of Biological Chemistry*, Vol.26, No.1(2), (February 2010), pp.21-30, ISSN 1949-8454.

Dos Santos A.L. (2010). Protease expression by microorganisms and its relevance to crucial physiological/pathological events. *World Journal of Biological Chemistry*, Vol.26, No.2, (March 2011), pp.48-58, ISSN 1949-8454.

Elias, C.G.; Pereira, F.M.; Dias, F.A.; Silva, T.L.; Lopes, A.H.; d'Avila-Levy, C.M.; Branquinha, M.H.; Santos, A.L. (1951/52). Cysteine peptidases in the tomato trypanosomatid *Phytomonas serpens*: influence of growth conditions, similarities with cruzipain and secretion to the extracellular environment. *Experimental Parasitology*, Vol.120, No.4, (December 2008), pp.343-52, ISSN 0014-4894.

Fampa, P.; Lisboa, C.V.; Jansen, A.M.; Santos, A.L.S.; Ramirez, M.I. (1908). Protease expression analysis in recently field-isolated strains of *Trypanosoma cruzi*: a heterogeneous profile of cysteine protease activities between TC I and TC II major phylogenetic groups. *Parasitology*, Vol.135, No.9, (August 2008), pp.1093–1100, ISSN 0031-1820.

Fernandes, O. Análise da estrutura primária do gene de mini-exon em diferentes tripamosomatídeos e sua utilização como marcador molecular. (PhD Thesis, Instituto Oswaldo Cruz, 1996), 268p.

Fernandes, O.; Mangia, R.H.; Lisboa, C.V.; Pinho, A.P.; Morel, C.M.; Zingales, B.; Campbell, D.A. & Jansen, A.M. (1908). The complexity of the sylvatic cycle of *Trypanosoma cruzi* in Rio de Janeiro state (Brazil) revealed by the non-transcribed spacer of the mini-exon gene. *Parasitology*, Vol.118, No. Pt2, (February 1999), pp.161–166, ISSN 0031-1820.

Fernandes, O.; Santos, S.S.; Cupolillo, E.; Mendonça, B.; Derre, R.; Junqueira, A.C.; Santos, L.C.; Sturm, N.R.; Naiff, R.D.; Barret,T.V.; Campbell, D.A. & Coura, J.R. (1908). A mini-exon multiplex polymerase chain reaction to distinguish the major groups of *Trypanosoma cruzi* and *T. rangeli* in the Brazilian Amazon. *Transactions of the Royal Society of Tropical Medicine and Hygiene*, Vol.95, No.1, (January/February 2001), pp.97-99, ISSN 0035-9203.

Fernandes, O.; Souto, R.P.; Castro, J.A.; Pereira, J.B.; Fernandes, N.C.; Junqueira, A.C.; Naiff, R.D.; Barrett, T.V.; Degrave, W.; Zingales, B.; Campbell, D.A. & Coura, J.R. (1952). Brazilian isolates of *Trypanosoma cruzi* from humans and triatomines classified into two lineages using mini-exon and ribosomal RNA sequences. *American Journal of Tropical Medicine and Hygiene*, Vol.58, No.6, (June 1998), pp.807-811, ISSN 0002-9637.

Ferreira, E.; Souza, P,S.; Fonseca Filho, M. & Rocha, I. (1986). Nota sobre a distribuição do *Triatoma vitticeps* (Stal, 1859), (Hemiptera, Reduviidae), no Estado do Rio de Janeiro, Brasil. *Revista Brasileira de Malariologia e Doenças Tropicais*, Vol.38, (1986), pp.11-14, ISSN 0034-7256.

Ferreira, M.S.; Lopes, E.R.; Chapadeiro, E.; Dias, J.C.P. & Luquetti, A.O. (1996). Doença de Chagas. In: *Tratado de Infectologia* (9ª ed.), Veronesi, R. & Foccacia, R., pp.1175-1213, Atheneu, ISBN 85737, São Paulo, Brasil.

Freitas, J.M.; Augusto-Pinto, L.; Pimenta. J.R.; Bastos-Rodrigues, L.; Goncalves, V.F.; Teixeira, S.M.; Chiari, E.; Junqueira, A.C.; Fernandes, O.; Macedo, A.M.; Machado, C.R. & Pena, S.D. (2005). Ancestral genomes, sex, and the population structure of

Trypanosoma cruzi. *PLoS Pathogens*, Vol.2, No.3, (March 2006), pp.e24, ISSN 1553-7366.

Garcia, E.S. & Azambuja, P. (2000). Fisiologia de triatomíneos: desenvolvimento, reprodução e interação com o *Trypanosoma cruzi*. In: *Trypanosoma cruzi e Doença de Chagas* (2 ed.), Brener, Z.; Andrade, Z.A. & Barral-Neto, M., pp.41-47, Guanabara Koogan, ISBN 85-277-0563-X, Rio de Janeiro, Brasil.

Gardner, G. (1942). *Viagens no Brasil, principalmente nas províncias do norte e nos distritos do ouro e do diamante durante os anos de 1836-1842* (2 ed.), EDITORA Nacional, ISBN 978-85-232-0587-4, São Paulo, Brasil.

Gaunt, M.W.; Yeo, M.; Frame, I.A.; Stothard, J.R.; Carrasco, H.J.; Taylor, M.C.; Solis Mena, S.; Veazy, P.; Miles, G.A.J.; Acosta, N.; Rojas de Arias, A. & Miles, M.A. (1869). Mechanism of genetic exchange in American trypanosomes. *Nature*, Vol.421, (February 2003), pp. 936–939, ISSN 0028-0836.

Gomes, S.A.O.; Fonseca de Souza, A.L.; Silva, B.A.; Kiffer-Moreira, T.; Santos-Mallet J.R.; Santos A.L.S.; Meyer-Fernandes J.R. (1951/52). *Trypanosoma rangeli*: differential expression of cell surface polypeptides and ecto-phosphatase activity in short and long epimastigote forms. *Experimental Parasitology*, Vol.112, No.4, (April 2006), pp. 253–262, ISSN 0014-4894.

Gomes, S.A.O.; Misael, D.; Silva, B.A., Feder, D.; Silva, C.S; Gonçalves, T.C.M.; Santos, A.L.S. and Santos-Mallet, J.R. (1987). Major cysteine protease (cruzipain) in Z3 sylvatic isolates of *Trypanosoma cruzi* from Rio de Janeiro, Brazil. *Parasitology Research*, Vol.105, No.3, (September 2009), ISSN 1432-1955.

Gonçalves, T.C.M. Aspectos ecológicos de *Triatoma vitticeps* (Stal, 1859) (Hemiptera, Reduviidae), com caracterização das amostras de *Trypanosoma cruzi* Chagas, 1909 (Kinetoplastida, Trypanosomatidae) isoladas desse triatomíneo, no município de Santa Maria Madalena, Estado do Rio de Janeiro. (PhD Thesis, Instituto Oswaldo Cruz, 2000), 125 p.

Gonçalves, T.C.M.; Oliveira, Edson.; Dias, L.S.; Almeida, M.D.; Nogueira, W.O. & Ávila-Pires, F.D. (1909). An Investigation on the Ecology of *Triatoma vitticeps* (Stal, 1859) and its Possible Role in the Transmission of *Trypanosoma cruzi*, in the Locality of Triunfo, Santa Maria Madalena Municipal District, State of Rio de Janeiro, Brazil. *Memórias do Instituto Oswaldo Cruz*, Vol.93, No.6, (November/December 1998), pp.711-717, ISSN 0074-0276.

Gonçalves, T.C.M.; Victorio, V.M.N.; Jurberg, J. & Cunha, V. (1909). Biologia do *Triatoma vitticeps* (Stal, 1859) em condições de laboratório (Hemiptera: Reduviidae: Triatominae). I. Ciclo evolutivo. *Memórias do Instituto Oswaldo Cruz*, Vol.83, No.4, (October/December 1988), pp.519:523, ISSN 0074-0276.

Gonçalves, T.C.M.; Victorio, V.M.N.; Jurberg, J. & Cunha, V. (1909). Biologia do *Triatoma vitticeps* (Stal, 1859) em condições de laboratório (Hemiptera: Reduviidae: Triatominae). II. Resistência ao jejum. *Memórias do Instituto Oswaldo Cruz*, Vol.84, No.1, (January/March 1989), pp.131-134, ISSN 0074-0276.

Guimarães, F.N. & Jansen, G. (1909). Um foco potencial de tripanosomiase americana na cidade do Rio de Janeiro (Distrito Federal). *Memórias do Instituto Oswaldo Cruz*, Vol.39, No.3, (December 1943), pp.405-420, ISSN 0074-0276.

Heitzmann-Fontenelle, T.J. (1976). Bionomia comparativa de triatomíneos. IV. *Triatoma vitticeps* (Stal, 1859) (Hemiptera, Reduviidae). *Ecossistema*, Vol.5, No.1, (August 1980), pp.39-46, ISSN: 0100-4107.

Hoare, C.A. (1964). Morphological and taxonomic studies on mammalian trypanosomas. X: Revision of the systematics. *Journal of Protozoology*, Vol.11, (May 1964), pp.200-207, ISSN 0022-3921.

Ienne, S.; Pedroso, A.; Carmona e Ferreira, R.; Briones, M.R.S. & Zingales, B. (2001). Network genealogy of 195-bp satellite DNA supports the superimposed hybridization hypothesis of *Trypanosoma cruzi* evolutionary pattern. *Infection, Genetics and Evolution*, Vol.10, No.5, (July 2010), pp.601–606, ISSN 1567-1348.

Jansen, A.M.; Pinho, A.P.S.; Lisboa, C.V.; Cupolillo, E.; Mangia, R.H. & Fernandes, O. (1909). The sylvatic cycle of *Trypanosoma cruzi*: a still unsolved puzzle. *Memórias do Instituto Oswaldo Cruz*, Vol.94, Suppl I, (September 1999), pp.203-204, ISSN 0074-0276.

Lainson, R.; Shaw, J.J.; Fraiha, H.; Miles, M.A. & Drapes, C.C. (1908). Chagas' disease in the Amazon basin: I. *Trypanosoma cruzi* infections in silvatic mammals, triatomine bugs and man in state of Pará, north Brazil. *Transactions of the Royal Society of Tropical Medicine and Hygiene*, Vol.73, No.2, (March 1979), pp.193-204, ISSN 0035-9203.

Lent, H. & Wygodzinsky, P. (1979). Revision of the Triatominae (Hemiptera: Reduviidae), and their significance as vectors of Chagas' disease. *Bulletin of the American Museum of Natural History*, Vol.163, No.3, (July 1979) pp.123-520, ISSN 0003-0090.

Lent, H. (1942). Transmissores da moléstia de Chagas no estado do Rio de Janeiro. *Rev F de Medicina*, Vol.6, pp.3-13 ISSN 0036-4665.

León, L.A. (1962). Contribución a la historia de los transmissores de la enfermedad de Chagas (Del siglo XVI a XIX).In: *Anais Congresso Internacional sobre doença de Chagas* 3, Eds. pp.761- 770, Editora, ISBN 00029637, Rio de Janeiro, Brasil.

Lewis, M.D.; Llewellyn, M.S.; Gaunt, M.W.; Yeo, M.; Carrasco, H.J., & Miles, M.A. (1971). Flow cytometric analysis and microsatellite genotyping reveal extensive DNA content variation in *Trypanosoma cruzi* populations and expose contrasts between natural and experimental hybrids. *International Journal for Parasitology*, Vol.39, No.12, (October 2009) pp.1305–1317, ISSN 0020-7519.

Lisbôa, C.V.; Mangia, R.H.; Menezes-Trajano, V.; Ivo, A.; Nehme, N.S.; Morel, C.M. & Jansen AM (1909). Ecological aspects of the circulation of *Trypanosoma cruzi* in the sylvan environment. *Memórias do Instituto Oswaldo Cruz*, Vol.91, Suppl, pp.279, (November 1996) ISSN 0074-0276.

Lopes, C.M.; Mallet, J.R.S.; Ramos, L.B.; Giordano, C.; Silva, J.L.; OLIVEIRA, M. L. R.; Misael, D.S.; Silva, C.S.; Filho, M.F.; Rodrigues, M.L.J. & Gonçalves, T.C.M. (2009). Visitação e Colonização do Ambiente Domiciliar, por populações de Triatomineos em Regiões do Estado do Rio de Janeiro. In: *XXV Reunião de Pesquisa Aplicada em Doença de Chagas*. pp. 09, Editora UFTM- Uberaba, MG, Brasil.

Martinez, J.; Campetella O.; Frasch, A.C.; Cazzulo J.J. (1970). The major cysteine proteinase (cruzipain) from *Trypanosoma cruzi* is antigenic in human infections. *Infection and immunity*, Vol.59, No.11, (November 1991), pp.4275-4277, ISSN 0019-9567.

Martins, A.; Versiani, V. & Tupinambá, A. (1909). Estudos sobre a tripanosomíase americana em Minas Gerais. *Memórias do Instituto Oswaldo Cruz*, Vol.35, No.2, (July 1940), pp.286-301, ISSN 0074-0276.

Mendonça, M.B.A.; Nehme, N.S.; Santos, S.S.; Cupolillo, E.; Vargas, N.; Junqueira, A.; Naiff, R.D.; Barrett, T.V.; Coura, J.R.; Zingales, B. & Fernandes, O. (1908). Two main clusters within *Trypanosoma cruzi* zymodeme III are defined by distinct regions of the ribosomal RNA cistron. *Parasitology*, Vol.124, No.Pt2, (February 2002), pp.177–184, ISSN 0031-1820.

Miles, M.A. & Cibulkis, R.E. (1985). Zymodeme characterization of *Trypanosoma cruzi*. *Parasitology Today*, Vol.2, No.4, (January 1986), pp.94-97, ISSN 0169-4758.

Miles, M.A. (1985). *Trypanosoma cruzi*: analysis of isozymes and antigenic expression. *Annales de la Societe Belge de Medecine Tropical*, Vol.65, Suppl I, (March-April 1985), pp.67-69, ISSN 0365-6527.

Miles, M.A., Souza, A.A., Póvoa, M., Shaw, J.J., Lainson, R., Toye, P.J. (1869). Isozymic heterogeneity of *Trypanosoma cruzi* in the first autochthonous patients with Chagas' disease in Amazonian Brazil. *Nature*, Vol.272, (April 1978), pp.819-821, ISSN 0028-0836.

Miles, M.A.; Cedillos, R.A.; Povoa, M.M.; de Souza, A.A.; Prata, A. & Macedo, V. (1823). Do Radically dissimilar *Trypanosoma cruzi* strains (zymodemes) cause Venezuelan and Brazilian forms of Chagas' disease? *Lancet*, Vol.20, No.1, (June 1981b), pp.1338-1340, ISSN 0140-6736.

Miles, M.A.; Lanham, S.M.; Souza, A.A. & Póvoa, M. (1980). Further enzymic characters of *Trypanosoma cruzi* and their evalution for strain identification. *Transactions of the Royal Society of Tropical Medicine and Hygiene*, Vol.74, No.2, pp.221-237, ISSN 0035-9203.

Miles, M.A.; Llewellyn, M.S.; Lewis, M.D.; Yeo, M.; Baleela, R.; Fitzpatrick, S.; Gaunt, M.W. & Mauricio, I.L. (1908). The molecular epidemiology and phylogeography of *Trypanosoma cruzi* and parallel research on *Leishmania*: looking back and to the future. *Parasitology*, Vol.136, No.12, (October 2009), pp.1509-1528, ISSN 0031-1820.

Miles, M.A.; Póvoa, M.; Souza, A.A.; Lainson, R.; Shaw, J.J. & Ketteridge, D.S. (1908). Chagas'disease in the Amazon Basin: II. The distribuition of *Trypanosoma cruzi* zymodemes 1 and 3 in Pará State, north Brazil. *Transactions of the Royal Society of Tropical Medicine and Hygiene*, Vol.75, No.5, (May 1981a), pp.667-674, ISSN 0035-9203.

Miles, M.A.; Toye, P.J.; Oswald, S.C. & Godfrey, D.G. (1908). The identification by isoenzyme patterns of two district strain-groups of *Trypanosoma cruzi*, circulating independently in a rural area of Brazil. *Transactions of the Royal Society of Tropical Medicine and Hygiene*, Vol.71, No.3, (July 1977), pp.217-225, ISSN 0035-9203.

Monteiro, F.A.; Barrett, T.V.; Fitzpatrick S.; Cordón-Rosales, C.; Feliciangeli, D. & Beard CB. (1952). Molecular phylogeography of the Amazonian Chagas disease vectors *Rhodnius prolixus* and *R. robustus*. *Molecular Ecology*, Vol.12, No.4, (April 2003), pp.997-1006.

Monteiro, F.A.; Wesson, D.M.; Dotson, E.M.; Schofield, C.J. & Beard CB. (2000). Phylogeny and molecular taxonomy of the Rhodniini derived from mitochondrial and nuclear DNA sequences. *American Journal of Tropical Medicine and Hygiene*, Vol.62, No.4, (April 2000), pp.460-465.

Neiva, A. (1914). Presença em uma localidade do Estado do Rio de um novo transmissor da moléstia de Chagas encontrado infectado em condições naturais (Nota prévia). *Brasil Med*, Vol.28, pp.333-335.

Noireau, F.; Diosque, P.; Jansen A.M. (2009). Trypanosome cruzi: adaptation to its vectors and its hosts. Vet. Res. Vol.40, No.26 (February 2009), pp. 1 to 23, ISSN 0928 4249

Oliveira, M.L.R.; Lopes, C.M.; Gonçalves, T.C.M.; Mallet, J.R.S.; Misael, D.S.; Silva, A.H.R. & Duarte, R. (2010). Determinação de Fontes Alimentares de *Triatoma vitticeps* nas regiões serrana, norte e noroeste do estado do Rio de Janeiro. In: *XXVI Reunião de Pesquisa Aplicada em Doença de Chagas.* pp. 12, Editora UFTM, Uberaba, Brasil.

Pinho, A.P.; Gonçalves, T.C.M.; Mangia, R.H.; Russel, N.N. & Jansen, A.M. (1909). The occurrence of *Rhodnius prolixus* Stal, 1859, naturally infected by *Trypanosoma cruzi* in the State of Rio de Janeiro, Brazil (Hemiptera, Reduviidae, Triatominae). *Memórias do Instituto Oswaldo Cruz,* Vol.93, No.2, (March/April 1998), pp.141-143, ISSN 0074-0276.

Pinto, C. (1931). Valor do rostro e antenas na caracterização dos gêneros de Triatomídeos. Hemíptera, Reduvioidea. *Boletim de Biologia,* Vol.19, pp.45-136, ISSN 00063185.

Rao, M.B.; Tanksale, A.M.; Ghatge, M.S.; Deshpande, V.V. (1997). Molecular and biotechnological aspects of microbial proteases. Microbiology and Molecular Biology Reviews, Vol.62, No.3, (September 1998), pp. 597-635, ISSN 1098-5557.

Ready, P.D. & Miles, M.A. (1908). Delimitation of *Trypanosoma cruzi* zymodemes by numerical taxonomy. *Transactions of the Royal Society of Tropical Medicine and Hygiene,* Vol.74, No.2, (June 1980), pp.38–242, ISSN 0035-9203.

Santos, A. L. S. (2010). Aspartic peptidase inhibitors as potential bioctive pharmacological compounds against human fungal pathogens. In: Combating Fungal Infection: Problems and Remedy, pp.289, ed Springer Heidelberg Dordrechithondon, ISBN 978-3-642-12172-2, e-ISBN 978-3-642-12173-9, New York.

Santos, U.M.; Pinto, A.F.S.; Almeida, A.Z.; Zaganelli, F.L.; Carrancho, P.V.; Netto, N.A. (1967). Doença de Chagas no estado do Espírito Santo. III: Vetores do *Trypanosoma.* *Revista da Sociedade Brasileira de Medicina Tropical,* Vol.3, No.1, (December 1969), pp.51-52, ISSN 0037-8682.

Santos-Mallet, JR, Silva CS, Gomes SAO, Oliveira DL, Santos CL, Sousa DM, Oliveira LR, Pinheiro NL, Gonçalves TCM (2008). Molecular characterization of *Trypanosoma cruzi* sylvatic isolates from Rio de Janeiro, Brazil. Parasitology Research (June 2008) volume 103, No. 5, pp.1041-1045, ISSN: 1432-1955.

Sessa, P.A. & Carias, V.R.D. (1967). Infecção natural de triatomíneos do Espírito Santo por flagelados morfologicamente semelhantes ao *Trypanosoma cruzi.* *Revista da Sociedade Brasileira de Medicina Tropical,* Vol.19, No.2, (April/June 1986), pp.99-100, ISSN 0037-8682.

Silva C.S. Estudo morfobiológico e histopatológico de amostras silvestres de *Trypanosoma cruzi* isoladas de *Triatoma vitticeps* (Stal, 1959) no estado do Rio de Janeiro. (Master Thesis, Universidade Federal Rural do Rio de Janeiro, 2006), 85p.

Silva, I.G. Influência da temperatura na biologia de 18 espécies de triatomíneos (Hemiptera: Reduviidae) e no xenodiagnóstico. (MSC Thesis, Universidade Federal do Paraná, Curitiba, 1985), 169p.

Silveira, A.C.; Alencar, T.A. & Máximo, M.H.C. (1983). Sobre o *Triatoma vitticeps* (Stål, 1859), no estado do Espírito Santo, Brasil. In: *X Reunião Anual Pesquisa Básica doença de Chagas,* Caxambu, Brasil. p.58.

Silveira, A.C.; Sakamoto, T.; Faria Filho, O.F. & Gil, H.S.G. (1949). Sobre o foco de triatomíneos domiciliados na baixada fluminense. *Revista Brasileira de Malariologia e Doenças Tropicais*, Vol.34, (mar 1982), pp.50-58.

Soares, M.J.; Souto-Padrón, T.; De Souza W. (1966). Identification of a large pre-lysosomal compartment in the pathogenic protozoon *Trypanosoma cruzi*. *Journal of cell science*, Vol.102, No.Pt 1, (May 1992), pp.157–167, ISSN 0021-9533.

Sousa D.M. Aspectos morfobiológicos, bioquímicos e susceptibilidade ao benznidazol de isolados silvestres de *Trypanosoma cruzi*. (Master Thesis, Instituto Oswaldo Cruz, 2009), 85p.

Sousa, M.A. Tripanosomatídeos de insetos e plantas: análise do crescimento, diferenciação e divisão celular, biometria e fenômenos sugestivos de sexualidade. Valor taxonômico. (PhD Thesis, Instituto Oswaldo Cruz, 2000), 161 pp.

Souto, R. P.; Fernandes, O.; Macedo, A. M.; Campbell, D. A.; Zingales, B. (1980). DNA markers define two major phylogenetic lineages of *Trypanosoma cruzi*. *Molecular and Biochemical Parasitology*. Vol.83, No.2, (December 1996), pp.141-152, ISSN 0166-6851.

Stothard, J.R.; Frame, I.A.; Carrasco, H.J. & Miles, M.A. (1908). On the molecular taxonomy of *Trypanosoma cruzi* using riboprinting. *Parasitology*, Vol.117, Pt3, (September 1998), pp.243–247, ISSN 0031-1820.

Sturm, N.R., Campbell, D.A. (2009). Alternative lifestyles: The population structure of *Trypanosoma cruzi*. *Acta Tropica*;Vol.115, No.1-2, (July/August 2010), pp.35-43, ISSN 0001-706X.

Talice, R.V.; Costa, R.S.; Rial, B. & Osimani, J.J. (1940). Los 100 primeros casos agudos confirmados de Enfermedad de Chagas (Triapanosomiasis americana) en el Uruguay. A. Monteverde y Cia., ISBN 0066-4170, Montevideo.

Tibayrenc, M. & Ayala, F.J. (1947). Isoenzyme variability in *Trypanosoma cruzi*, the agent of Chagas disease: genetical, taxonomical and epidemiological significance. *Evolution*, Vol.42, No.2, (March 1988), pp.277-292, ISSN 0014-3820.

Tibayrenc, M. (1971). Genetic epidemiology of parasitic protozoa and other infectious agents: the need for an integrated approach. *International Journal for Parasitology*, Vol.28, No.1, (January 1998), pp.85-104, ISSN 0020-7519.

Tibayrenc, M.; Ward, P.; Moya, A. & Ayala, F.J. (1915). Natural populations of *Trypanosoma cruzi*, the agent of Chagas disease, have a complex multiclonal structure. *Proceedings of the national academy of sciences of the United States of America (PNAS)*, Vol.83, No.1, (January 1986), pp.115-119, ISSN 0027-8424.

Tomazi, L.; Kawashita, S. Y.; Pereira, P. M.; Zingales, B. & Briones, M.R.S. (2002). Haplotype distribution of five nuclear genes based on network genealogies and Bayesian inference indicates that *Trypanosoma cruzi* hybrid strains are polyphyletic. *Genetics and Molecular Research*, Vol.8, No.2, (April 2009), pp.458-476, ISSN 1676-5680.

Urbina, J.A. (1909). Parasitological cure of Chagas disease: is it possible? Is it relevant? *Memórias do Instituto Oswaldo Cruz*, Vol.94, Suppl I, (September 1999), pp.49-355, ISSN 0074-0276.

Vikerman, K. (1943/44). Development cycle and biology of pathogenic trypanosomes. *British Medical Bulletin*, Vol.41, No.2, (April 1985), pp.105-114, ISSN 0007-1420

Wallace, F.G. (1951/52). The trypanosomatid parasites of insects and arachnids. *Experimental Parasitology*, Vol.18, No.1, (February 1966), pp.124-193, ISSN 0014-4894.

WHO. Expert Commitee on the Control of Chagas disease (2000: Brasilia, Brazil). Control of Chagas disease: Second Report of the WHO expert committee (WHO technical report series; 905) (2002), pp. 1-120, ISBN 9241209054.

Zeledón, R.; Alvarenga, N.J. & Schosinsky. (1977). Ecology of *Trypanosoma cruzi* in the insect vector. In new approaches in American trypanosomiasis research. *Pan American Health Organization Scientific Publication*, Vol. 347, (july 1977)pp.59-70, ISSN 1020-4989.

Zingales, B (2011). *Trypanosoma cruzi*: um parasita, dois parasitas ou vários parasitas da doença de Chagas? *Trypanosoma cruzi*: one parasite, two parasites or several parasites of Chagas disease? Revision. *Revista da Biologia*, Vol.6b, (June 2011) pp.44-48, ISSN 1984-5154.

Zingales, B.; Andrade, S.G.; Briones, M.R.S.; Campbell, D.A.; Chiari, E.; Fernandes, O.; Guhl, F.; Lages-Silva, E.; Macedo, AM.; Machado, CR.; Miles, M.A.; Romanha, A.J.; Sturm, N.R.; Tibayrenc, M. & Schijman, A.G. (1909). A new consensus for *Trypanosoma cruzi* intraspecific nomenclature: second revision meeting recommends TcI to TcVI. *Memórias do Instituto Oswaldo Cruz*, Vol.104, No.7, (November 2009), pp.1051-1054, ISSN 0074-0276.

Zingales, B.; Stolf, B.S.; Souto, R.P.; Fernandes, O. & Briones, M.R. (1909) Epidemiology, biochemistry and evolution of *Trypanosoma cruzi* lineages based on ribosomal RNA sequences. *Memórias do Instituto Oswaldo Cruz*, Vol.94, Suppl I, (September 1999), pp.159-164, ISSN 0074-0276.

9

Molecular Microbiology Applied to the Study of Phytopathogenic Fungi

Carlos Garrido, Francisco J. Fernández-Acero, María Carbú,
Victoria E. González-Rodríguez, Eva Liñeiro, and Jesús M. Cantoral
Microbiology Laboratory, Faculty of Marine and Environmental Sciences
University of Cádiz, Puerto Real
Spain

1. Introduction

Fungi is an extensive group of eukaryotic microorganisms, generally they are microscopic and usually filamentous. It is estimated that there are between 70,000 and 1.5 millions species of fungi, most of them are being discovering and describing (Agrios, 2005). Most of the known hundred thousand fungal species are strictly saprophytic, living on decomposing dead organic matter. About fifty species cause disease in human, and more than ten thousand species can cause disease in one (obligate parasite) or many kinds of plants (non-obligate parasites) (Fernández-Acero et al., 2007a).

Phytopathogenic fungi are able to infect any tissue at any stage of plant growth. Plant pathogenic fungi show a complex life cycles, including both sexual and asexual reproduction stages (Agrios, 2005). Moreover, complex infection cycles and carbon assimilation is displayed (Garrido et al., 2010). These biological variability give them the possibility to develop its biological role from very climatologically different environments, since dry and desert zones until wet and hot regions in the tropic and equatorial area to the capacity to attack all plant tissues, from leaves to roots (Agrios, 2005).

During the last decades, the development of molecular methods has lead the Scientifics community to accumulate a high quantity of information from different molecular approaches (Fernández-Acero et al., 2011; Garrido et al., 2009b). Advances into Genomics, Transcriptomics, Proteomics, and more recently, Metabolomics are transforming research into fungal plant pathology, providing better and more accurate knowledge about the molecular biology and infection mechanisms showed by these fungi (Garrido et al., 2010).

Since 1992, our research group has been working with two of the most aggressive plant pathogens, which have been established such a model organisms for molecular and phytaphology studies: *Botrytis cinerea* and *Colletotrichum acutatum* (Fernández-Acero et al., 2006b, 2007a; Garrido et al., 2009b, 2010; Perfect et al., 1999). These genera include some of the most destructive plant pathogen species known. They induce worldwide diseases as, between others, the grey mould on grapes and the anthracnose on strawberries, respectively (Coley-Smith et al., 1980; Elad et al., 2004; Sutton, 1992). The losses caused by the phytopathogenic fungi *Botrytis cinerea* and *Colletotrichum acutatum* have been quantified between 10 and 100 million of Euros per year in Europe (Fernández-Acero et al., 2007a). Losses caused by *B. cinerea* in French vineyards oscillate between 15% and 40%; in Holland, *B. cinerea* generates losses of about 20% of the flower crop; and, in Spain, the losses fluctuate between 20% and 25% of the strawberry crops (Fernández-Acero et al., 2007a). *Colletotrichum* spp. causes up to 80% plant death in nurseries and yield losses of >50%, being a major disease of cultivated strawberry (Denoyes-Rothan et al., 2004; Garrido et al., 2009a).

Our group has carried out an intense research activity of the molecular microbiology of these plant pathogens. These studies involve several molecular approaches in which the gel electrophoresis plays an important role. In this chapter, we will summarize the results obtained, and the molecular methods used for the study and characterization of the phytopathogen fungi *Botrytis cinerea* and *Colletotrichum* spp., all of them strongly related with different types of gel electrophoresis approaches and downstream protocols, including, between others, Pulse Field Gel Electrophoresis, agarose gel electrophoresis of DNA, Restriction Fragment Polymorphism Analyses, Southern-blot, Polyacrylamide Gel Electrophoresis and Two dimensional gel electrophoresis of proteins. These electrophoretic methods will be used to structure the development of chapter, describing the technical bases of each method and showing the approaches carried out and the results obtained.

2. Chromosomal polymorphism and genome organization in *Botrytis cinerea* and *Colletotrichum* spp.

Botrytis cinerea and *Colletotrichum acutatum* are two species of phytopathogenic fungi that show a very high level of phenotypic diversity among isolates. These fungi show complex cycles of life and infection, including both sexual and asexual forms (Garrido et al., 2008; Vallejo et al., 2002). Also high levels of somatic variability appear when the fungi are grown *"in vitro"*, depending on the medium, temperature, light and other factors, which even determine differences in cultural characteristics, production of reproductive structures and pathogenicty between others (Bailey & Jeger, 1992; Carbu, 2006; Garrido et al., 2009b; Rebordinos et al., 1997, 2000; Vallejo et al., 1996, 2001). These fungi do not show a high level of host specificity and they infect many different genera of hosts, adapting their infection strategy and metabolism to the environment conditions and kind of plant colonized. They are notoriously variable genera about which many fundamental questions relating to taxonomy, evolution, origin of variation, host specificity and mechanisms of pathogenesis remain to be answered (Bailey & Jeger, 1992; Elad et al., 2004).

Many research projects are aimed to study the genome organization and chromosomal polymorphism trying to find the origin of phenotypic variability showed by these fungi. In the past decades, several strategies have been tested on lower fungi such as *B. cinerea* and *Colletotrichum* spp., i.e. cytological karyotyping, analysis of progeny from crosses between strains, sexual hybridizations, etc. These assays looked for a relation between molecular and phenotypic variability (Carbu, 2006; Faretra & Antonacci, 1987; Faretra et al., 1988; Vallejo et al., 1996). Cytological studies showed a very high level of difficulty in this group of microorganisms due to small size and/or the difficulty to condense sufficiently the chromosomes to make them visible by microscope. These characteristics made difficult to obtain reliable information about the genome organization of these fungi, and the obtaining of conclusive results about their biological mechanisms of recombination and chromosomal polymorphisms.

The development of Pulse- field gel electrophoresis (PFGE) resolved many problems found with cytogenetic studies in filamentous fungi. This technique has been widely used since the 90s for genomic characterization into fungal plant pathogens. PFGE allows the separation of large DNA molecules (DNAs from 100 bases to over 10 megabases (Mb) may be effectively resolved) which would all co-migrate in conventional agarose gels. This technique has proved to be a very useful tool to study aspects of genome organization in several yeast and fungi. It has led to the discovery that most species exhibit chromosome-length polymorphisms (CLPs), revealing a high level of intraspecific, and even population-level variability (Vallejo et al., 2002).

Technically, PFGE resolves chromosome-sized DNAs by alternating the electric field between spatially distinct pairs of electrodes. The electrophoresis cell consists of an array with 24 horizontal electrodes arranged in a hexagon. Agarose gels are electrophoresed horizontally, submerged under recirculated buffer. The system (CHEF-"Clamped Homogeneous Electric Field" and PACE "Programmable Autonomously Controlled Electrodes", from BIO-RAD) provides highly uniform, or homogenous, electric fields within the gel, using an array of 24 electrodes, which are held to intermediate potentials to eliminate lane distortion. Thus, lanes are straight. The system maintains uniform field using patented Dynamic Regulation. The electrodes sense changes in local buffer conductivity due to buffer breakdown, change in buffer type, gel thickness, or temperature, and potentials.

The preparation of samples for resolving chromosomal karyotypes by PFGE is not exempt of difficulty due to the biological characteristic of fungal cells. Fungus has to be growth in an optimal culture medium and mycelium harvest after determinate time which depends of the fungal species. This time is very important because is necessary to obtain the highest number of fungal cells in metaphase stage (Carbu, 2006; Garrido et al., 2009b). Chromosomes are condensed and highly coiled in metaphase, which makes them most suitable for visual analysis. After young mycelium is harvested, it is necessary to produce protoplasts using different mixes of lysing enzymes, which digest the fungal cell wall after incubation. Protoplast suspensions are mixed with low melting point agarose, adjusted to final concentration of 1×10^8 protoplast ml^{-1}, and solidified plugs of agarose containing protoplast are digested with proteinase K. The digestion produces pores in the plasma membrane, providing the possibility to extract the chromosomal by PFGE (Garrido et al., 2009b).

Gels are prepared with a special type of agarose. It depends of the DNA molecules sizes because there are different commercial preparations, some of them for DNA molecules higher than 10 Mb, i.e. PFGE ™Megabase agarose (Bio-Rad). Plugs are cast in the gel, and this is placed in the center of the hexagon formed by the 24 electrodes. Many parameters of the electrophoresis have to be optimized, since the type and concentration of running buffer, temperature of buffer, voltage and time of pulses, angles of electric fields. Depending of instrument setup, we can resolve the electrophoretic karyotype (EK) only with one experiment, like in the case of *Botrytis cinerea*; or even it could be necessary two different steps/running conditions, due to the high differences in sizes of the chromosomal DNA molecules. After electrophoresis, gels are stained using i.e. ethidium bromide and visualized using a UV light system.

PFGE has been widely used by our group to study the genome organization and Chromosomal Polymorphisms (CPL) in *B. cinerea* and *C. acutatum*. We have determined the number and sizes of chromosomes in both species, and therefore we have estimated the genome size for these fungi; the high level of CPL displayed by them, represented in the different EK profiles showed by the strains; and PFGE has made possible downstream applications such as Southern-blot analysis using different probes. All the results accumulated during the last years have provided a better understanding about the genome organization and the molecular bases of asexual and sexual reproduction of these fungi. They proved that polymorphism has been observed in both asexual and sexual fungi and most likely results from both mitotic and meiotic processes, especially in the case of *Botrytis cinerea* (Vallejo et al., 2002).

When a study of PFGE has made, it is usual to find chromosomal bands of different intensity and therefore it is important to consider several technical aspects that can have influence in the interpretation of the final results, and the conclusions obtained: i) a double band could be composed of two coumpounds of a couple of homologous chromosomes or of two heterologous chromosomes of similar size, and ii) two homologous chromosomes can differ in size and appear like two heterologous ones. Due to this fact, depending on the aims of the study, sometimes further hybridization studies are necessary in order to determine the linkage groups of each of the bands (Carbu, 2006; Vallejo et al., 1996).

Botrytis cinerea strains studied by our group were isolated from different hosts and geographical origins. We found different EK profiles between isolates, which did not follow any correlation with the host, year of isolation, or phenotypical characteristics. We have found that the number of chromosomal bands varied between 5 and 12, and they ranged between 1.80 and 3.8 Mb. These results made possible to estimate the minimal genome size of *B. cinerea* genome, found between 14.5 and 22.7 Mbp (Carbu, 2006; Vallejo et al., 1996, 2002) (Fig. 1a).

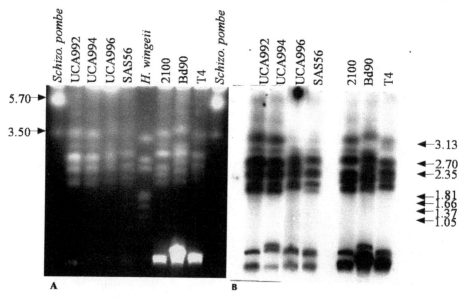

Fig. 1. A.- PFGE chromosomal separation of selected *B. cinerea* isolates. The molecular sizes were estimated using *Schizosaccharomyces pombe* (line 1 and 10), and *Hansenula wingeii* (line 6) chromosomes as reference molecular markers (Bio-Rad). B.- Southern-blot hybridization using a telomeric DNA probe to hybridise the PFGE separated chromosomal bands.

The *B. cinerea* strains showed a high level of CLPs, revealing the facility to support chromosomal rearrangements in this species, and could be the basis of the high degree of adaptability to the environmental conditions. Our group has also studied crosses between strains with different EK profiles. This study had as main aim to analyze the chromosomal rearrangements and chromosomal segregation in the crossed strains, in order to clarify the controversy appeared about the possibility that a high level of CLPs between strains, could inhibit meiosis (Zeigler, 1998), and therefore to be one possible reason to explain the low level of sexual reproduction that take place in *B. cinerea* under natural conditions (Carbu, 2006; Giraud et al., 1997).

The crosses between strains produced fertile strains (more than 100 ascospores studied) and our results demonstrated that chromosomal rearrangements did not affect the capacity to reproduce sexually in *B. cinerea*. It was observed than only several isolates recovered the parental EKs. New chromosomes sizes were identified and some bands were lost from the parental to descendants EKs. All these results, along with a segregation analyses carried out in the decendants, represented strong evidence that some strains might not be haploid, and that aneuploidy and differences in ploidy levels are present in this species (Vallejo et al., 2002). Our group has also studied how during a long period of time, reproducing the fungus "*in vitro*", there were not detected changes in the EK of a given strain. All results together, proved that mitotic growth does not provide EK variability in this fungus, being the chromosomal rearrangements generated after meiotic recombination the causal agent of EK variability in *B. cinerea* (Carbu, 2006).

In the case of the species *C. acutatum*, there were not data published about the EKs and CLPs among isolates until the last 2009 (Garrido et al., 2009b). PFGE had been used with other species of this genus, like *C. gloeosporioides* (Masel et al., 1993) and *C. lindemouthianum* (O'Sullivan et al., 1998). *Colletotrichum* spp. displayed an estimated genome sizes higher than *B. cinerea*. Protocols to separate the chromosomes molecules were carried out in two different experimental setups, including variations in the pulse of electric field, percentage of agarose gels and duration of the assays (Masel et al., 1993), i.e. for separation of larger chromosomal molecules in *C. gloeosporioides*, Masel et al. (1993) optimized an PFGE approach running a electrophoresis of seven days long. During this experiment, it was necessary to replace the running buffer each two days to obtain a better resolution in the final image. Similar protocols were used to resolve EK from *C. lindemouthianum* strains(O'Sullivan et al., 1998).

The karyotype of *C. acutatum* was studied by our group in several strains isolated from different geographical origins. They had showed differences in the morphological characteristics in relation to the color and texture of mycelium, ratio of growth in different medium, pathogenicity and level of conidia production (Garrido et al., 2008, 2009b). Protocol to obtain *C. acutatum* protoplasts and the PFGE conditions to separate chromosomes were optimized based on our previous experience with *B. cinerea*. We optimized a PFGE running conditions to separate chromosomes between approximatele 0.1 and 9 Mb after only 72 h. of running. This protocol improved substantially those previously described for *Colletotrichum* spp., which took longer due to the two steps needed to resolve the complete karytopyes. Those longer protocols (Masel et al., 1993) were also tested, and we got the same number and sizes of chromosomal bands, proving the improvement of our optimized 72-hours protocol (Garrido et al., 2009b).

C. acutatum strains showed EK profiles containing between six and nine chromosomal bands with different sizes ranging from 0.1 and 8 Mb. The total minimal genome size estimated for *C. acutatum* ranged between 29 and 36 Mb, which is similar to that previously described for other species of *Colletotrichum* (Masel et al., 1993; O'Sullivan et al., 1998). We observed CLPs between strains studies but further analyses with a high number of isolates could be necessary in order to obtain strong conclusions about the CLPs showed by the species and how this variability could affect the sexual and asexual reproduction of this species in the environment (Garrido et al., 2009b).

PFGE gels from *B. cinerea* and *C. acutatum* were used in downstream applications, like Southern-blot analyses. Gels were transferred to Hybond-N membranes and they hybridised with a telomeric probe confirming that all the bands represented chromosomes. The description of Southern-blot analyses will be described in the next section, but it proved how PFGE, not only provides the possibility to obtain interesting conclusions about the biology and genome organization of these fungi, but also gel electrophoresis techniques are often the starting point for interesting downstream applications that provide more information in the researches of these fungi (Fig. 1b).

In our PFGE studies in *B. cinerea* and *C. acutatum*, it has not been observed a higher EKs variability that showed by phenotypic characteristics among strains (Carbu, 2006; Garrido et al., 2008, 2009a, 2009b; Rebordinos et al., 2000; Vallejo et al., 1996). Phenotypic features were very highly variable between strains with the same EKs. Therefore, we cannot conclude that there is a direct relation between morphological, physiological and pathogenic variability directly related with heterokaryosis, aneuploidy and a variable level of ploidy among strains. New proteomics approaches to *B. cinerea* and *Colletotrichum* spp., which will be described during next pages, is contributing with very interesting data, that in conjunction with genomic information, disclose that phenotypic variation is more related with the synthesis of proteins and their post-transductional modifications, and not only by genotypes encoding them (Fernández-Acero et al., 2011).

3. Phylogenetic relationships between strains of *Colletotrichum* spp. using telomeric fingerprinting

Colletotrichum acutatum is a widely spread species that can be found throughout the world (Whitelaw-Weckert et al., 2007). *C. acutatum* causes anthracnose on a number of economically important crops, including woody and herbaceous crops, ornamentals, fruits, conifers and forage plants (Sreenivasaprasad & Talhinhas, 2005). It was classified as an organism of quarantine significance in Canada from 1991 to 1997, in the UK and the EU since 1993, and it can be found widely spread in the southwest region of USA (EPPO/CABI, 1997; Garrido et al., 2009a; Mertely and Legard, 2004). Investigations of *C. acutatum* were focused in two main aspects of the pathogen: i) cultural and morphological studies (Afanador-Kafuri et al., 2003; Denoyes-Rothan & Baudry, 1995; Garrido et al., 2008;) and ii) molecular approaches using molecular techniques including isoenzyme comparisons, Restriction Fragment Length Polymorphism (RFLP) analyses of mitochondrial DNA, Amplified Fragment Length Polymorphism (AFLP), AT rich analyses, Random Amplified Polymorphic DNA (RAPD), and ITS sequences analyses for specific PCR sequencing and identification (Buddie et al., 1999; Freeman et al., 1993; Garrido et al., 2009a, 2009b; Sreenivasaprasad et al., 1996; Talhinhas et al., 2005).

Sreenivasaprasad & Talhinhas (2005) studied *C. acutatum* populations from several hosts and different geographical origins. They established molecular groups based on sequences analyses of the internal transcribed spacers (ITS) of ribosomal DNA polymorphic regions (Sreenivasaprasad & Talhinhas, 2005). ITS regions have been widely used on molecular approaches for studying relationship between microorganisms, and it is also very useful regions for designing molecular approaches to identification and diagnostic protocols, due to the high variability showed by the sequences among species and even strains (Garrido et al., 2009a). The classification carried out by Sreenivasaprasad & Talhinhas (2005), established eight molecular groups for *C. acutatum* species. These molecular groups have been widely used to study the genotypic and phenotypic diversity of this fungus, and to classify isolates from different origin (Whitelaw-Weckert et al., 2007).

During the last years, we carried out a study to classify a worldwide collection of *C. acutatum* strains isolated from thirteen countries (Australia, Canada, France, Germany, Japan, The Netherlands, New Zealand, Norway, Portugal, Spain, Switzerland, USA and UK). For this purpose we used two different molecular approaches in order to study the phylogenetic relationship between strains: i) a sequencing analysis of the internal transcribed spacers (ITS) of the 5.8S ribosomal DNA polymorphic regions; ii) a telomeric fingerprinting study by Southern-blot hybridization, using a telomeric probe after RFLP digestions of genomic DNA (Garrido et al., 2009b).

In total, eighty-one 5.8S-ITS sequences were studied, several strains were sequenced by our group, and other ones used from databases such as reference sequences for allocating our strains in the previously established molecular groups for *C. acutatum*. ITS regions, including 5.8S rDNA, were amplified by conventional PCR using universal primers ITS1 and ITS4 (White et al., 1990). After PCR amplification, products were loaded in a conventional 1% agarose gel for conventional DNA electrophoresis. Products were cut from the gels using a purification kit, DNA was quantified, and subsequently sequenced in both directions (Garrido et al., 2009b).

The phylogenetic study carried out with the sequences allowed us to allocate the strains into *C. acutatum* molecular groups described by Sreenivasaprasad & Talhinhas (2005), but the analysis of bootstrap in the neighbout-joining phylogenetic tree, published by Garrido et al. (2009), showed interesting data about the molecular groups. In base of that analyses, the nine molecular groups previously described (Whitelaw-Weckert et al., 2007), could be grouped in only four groups. Our results proved that A1, A2, A5 A8 and A9 subgroups showed a bootstrap support of 90%, and therefore could be considered such as large group in base to the analyses of the sequences of ITS regions (Garrido et al., 2009b). The same result was observed for subgroups A6 and A4, since these subgroups clustered together with a strong bootstrap support of 91% (Garrido et al. 2009). Our results supported a new classification into four molecular groups instead the nine previously described for this species in base to the ITS sequences (Garrido et al., 2009b).

The phylogenetic analyses showed that the majority of the strains studied grouped in the group A2. This happened because many strains from Spain were included in the analyses. The results proved the high level of similarity between *C. acutatum* strains isolated from Spain. It is also interesting that the A2 group included, principally, isolates from Spain, Portugal, France, UK and USA. *C. acutatum* was first described in the southwest region of the USA, and then it was observed in France and UK at the beginning of the 80s. It is not clear how the pathogen was introduced into production fields in Europe, but it is thought that the pathogen could have arrived since the American nurseries to the EU (Freeman & Katan, 1997). It should have arrived to France, UK and the Iberian Peninsula fields. The arrival of the pathogen was facilitated by the intense international trade between these countries related with strawberry crop. Therefore the fungus could be introduced by infected plants, contaminated soil associated with strawberry crowns at planting, and quiescent infections on strawberry leaves or fruits (Garrido et al., 2008, 2009b; Leandro et al., 2001, 2003).

In order to complete the phylogenetic classification of our *C. acutatum* strain collection, a different molecular approach was carried out. The results obtained were compared with those from the ITS sequences analyses. We used the profiles obtained after restriction enzymes digestions of genomic DNA, and then hybridized with a telomeric probe by Southern-blot hybrisization. Genomic DNA of *C. acutatum* strains were digested to completion with several restriction enzymes in independent experiments (*Bam*HI, *Eco*RI, *Hind*III and *Pst*I). Gel electrophoresis is an intermediate point of the complete protocol. It make possible to separate the DNA fragments obtained after the restriction enzymes digestions. In this case, we used a 1.5% agarose gel, and electrophoresis was carried out in a conventional horizontal tray for DNA electrophoresis (Bio-Rad). After separation of digested fragments, gels were blotting to Hybond-N membrans, being ready for subsequently hybridization (Garrido et al., 2009b).

For Southern-blot hybridization we used a telomeric probe to get hybridization in the telomeric regions. These regions are located at the end of the lineal chromosomes of most eukaryotic organisms, and they are named telomeres. Telomeres are regions of repetitive DNA sequences that protect the end of the chromosome from deterioration or from fusion with neighboring chromosomes. The repeated sequences is dependent of the species. For *C. acutatum* telomeres, we produced our telomeric probe, (TTAGGG)n, by PCR in the absence of a template using (TTAGGG)5 and (CCCTAA)5 primers as it was described by Ijdo et al. (Ijdo et al., 1991). The Hybond-N membranes were allowed to hybridize with the telomeric probe; films images were digitalized and telomeric profiles were analysed using Fingerprinting II software v3.0 (Bio-Rad).

The experimental setup described provided the possibility to obtain two different kinds of results/conclusions from the study: I) Selected restriction enzymes used for RFLP did not produce any cut in the telomeric regions of *C. acutatum* strains. Each band represents a physically distinct telomere extremity. Therefore, taking into consideration the higher number of telomeric extremities and then divided into two, we can determine the number of chromosomes among strains studied. II) The fingerprinting analyses of the telomeric profiles, carried out using Fingerprinting II software, make possible to produce phylogenetic trees based in the similarity of the profiles showed among the strains. Therefore, these results could be compared with those obtained from phylogenetic groups based on ITS sequences.

Among the fifty-two isolates analysed by telomeric fingerprinting, the number of band or telomeres oscillated between twelve and eighteen. Therefore, the minimum number of estimated chromosomes was from six to nine among *C. acutatum* isolates (Garrido et al.,

2009b). In this study the number of strains studied was higher than those studied by FPGE, and although fingerprinting analyses did not make possible to study the chromosomal length polymorphisms among the isolates, the minimum numbers of estimated chromosomes are coincident with those obtained from FPGE analyses, showed in the last section of this chapter.

The telomeric profiles obtained for each isolate of *C. acutatum* were analysed. UPGMA dendogram showed a representative grouping among the isolates, which was coincident with the grouping in the neighbuor-joining phylogenetic tree based on sequences of rDNA ITS regions (Garrido et al., 2009b). All the strains previously classified in the A2 molecular groups, also clustered in a large group with more than 70% of similarity based in this case in the telomeric fingerprinting profiles. These results proved the high level of similarity shows by these isolated, not only based in sequence similarity of one specific region but also in their genotypes and genome organization among *C. acutatum* strains, which suggests a common origin of the strains among the different molecular groups (Garrido et al., 2009b; Talhinhas et al., 2005).

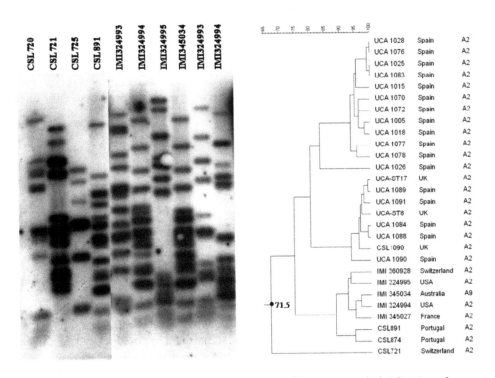

Fig. 2. Left.- Telomeric fingerprinting patterns obtained by telomeric hybridisation of Southern blots from HindIII-DNA digestions. Right.- Combined UPGMA dendograms with the *C. acutatum* isolates belonging to A2 group, based on Dice coefficients generated using a composite data set from individual experiments of each enzyme digestion (*Bam*HI, *Eco*RI, *Hind*III and *Pst*I) hybridised with a telomeric probe.

4. Develop of molecular methods for detection and identification of phytophatogenic fungi – Monitoring of the diseases causing by *Colletotrichum* spp.

Many fungal plant pathogens produce similar symptoms when they develop diseases among different hosts. Currently, the ability to detect, identify and quantify plant pathogens accurately is the cornerstone of plant pathology (Garrido et al., 2011). The reliable identification of the organism(s) responsible for a crop disease is an essential prerequisite to apply the correct disease management strategies and the most appropriate control measures to take. Besides, many pathogens are subjected to special regulation through quarantine programs agreed among producer countries. For all these reasons, pathogen identification is crucial to all aspect of fungal diagnostics and epidemiology in the field of plant pathology, but also in medical science, environmental studies and biological control (Alastair McCartney et al., 2003; Atkins et al., 2003).

Since the 1990's, new methods based on molecular biology have provided new tools for more accurate and reliable detection, identification and quantification of plant pathogens. These methods are based on immunological and DNA/RNA study strategies, including, amongst others: RFLP analyses of mitochondrial DNA (Garrido et al., 2008; Sreenivasaprasad et al., 1992), AFLP, AT-rich analyses (Freeman et al., 2000a, 2000b), RAPD-DNA (Whitelaw-Weckert et al., 2007), genus-specific and species-specific PCR primers (Garrido et al., 2008; Martínez-Culebras et al., 2003; Mills et al., 1992; Sreenivasaprasad et al., 1996), real-time PCR studies (Garrido et al., 2009a), and ELISA assays (Hughes et al., 1997). Diagnosis time can be reduced from a period of weeks, typically experienced with culture plating, to only a few days, thus allowing the appropriate control methods to implemented much sooner and more effectively (Atkins et al., 2003).

Advances in polymerase chain reaction technology have opened alternative approaches to the detection and identification of fungal pathogens. The development of PCR technology relies on three fundamental steps: i) the selection of a specific target region of DNA/RNA to identify the fungus; ii) extraction of total community DNA/RNA from the environmental sample; iii) a method for identifying the presence of the target DNA/RNA region in the sample (Garrido et al., 2011). Our group have optimised a very high sensitive protocol for diagnosis and identification of the fungal genus *Colletotrichum*, and the species *C. acutatum* and *C. gloeosporioides* (Garrido et al., 2009a).

The sensitivity of PCR-based protocols depends mainly on the instrumentation and technique used (i.e. conventional PCR *vs.* real-time PCR), but in a high proportion of cases this sensitivity depends on the quality of the total community DNA/RNA extracted from the environmental samples. Garrido et al. (2009) optimized a DNA extraction protocol that can be used for samples of strawberry plant material directly, or from fungal colonies removed from an agar plate. This method uses sample material physically ground using a grinding machine, in the presence of CTAB lysis buffer. The lysated samples are washed in various chemical products (chloroform, isopropanol, ethanol, etc.) and then the final step involves using Magnesil® beads and GITC lysis buffer (guanidinium thiocyanate buffer) in a Kingfisher robotic processor (Kingfisher ML, Thermo Scientific). The new method was tested with roots, crowns, petioles, leaves and fruits and the extraction methods always showed very high yields of DNA in both quantity and quality. Although, a wide range of commercial kits are available for extraction of fungal DNA, they can represent a high cost per sample analysed, and they are not always totally reliable in not co-extracting PCR inhibitors, needed a dilution of samples prior to PCR reactions. The optimised protocol did not co-extracted PCR-inhibitors from any samples, and therefore, the sensitivity of the detection protocol is improved using this DNA extraction protocols (Garrido et al., 2011).

To date, conventional PCR has been a fundamental part of fungal molecular diagnosis, but it shows several limitations: i.e. gel-based methods, possibility of quantification, sensitivity, etc. The development of real-time PCR has been a valuable response to these limitations (Garrido et al., 2011). This technology improve the sensitivity, accuracy and it is less time-consuming that conventional end-point PCR. For development and optimization of *Colletotrichum* diagnosis protocols, the commonly-used ribosomal RNA genes were used, because of the highly variable sequences of the internal transcribed spacers ITS1 and ITS2, which separate the 18S/5,8S and 5,8S/28S ribosomal RNA genes, respectively (Garrido et al., 2009a). Specific genus and species sets of primers and probes were designed for real-time PCR amplifications using TaqMan® chemistry technology. This system consists of a fluorogenic probe specific to the DNA target, which anneals to the target between the PCR primers; TaqMan® tends to be the most sensitive and simply methods for real-time PCR detection (Garrido et al., 2009a, 2011).

The specificity of all assays was tested using DNA from isolates of six species of *Colletotrichum* and from DNA of another nine fungal species commonly found associated with strawberry material. All the new assays were highly specific for *Colletotrichum* spp., *C. acutatum* and *C. gloeosporioides*, no cross-reactions were observed with either related plant pathogens or healthy strawberry plant material. The sensitivity of the new real-time PCR assays was compared with that of previously published conventional PCR assays; they were confirmed to be 100 times more sensitive than the latter. The *C. acutatum*-specific real-time PCR assay was also compared with an existing ELISA assay for the diagnosis of this pathogen. Real-time PCR permitted the detection of the pathogen in samples that gave negative results for *C. acutatum* using ELISA. The real-time PCR assay detected the equivalent of 7.2 conidia per plant inoculated with a serial dilution of *C. acutatum* spores, demonstrating the high degree of sensitivity of the method (Garrido et al., 2009a).

The new protocols were tested for monitoring the development of anthracnose disease in strawberry in the field in the south of Spain. The real-time PCR results showed a progressive increase of target DNA between January and June. The results showed that an increase in lesion development was accompanied by an increase in the amount and incidence of the pathogen as the season progressed. These results showed that new methods are suitable for diagnosis, identification and monitoring of the disease using field samples of strawberry and also, they permitted the detection of the pathogens from artificially infected symptomless plant material. Therefore, the methods described, based on real-time PCR, proved useful for studying the epidemiological routes of these strawberry pathogens in fields and nurseries (Garrido et al., 2009a, 2011).

5. Proteomics approaches of phytopathogenic fungi

In spite of the advances done by the described techniques above, nowadays proteomics is the most realistic and effective set of tools to unravel complex mixtures of proteins, describing the current molecular biology age as "post-genomic era". The term proteome was coined in 1995 by Wilkins et al (Wilkins et al., 1995), later the term proteomics appeared by James et al. (James, 1997). Proteome is defined as the complete set of proteins expressed by an organism, in a particular biological state. Proteomics may be introduced as a set of techniques that allow to study and to describe the proteome. The impact of the proteomic approaches is mainly based in a group of widely used techniques such as liquid chromatography or two dimensional gel electrophoresis, to separate complex protein mixtures, defining the proteome. However, the increasing relevance of these studies has been pushed by the improvements done in mass spectrometry system, allowing the analysis of peptides and proteins and/or by the increase number of proteins entries in the databases, making easier protein analysis and identification.

Main proteome characteristic is that it is a high dynamic system. It is even more complex than genomics, due to while the genome of an organism is more or less constant, the number of obtained proteomes from a specific genome is almost infinite. It depends of the assayed cell, tissue, culture conditions, etc. Each change produces a modification in the observed proteome. An additional factor of complexity is that there are changes that occur in proteome that are not encoded in the genome. These changes are mainly based on two sources, (i) the editing of the mRNA and (ii) post-translational modifications (PTMs) that normally serve to modify or modulate the activity, function or location of a protein in different contexts physiological or metabolic. There are more than 200 different described PTMs (phosphorylation, methylation, acetylation, etc.). They transform each single gene into tens or hundreds of different biological functions. Before proteomics achievements, the differential analysis of the genes, that were expressed in different cell types and tissues in different physiological contexts, was done mainly through analysis of mRNA. However, for wine yeast it has been proved that there is no direct correlation between mRNA transcripts and protein content (Rossignol et al., 2006). It is known that mRNA is not always translated into protein, and the amount of protein produced by a given amount of mRNA depends on the physiological state of the cell. Proteomics confirms the presence of the protein and provides a direct measure of its abundance and diversity.

In terms of methodology, proteomics approaches are classified in two groups, (i) gel free systems, based in the use of different chromatography methods, and (ii) gel based methods, using mainly two dimensional polyacrylamide gel electrophoresis (2DE), that will be the core of our discussion. As a schematic summary, the typical workflow of a proteomic experiment begins with the experimental design. It must be deeply studied, and it will delimit the obtained conclusions, even more when comparison between two strains, cultures or physiological stages between others, are done. From an optimal point of view, only one factor must change between the different assayed conditions (Fernández-Acero et al., 2007a, 2007b). It must contain the use of different biological replicates depending of the used strategy, usually from 3 to 5. The next key step is to obtain a protein extract with enough quality to separate the complex mixture of proteins. Usually, the protein extraction is done in sequential steps (Garrido et al., 2010). First, the biological sample is disrupted using mechanical or chemical techniques. Then, proteins are precipitated and cleaned. Most of the protocol use acetone and trichloroacetic acid. During the next step the proteome is defined and visualized using electrophoretic techniques. 2DE has been widely used for this purpose. Using this technique proteins are separated using two different parameters. During the first dimension, proteins are separated by their isoelectric point using an isoelectrofocusing (IEF) device. Then, the focused strips are used to load in a polyacrylamide gel, where the proteins are separated by their molecular weight. This system allows the separation of hundreds of proteins from a complex mixture. The gels are visualized with unspecific protein stains (those that stain total proteins, such as Coomassie, Sypro, Silver, etc.), or specific ones (those staining solution prepared to detect specific groups of proteins, mainly post-translationals modifications, i.e. Phospho ProQ diamond). The gels are digitalized and analyzed with specific software to reveal the significant spots. Those spots are identified using mass spectrometry. MALDI TOF/TOF is commonly used for 2DE approaches. The huge list of identified proteins obtained is studied to reveal the biological relevance of each identification.

Unfortunately, the number of papers related to fungal proteomics is still poor compared with the application of this technology to other biological sources. As an example, a simple search in WOK website (web of knowledge, http://www.accesowok.fecyt.es/) get 809 entries when the terms "proteom*"and "fung*" are used, whereas 51237 entries are displayed when "proteom*" is used alone. In spite of the numerical results obtained may vary depending of the used keywords and web resource, the fact is that there is a lot of work to do to bring fungal proteomic information at the same level that is obtained with other biological sources. This lack is mainly caused by (i) the difficulties to obtain proteins with enough quality to 2DE separations and (ii) the lack of protein sequences listed in the databases. Our research group was pioneer solving these problems and preparing the first proteomic approaches to the phytopathogenic fungi Botrytis cinerea (Fernández-Acero et al. 2006).

Fungi posse strong cell walls. This makes difficult the cell breakage using standard protocols. Moreover, fungal proteins extract are characterised by its high concentration of glycosylated proteins that produces dense extracts, dragging a lot of impurities that disturb protein electrofocusing. We optimized a protocol based on a first phosphate buffer solubilisation followed by a typical TCA/Acetone precipitation. Using this protocol we developed the first proteomic map of Botrytis cinerea (Fernández-Acero et al., 2006b). Using this optimized approach we prepared a differential proteomics approach based on 2DE, comparing the proteomes of two B. cinerea strains differing in virulence (Fernández-Acero et al., 2007b). In spite of this protocol has been widely cited and used (Cobos et al., 2010; Fernández-Acero et al., 2010, 2011; Michielse et al., 2011; Moreira et al., 2011; Sharma et al., 2010; Yang et al., 2011), our recent data suggest that the phosphate buffer solubilisation produces an artificial enrichment of soluble proteins in our assayed extracts. For this reason, we improved our method using a phenol based protocol preparing a Botrytis cinerea map during cellulose degradation (Fernández-Acero et al., 2010). Based on this protocol, adding a previous step of precipitation with DOC, we developed the analysis of the main fungal subproteome, the secretome. We identified 76 secreted proteins from cultures where the virulence was induced with different plant-based elicitors (Fernández-Acero et al., 2010). New projects to unravel proteome content of Botrytis cinerea and Colletotrichum acutatum are running.

All the proteomic approaches developed on B. cinerea has been facilitated by the availability of fungal genome sequence (Amselem et al., 2011) (http://urgi.versailles.inra.fr/ Species/Botrytis, and http://www.broadinstitute.org/annotation/genome/botrytis_ cinerea/Home.html). Summarizing all our identified spots, we do not get the 3% of the predicted genome. The method to capture new fungal proteins, its identification by mass spectrometry and to determine their biological relevance needs to be determined yet. By using our previous experience with *B. cinerea*, we are developing proteomic approaches to *C. acutatum*. Its conidial germination, mycelia dataset and secretome are characterized by 2DE. The key challenge is in our opinion, the use of the collected information to develop new methodologies to fight against plants pathogens. As a future prospect, the development of new environmental friendly proteomics-based fungicides has been discussed (Fernández-Acero et al., 2011).

6. Acknowledgements

This research has been financed by the Spanish Government DGICYT - AGL2009-13359-CO2/AGR, by the Andalusian Government: Junta de Andalucía, PO7-FQM-002689, http://www.juntadeandalucia.es/innovacioncienciayempresa; Programa Operativo 2007-2013 (FEDER-FSE) (18INSV2407, 18INSV2610), and by the CeiA3 International Campus of excellence in Agrifood (18INACO177.002AA, http://www.uco.es/cei-A3/). Victoria E. González-Rodríguez was financed by the grant FPU of the Ministerio de Educación, Government of Spain, Ref. AP2009-1309; Eva Liñeiro was financed by the grant of the University of Cádiz Ref. 2010-152.

7. References

Afanador-Kafuri, L., Minz, D., Maymon, M., Freeman, S. (2003). Characterization of *Colletotrichum* isolates from Tamarillo, Passiflora, and Mango in Colombia and identification of a unique species from the genus. *Phytopathology*, Vol. 93, pp. 579-587, 0031-949X

Agrios, G.N. (2005). *Plant Pathology*. (5th Ed). Elsevier Academic Press, ISBN: 13: 978-0120445653, San Diego, CA

Alastair McCartney, H., Foster, S.J., Fraaije, B.A., Ward, E. (2003). Molecular diagnostics for fungal plant pathogens. *Pest Management Science*, Vol. 59, pp. 129-142, 1526-498X

Amselem, J., Cuomo, C.A., van Kan, J.A.L., Viaud, M., Benito, E.P., Couloux, A., Coutinho, P.M., de Vries, R.P., Dyer, P.S., Fillinger, S., Fournier, E., Gout, L., Hahn, M., Kohn, L., Lapalu, N., Plummer, K.M., Pradier, J.-., Quévillon, E., Sharon, A., Simon, A., Have, A., Tudzynski, B., Tudzynski, P., Wincker, P., Andrew, M., Anthouard, V., Beever, R.E., Beffa, R., Benoit, I., Bouzid, O., Brault, B., Chen, Z., Choquer, M., Collémare, J., Cotton, P., Danchin, E.G., Da Silva, C., Gautier, A., Giraud, C., Giraud, T., Gonzalez, C., Grossetete, S., Güldener, U., Henrissat, B., Howlett, B.J., Kodira, C., Kretschmer, M., Lappartient, A., Leroch, M., Levis, C., Mauceli, E., Neuvéglise, C., Oeser, B., Pearson, M., Poulain, J., Poussereau, N., Quesneville, H., Rascle, C., Schumacher, J., Ségurens, B., Sexton, A., Silva, E., Sirven, C., Soanes, D.M., Talbot, N.J., Templeton, M., Yandava, C., Yarden, O., Zeng, Q., Rollins, J.A., Lebrun, M.-., Dickman, M. (2011). Genomic analysis of the necrotrophic fungal pathogens *Sclerotinia sclerotiorum* and *Botrytis cinerea*. *PLoS Genetics*, Vol. 7, No. 8, e1002230. doi:10.1371/journal.pgen.1002230, 1553-7390

Atkins, S.D., Clark, I.M., Sosnowska, D., Hirsch, P.R., Kerry, B.R. (2003). Detection and quantification of *Plectosphaerella cucumerina*, a potential biological control agent of potato cyst nematodes, by using conventional PCR, Real-Time PCR, Selective Media, and Baiting. *Applied and Environmental Microbiology*, Vol. 69, pp. 4788-4793, 0099-2240

Bailey, J.A., Jeger, M.J. (1992). *Colletotrichum: biology, pathology and control*. CAB international, ISBN: 0-85198-756-7, Wallingford

Buddie, A.G., Martínez-Culebras, P.V., Bridge, P., García-López, M.D., Querol, A., Cannon, P.F., Monte, E. (1999). Molecular characterization of *Colletotrichum* strains derived from strawberry. *Mycological Research*, Vol. 103, pp. 385-394, 0953-7562

Carbu, M. (2006). *Estudio de la variabilidad genetica y organizacion cromosomica en el hongo fitopatogeno "Botrytis cinerea"*. Thesis doctoral. University of Cádiz, Cádiz

Cobos, R., Barreiro, C., Mateos, R.M., Coque, J.R. (2010). Cytoplasmic- and extracellular-proteome analysis of *Diplodia seriata*: A phytopathogenic fungus involved in grapevine decline. *Proteome Sciences*, Vol. 8, doi: 10.1186/1477-5956-8-46, 1477-5956

Coley-Smith, J.R., Verhoheff, K., Jarvis, W.R. (1980). *The Biology of Botrytis*. Academic Press, ISBN 0-1217 98508X, London, UK

Denoyes-Rothan, B., Guerin, G., Lerceteau-Kohler, E., Risser, G. (2004). Inheritance of resistance to *Colletotrichum acutatum* in *Fragaria x ananassa* . *Phytopathology*, Vol. 95, pp. 405-412, 0031-949X

Denoyes-Rothan, B., Baudry, A. (1995). Species identification and pathogenicity study of French *Colletotrichum* strains isolated from strawberry using mophological and cultural characteristics. *Phytopathology*, Vol. 85, pp. 53-57, 0031-949X

Elad, Y., Williamson, B., Tudzynski, P., Delen, N. (2004). *Botrytis: Biology, Pathology and Control*. Springer, ISBN: 978-1-4020-6586-1, Dordrecht, The Netherland.

EPPO/CABI (1997). *Quarantine Pests for Europe* (2nd edition). Edited by Smith IM, McNamara DG, Scott PR, Holderness M. CABI International, Wallingford, UK, 1425 pp.

Faretra, F., Antonacci, E., Pollastro, S. (1988). Sexual bahaviour and matting system of *Botryotinia fuckeliana* (*Botrytis cinerea*). *Microbiology*, Vol. 134, No. 9, pp. 2543-2550, 1350-0872

Faretra, F., Antonacci, E. (1987). Production of apothecia of *Botryotinia fuckeliana* (de Bary) Whetz under controled environmental conditions. *Phytopathologia Mediterranea*, Vol. 26, pp. 29-35, 0031-9465

Fernández-Acero, F.J., Colby, T., Harzen, A., Carbú, M., Wieneke, U., Cantoral, J.M., Schmidt, J. (2010). 2-DE proteomic approach to the *Botrytis cinerea* secretome induced with different carbon sources and plant-based elicitors. *Proteomics*, Vol. 10, pp. 2270-2280, 1615-9853

Fernández-Acero, F.J., Carbú, M., Garrido, C., Vallejo, I., Cantoral, J.M. (2007a). Proteomic Advances in Phytopathogenic Fungi. *Current Proteomics*, Vol. 4, pp. 79-88, 1570-1646

Fernández-Acero, F.J., Jorge, I., Calvo, E., Vallejo, I., Carbú, M., Camafeita, L.E., Garrido, C., López, J.A., Cantoral, J.M., Jorrin, J. (2007b). Proteomic analysis of phytopathogenic fungus *Botrytis cinerea* as a potential tool for identifying pathogenicity factors, therapeutic targets and for basic research. *Archives of Microbiology*, Vol. 187, pp. 207-215, 1432-072X

Fernández-Acero, F.J., Carbú, M., Garrido, C., Collado, I.G., Cantoral, J.M., Vallejo, I. (2006a). Screening Study of Potential Lead Compounds for Natural Product-based Fungicides Against *Phytophthora* Species. *Journal of Phytopathology*, Vol. 154, pp. 616-621, 0931-1785

Fernández-Acero, F.J., Jorge, I., Calvo, E., Vallejo, I., Carbú, M., Camafeita, E., Lopez, J.A., Cantoral, J.M., Jorrin, J. (2006b). Two-dimensional electrophoresis protein profile of the phytopathogenic fungus *Botrytis cinerea*. *Proteomics*, Vol. 6, No. 1, pp. 88-96, 1615-9853

Fernández-Acero, F.J., Carbú, M., El-Akhal, M.R., Garrido, C., González-Rodríguez, V.E., Cantoral, J.M. (2011). Development of proteomics-based fungicides: New strategies for environmentally friendly control of fungal plant diseases. *International Journal of Molecular Sciences*, Vol. 12, pp. 795-816, 1422-0067

Freeman, S., Katan, T. (1997). Identification of *Colletotrichum* species responsible for anthracnose and root necrosis of strawberry in Israel. *Phytopathology*, Vol. 87, pp. 516-521, 0031-949X

Freeman, S., Minz, D., Jurkevitch, E., Maymon, M., Shabi, E. (2000a). Molecular analyses of *Colletotrichum* species from Almond and other fruit. *Phytopathology*, Vol. 90, pp. 608–614, 0031-949X

Freeman, S., Shabi, E., Katan, T. (2000b). Characterization of *Colletotrichum acutatum* causing anthracnose of Anemone (*Anemone coronaria* L.). *Phytopathology*, Vol. 66, No. 12, pp. 5267-5272, 0031-949X

Freeman, S., Pham, M., Rodríguez, R.J. (1993). Molecular genotyping of *Colletotrichum* species based on arbitrarily primed PCR, A+T-rich DNA, and nuclear DNA analyses. *Experimental Mycology*, Vol. 17, pp. 309-322, 0147-5975

Garrido, C., Carbú, M., Fernández-Acero, F.J., González-Rodríguez, V.E., Cantoral, J.M. (2011). New Insight in the Study of Strawberry Fungal Pathogens. *Genes, Genomes and Genomics*, Vol. 5, No. 1, pp. 24-39, 1749-0383

Garrido, C., Cantoral, J.M., Carbú, M., González-Rodríguez, V.E., Fernández-Acero, F.J. (2010). New proteomic approaches to plant pathogenic fungi. *Current Proteomics*, Vol. 7, pp. 306-315, 1570-1646

Garrido, C., Carbú, M., Fernández-Acero, F.J., Boonham, N., Colyer, A., Cantoral, J.M., Budge, G. (2009a). Development of protocols for detection of *Colletotrichum acutatum* and monitoring of strawberry anthracnose using real-time PCR. *Plant Pathology*, Vol. 58, pp. 43-51, 1365-3059

Garrido, C., Carbú, M., Fernández-Acero, F.J., Vallejo, I., Cantoral, J.M. (2009b). Phylogenetic relationships and genome organization of *Colletotrichum acutatum* causing anthracnose in strawberry. *European Journal of Plant Pathology*, Vol. 125, pp. 397-411, 0929-1873

Garrido, C., Carbú, M., Fernández-Acero, F.J., Budge, G., Vallejo, I., Colyer, A., Cantoral, J.M. (2008). Isolation and pathogenicity of *Colletotrichum* spp. causing anthracnose of strawberry in south west Spain. *European Journal of Plant Pathology*, Vol. 120, pp. 409-415, 0929-1873

Giraud, T., Fortini, D., Levis, C., Leroux, P., Brygoo, Y. (1997). RFLP markers show genetic recombination in *Botryotinia fuckeliana* (*Botrytis cinerea*) and transposable elements reveal two sympatric species. *Molecular Biology and Evolution*, Vol. 14, pp. 1177-1185, 1537-1719

Hughes, K.J.D., Lane, C.R., Cook, R.T.A. (1997). Development of a rapid method for the detection and identification of *Colletotrichum acutatum*, In: *Diagnosis and Identification of Plant Pathogens*, Dehne HW, Adam G, Diekmann M, Frahm J, Mauler-Machnik A & van Halteren P), pp. 113-116. Kluwer Academic, ISBN: 0792347714, Dordrecht (NL).

Ijdo, J.W., Wells, R.A., Baldini, A., Reeders, S.T. (1991). Improved telomere detection using a telomere repeat probe (TTAGGG)n generated by PCR. *Nucleic Acids Research*, Vol. 19, pp. 4780, 0305-1048

James, P. (1997). Protein identification in the post-genome era: the rapid rise of proteomics. *Quarterly Reviews of Biophysics*, Vol. 30, pp. 279-331, 0033-5835

Kim, Y., Nandakumar, M.P., Marten, M.R. (2007). Proteomics of filamentous fungi. *Trends in Biotechnology*, Vol. 25, pp. 395-400, 0167-7799

Leandro, L.F.S., Gleason, M.L., Nutter, F.W., Wegulo, S.N., Dixon, P.M. (2003). Strawberry plant extracts stimulate secondary conidiation by *Colletotrichum acutatum* on symptomless leaves. *Phytopathology*, Vol. 93, pp. 1285-1291, 0031-949X

Leandro, L.F.S., Gleason, M.L., Nutter, F.W., Wegulo, S.N., Dixon, P.M. (2001). Germination and Sporulation of *Colletotrichum acutatum* on symptomless strawberry leaves. *Phytopathology*, Vol. 91, pp. 659-664, 0031-949X

Martínez-Culebras, P.V., Querol, A., Suarez-Fernández, M.B., García-López, M.D., Barrio, E. (2003). Phylogenetic relationships among *Colletotrichum* pathogens of strawberry and design of PCR primers for their identification. *Journal of Phytopathology*, Vol. 151, pp. 135-143, 0931-1785

Masel, A., Irwin, J.A.G., Manners, J.M. (1993). DNA addition or deletion is associated with a major karyotype polymorphism in the fungal phytopathogen *Colletotrichum gloeosporioides*. *Molecular Genetics and Genomics*, Vol. 237, pp. 73-80, 1617-4615

Mertely, J.C., Legard, D.E. (2004). Detection, isolation, and pathogenicity of *Colletotrichum* spp. from strawberry petioles. *Plant Disease*, Vol. 88, pp. 407-412, 0191-2917

Michielse, C.B., Becker, M., Heller, J., Moraga, J., Collado, I.G., Tudzynski, P. (2011). The *Botrytis cinerea* Reg1 protein, a putative transcriptional regulator, is required for pathogenicity, conidiogenesis and for the production of secondary metabolites. *Molecular Plant-Microbe Interaction*, Vol 24, No. 9, pp. 1074-1085, 0894-0282

Mills, P.R., Sreenivasaprasad, S., Brown, A.E. (1992). Detection and differentiation of *Colletotrichum gloeosporioides* isolates using PCR. *FEMS Microbiology Letters*, Vol. 98, pp. 137-144, 0378-1097

Moreira, J., Almeida, R., Tavares, L., Santos, M., Viccini, L., Vasconcelos, I., Oliveira, J., Raposo, N., Dias, S., Franco, O. (2011). Identification of Botryticidal Proteins with Similarity to NBS–LRR Proteins in Rosemary Pepper (*Lippia sidoides* Cham.) Flowers. *The Protein Journal*, Vol. 30, pp. 32-38, 1572-3887

O'Sullivan, D., Tosi, P., Creusot, F., Cooke, M., Phan, T., Dron, M., Langin, T. (1998). Variation in genome organization of the plant pathogenic fungus *Colletotrichum lindemuthianum*. *Current Genetic*, Vol. 33, pp. 291-298, 0172-8083

Perfect, S.E., Bleddyn Hughes, H., O'Connell, R.J., Green, J.R. (1999). REVIEW: *Colletotrichum*: A model genus for studies on pathology and fungal-plant interactions. *Fungal Genetics and Biology*, Vol. 27, pp. 186-198, 1087-1845

Rebordinos, L., Vallejo, I., Santos, M., Collado, I.G., Carbú, M., Cantoral, J.M. (2000). Análisis genético y relación con patogenicidad en *Botrytis cinerea*. *Revista Iberoamericana de Micología*, Vol. 17, pp. 37-42, 1130-1406

Rebordinos, L., Santos, M., Vallejo, I., Collado, I.G., Cantoral, J.M. (1997). Molecular characterization of the phytopathogenic fungus *Botrytis cinerea*. *Recent Research Developments in Phytochemistry*, Vol. 1, pp. 293-307, 81-308-0052-7

Rossignol, M., Peltier, J., Mock, H., Matros, A., Maldonado, A.M., Jorrín, J.V. (2006). Plant proteome analysis: A 2004-2006 update. *Proteomics*, Vol. 6, pp. 5529-5548, 1615-9853

Sharma, N., Rahman, M.H., Kav, N.N.V. (2010). A possible proteome-level explanation for differences in virulence of two isolates of a fungal pathogen *Alternaria brassicae*. *Journal of Plant Biochemistry and Biotechnology*, Vol. 19, pp. 40-49, 0971-7811

Sreenivasaprasad, S., Talhinhas, P. (2005). Genotypic and phenotypic diversity in *Colletotrichum acutatum*, a cosmopolitan pathogen causing anthracnose on a wide range of hosts. *Molecular Plant Pathology*, Vol. 6, pp. 361-378, 0885-5765

Sreenivasaprasad, S., Sharada, K., Brown, A.E., Mills, P.R. (1996). PCR-based detection of *Colletotrichum acutatum* on strawberry. *Plant Pathology*, Vol. 45, pp. 650-655, 1365-3059

Sreenivasaprasad, S., Brown, A.E., Mills, P.R. (1992). DNA sequence variation and interrelationships among *Colletotrichum* species causing strawberry anthracnose. *Physiological and Molecular Plant Pathology*, Vol. 41, pp. 265-281, 0885-5765

Sutton, B.C. (1992). The genus *Glomerella* and its anamorph *Colletotrichum*, In: *Colletotrichum: Biology, Pathology and Control*, Bailey, J.A. & Jeger, M.J., pp. 1-26, CBA International, ISBN: 0-85198-756-7, Wallingford, UK.

Talhinhas, P., Sreenivasaprasad, S., Neves-Martins, J., Oliveira, H. (2005). Molecular and phenotypic analyses reveal association of diverse *Colletotrichum acutatum* groups and a low level of *C. gloeosporioides* with olive anthracnose. *Applied and Environmental Microbiology*, Vol. 71, pp. 2987-2998, 0099-2240

Vallejo, I., Carbú, M., Muñoz, F., Rebordinos, L., Cantoral, J.M. (2002). Inheritance of chromosome-length polymorphisms in the phytopathogenic ascomycete *Botryotinia fuckeliana* (anam. *Botrytis cinerea*). *Mycological Research*, Vol. 106, pp. 1075-1085, 0953-7562

Vallejo, I., Rebordinos, L., Collado, I.G., Cantoral, J.M. (2001). Differential behaviour of mycelial growth of several *Botrytis cinerea* strains on either Patchoulol- or Globulol-amended media. *Journal of Phytopathology*, Vol. 149, pp. 113-118, 0931-1785

Vallejo, I., Santos, M., Cantoral, J.M., Collado, I.G., Rebordinos, L. (1996). Chromosomal polymorphism in *Botrytis cinerea* strains. *Hereditas*, Vol. 124, pp. 31-38, 0018-0661

White, T.J., Bruns, S.L., Taylor, J.W. (1990). Amplification and direct sequencing of fungal ribosomal RNA genes for phylogenetics, In: *PCR Protocols: A Guide to Methods and Applications*, Innis, M. A., Gelfand, D. H., Sninsky, J. J., and White, T. J., pp. 315, Academic Press, Inc., New York

Whitelaw-Weckert, M.A., Curtin, S.J., Huang, R., Steel, C.C., Blanchard, C.L., Roffey, P.E. (2007). Phylogenetic relationships and pathogenicity of *Colletotrichum acutatum* isolates from grape in subtropical Australia. *Plant Pathology*, Vol. 56, pp. 448-463, 1365-3059

Wilkins, M.R., Sanchez, J.C., Gooley, A.A., Appel, R.D., Humphery-Smith, I., Hochstrasser, D.F., Williams, K.L. (1995). Progress with proteome projects: why all proteins expressed by a genome should be identified and how to do it. *Biotechnology & Genetic Engineering Reviews*, Vol. 13, pp. 19-50, 0264-8725

Yang, J., - Wang, L., - Ji, X., - Feng, Y., - Li, X., - Zou, C., - Xu, J., - Ren, Y., - Mi, Q., - Wu, J., - Liu, S., - Liu, Y., - Huang, X., - Wang, H., - Niu, X., - Li, J., - Liang, L., - Luo, Y., - Ji, K., - Zhou, W., - Yu, Z., - Li, G., - Liu, Y., - Li, L., - Qiao, M., - Feng, L., - Zhang, K. (2011). Genomic and Proteomic Analyses of the Fungus *Arthrobotrys oligospora* Provide Insights into Nematode-Trap Formation. *PLoS Pathogens*, Vol. 7, e1002179, 1553-7366.

Zeigler, R.S. (1998). Recombination in *Magnaporte grisea*. *Annual review of phytpathology*, Vol. 36, pp. 249-275, 0066-4286

Pulsed Field Gel Electrophoresis in Molecular Typing and Epidemiological Detection of Methicillin Resistant *Staphylococcus aureus* (MRSA)

Velazquez-Meza Maria Elena[1]*, Vázquez-Larios Rosario[2],
Hernández Dueñas Ana Maria[2] and Rivera Martínez Eduardo[2]
[1]Instituto Nacional de Salud Pública, Cuernavaca Morelos
[2]Instituto Nacional de Cardiología "Dr. Ignacio Chávez"
México D. F.

1. Introduction

Methicillin-resistant *Staphylococcus aureus* (MRSA) is an important threat to hospitalized patients worldwide and is responsible for a wide range of human diseases, including septicemia, endocarditis, pneumonia, osteomyelitis, toxic shock syndrome, and bacteremia (Tenover & Gaynes, 2000). This species nevertheless represents a serious public health burden, particularly the clones which are resistant to methicillin and other classes of antibiotics; the emergence of penicillin-methicillin-, and recently high-level vancomycin-resistant strains emphasize the importance and urgency of such rational prescribing policy for the treatment of MRSA infections (Appelbaum, 2007; Goldstein 2007). Multiple studies have shown clonal spreads of epidemic MRSA strains within hospitals, between hospitals within a country (Breurec et al., 2011; Nübel et al., 2010), and also between countries and continents (Breurec et al., 2011; Deurenberg et al., 2009; Diekema et al. 2001). There are only a limited number of nosocomial MRSA clones spread worldwide (the Iberian [ST247-SCCmec I], the Brazilian [ST239-IIIA], the Hungarian [ST239-III], the New York/Japan [ST5-II], the Pediatric [ST5-VI], the Berlin [ST45-IV], EMRSA-15 [ST22-IV], and the EMRSA-16 [ST36-II] clones) (Enright et al., 2002; Oliveira et al., 2001).

Molecular typing of MRSA is used to support infection control measures. Although Pulsed-field gel electrophoresis (PFGE) is well known and considered as golden standard, for establishing clonal relationships at the local level, its detection capacity seems to make it also too discriminative for global comparisons (McDougal et al.2003; Murchan et al., 2003). Recently multilocus sequence typing (MLST) has been proven to be the most adequate method both for long-term and global epidemiologic studies and for population genetic studies. Typing methods based on sequencing of more stable housekeeping genes (MLST) allow the creation of Internet-based curate databases and inter-laboratory data exchange

---

* Corresponding Author

(Enright et al., 2000) The combination of these methods allows the unambiguous assignment of collections of MRSA isolates or new MRSA clones (Enright et al., 2000).

The prevalence of MRSA in Mexico differs widely from one hospital to another and according to different studies performed; an increasing frequency of MRSA (7% in 1989, 14% in 2001 and 36% in 2004) are documented by reports of routine oxacillin disk diffusion tests only (Alpuche et al., 1989; Calderón et al., 2002; Chávez, 2004). This is of great concern, because it is a common experience that once MRSA is introduced in a hospital it is difficult to eradicate it (Creamer et al., 2011; Rebmann & Aureden, 2011). However, reports from Mexico documenting the clonality of MRSA isolates are very scarce, Aires de Sousa at al. in 2001 (Aires de Sousa et al., 2001) reported dominant and unique MRSA clone designated the Mexican clone (I::NH::M), identified by PFGE among isolates collected in 1997, 1998 from a pediatric hospital in Mexico, which had a rather limited resistance profile. In more recent studies which involve strains collected for the period 1997 to 2003 in two Mexican hospitals, PFGE distributed the MRSA isolates into two types M (clone EMRSA-16-U.K) and C (clone New York/Japan) these two clones were distinguished by antibiogram and other molecular properties (Echaniz et al., 2006; Velazquez et al., 2004).

The aim of this study was to identify MRSA clones circulating in a tertiary care hospital in Mexico City and their prevalence in the course of time 2002-2009. For this purpose, we used a phenotypic characterization and a combination of different molecular typing methods, including PFGE, hybridization with a Tn554 and mecA probes, staphylococcal cassette chromosome mec (SCCmec) and MLST.

2. Material and methods

2.1 Hospital setting

The Instituto Nacional de Cardiologia "Dr. Ignacio Chavez" (CAR) is a tertiary-care cardiology hospital located, in Mexico City with 246 beds, distributed 10 wards: surgery, adults and pediatric cardiology, neumology, nephrology, coronary unit and others. In addition the hospital has 17 external services. The microbiology laboratory receives an average of 18,000 samples annually. The hospital has 5,800 admission and 5,700 discharges per year.

2.2 Bacterial isolates

We studied a total of   ninety single-patient clinical MRSA isolates, between January 2002 and December 2009. The strains were collected from several clinical sources: bronchial secretions (n=34); wound secretions (n= 25), blood (n= 16); catheter (n=3); pleural liquid (n=3); peritoneal fluid (n=1) and others (n=13). MRSA strains were collected from different wards: pediatric surgery, adult surgery, coronary unit, nephrology, surgery and cardiology. Of the 90 MRSA isolates, 24 were from children and 66 were from adults.

2.3 Antimicrobial susceptibility testing

Antimicrobial susceptibility testing for MRSA isolates was performed using the automated method of MicroScan® (DADE-BEHRING, Sacramento, CA) for: penicillin, oxacillin,

Pulsed Field Gel Electrophoresis in Molecular Typing and Epidemiological Detection of Methicillin Resistant
Staphylococcus aureus (MRSA)

181

amoxicillin, cefotaxime, cephalothin, cefazolin, imipenem, trimethoprim-sulfamethoxazole, ciprofloxacin, chloramphenicol, clindamycin, erythromycin, clarithromycin, gentamicin, rifampin, tetracycline and vancomycin, following the Clinical Laboratory Standards Institute guidelines (Clinical Laboratory Standards Institute [CLSI], 2009).

2.4 Molecular typing

The whole genomic DNA was prepared as described previously (Chung et al., 2000). After digestion with *SmaI* endonuclease, DNA was separated in a CHEF-DRII apparatus (Bio-Rad, Birmingham, U.K) (Chung et al., 2000). Strains HU25, HPV107, HDE288, BK2464, JP27 and 96/32010, representing the Brazilian, Iberian, Pediatric, New York/Japan-USA, New York/Japan-Japan and EMRSA-16-U.K clones, were included in the PFGE gels as controls. The control strains were kindly provided by Prof. Herminia de Lencastre from the Molecular Genetics laboratory Institute de Tecnología Química e Biologica da Universidad Nova de Lisboa. Criteria of Tenover were used to compare different clones (Tenover et al., 1995). Strains BK2464 and HDE288 were used as SCC*mec* controls. Hybridization of *SmaI* digests with *mecA* and Tn554 probes (de Lencastre et al., 1994), SCC*mec* typing (Oliveira & de Lencastre, 2002) and MLST (Enright et al., 2000) were performed as previously described. Briefly, MLST is based in internal fragments of seven housekeeping genes (*arcC, aroE, glpF, gmk, pta, tpi, yqiL*) for each isolates, the alleles at the seven loci defined the allelic profiles, which corresponded to a sequence type (ST). ST designations were those assigned the MLST data base (http://www.mlst.net). The SCC*mec* typing system is defined by combining the class of the *mec* gene complex with the cassette chromosome recombinase gene (*ccr*) allotypes. The polymorphism in the vicinity of the *mecA* gene detected by probe *ClaI*-digested DNAs with a *mecA* probe and transposon Tn554 insertion patterns detecting by probing *ClaI* digestion DNAs with a specific probe (de Lencastre et al., 1994, Enright et al., 2000; Oliveira & de Lencastre, 2002).

2.5 Computer-fingerprinting analysis

The computer analysis of the banding patterns obtained by PFGE was done using the NTSYSpc software version 2.0.2.11 (Applied Biostatistics Inc.) after visual inspection. Each gel included reference strain *S. aureus* NCTC 8325 to normalize the PFGE profiles. For clusters analyses, the Dice coefficients were calculated to compute the matrix similarity and were transformed into an agglomerative cluster by the unweighted pair group method with arithmetic average (UPGMA).

3. Results

3.1 Antimicrobial susceptibility

The 90 isolates showed resistance to penicillin (100%), oxacillin (99.3%), amoxicillin (100%) , cefotaxime (100%), cephalothin (100%), cefazolin (100%), chloramphenicol (100%), imipenem (99.3%), ciprofloxacin (87.7%); eleven strains (12.2%) showed low susceptibility for clindamycin, erythromycin, clarithromycin and were susceptible to ciprofloxacin; two strains (2.2%) showed low susceptibility for oxacillin (MIC 4µg/mL) and imipenem. All strains were susceptible to rifampin, tetracycline, gentamicin, trimethoprim-sulfamethoxazole and vancomycin.

3.2 Molecular typing

3.2.1 PFGE analysis

The PFGE analysis separated the MRSA strains into three types, A (5 subtypes), B (3 subtypes) and C (6 subtypes) (Figure 1). PFGE pattern C and subtypes were predominant in this isolates n=72 (80%), Clone A, n=11 (12.2%) and B, n=7 (7.8%) were only found in the isolates of 2002, and these two clones (A and B) were totally replaced by clone C in 2004 and continue until 2009. The results produced by a computer analysis of the banding patterns show clearly the division of the three clone groups (A, B and C); interestingly, the A and B clone isolates have very similar PFGE patterns (coefficient similarity 95%). Nevertheless, the three clones A, B and C could easily be distinguished by antibiograms and other molecular properties as well (Table 1). The three clones were multiresistant, however, each one of them showed a characteristic resistance pattern; clone A was resistant to ß-lactams and showed a low susceptibility to clarithromycin, clindamycin, erythromycin and was susceptible to ciprofloxacin; while clones B and C were resistant to ß-lactams, clarithromycin, clindamycin, erytromycin and ciprofloxacin; only the strains with subtypes B1 and B2 showed low susceptibility for oxacillin and imipenem.

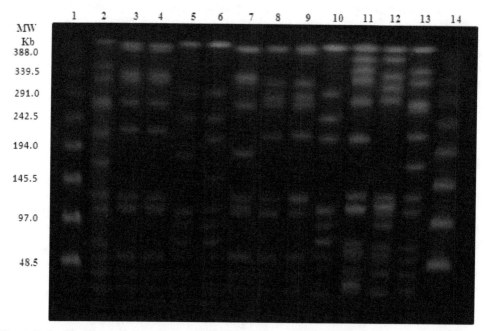

Fig. 1. Pulsed field gel electrophoresis profiles of MRSA clinical isolates from the Instituto Nacional de Cardiología "Dr. Ignacio Chávez", Mexico and representatives of international clones. Lanes: 1-14 lambda ladder used a molecular size (MW) markers; 2 and 13 reference strain NCTC8325; 3-4 (44CAR and 47CAR) pattern C; 5 (2CAR) pattern A; 6 (20CAR) pattern B; 7-12 (HDE288, BK2464, JP27, EMRSA16, HPV107 and HU25) control strains representative of Pediatric, New York/Japan-USA, New York/Japan-Japan, EMRSA16-U.K, Iberian and Brazilian clones.

Pulsed Field Gel Electrophoresis in Molecular Typing and Epidemiological Detection of Methicillin Resistant
Staphylococcus aureus (MRSA)

183

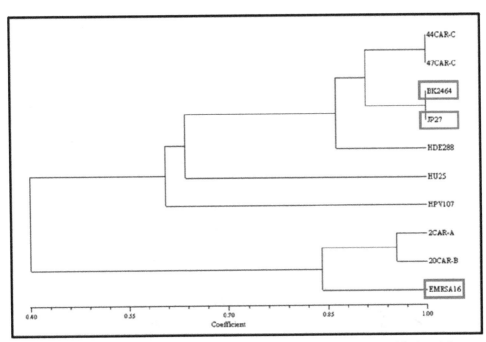

Fig. 2. Dendrogram comparing MRSA clones A, B and C from the Instituto Nacional de Cardiología "Dr. Ignacio Chávez", Mexico with different international MRSA clones: BK2464-New York/Japan-USA clone; JP27-New York/Japan-Japan clone; HDE288-Pediatric clone; HU25-Brazilian-clone; HPV107-Iberian clone; EMRSA-16-U.K. clone. For cluster analysis, Dice coefficients were calculated to compute matrix similarity a transformed into an agglomerative cluster with the unweighted pair group method with arithmetic average.

Property	Clone A	Clone B	Clone C
Antibiotype[1] (Resistance)	ß-lactams, (CLA,CD,ERY)[2]	ß-lactams, (CIP, CLA,CD, ERY)	ß-lactams, (CIP, CLA,CD, ERY)
Number of subtypes	5	3	6
SSCmec type[3]	IV	II	II
Hybridization bands(Kb) SmaI-mecA	~180	~211[4]	~211
Hybridization bands(Kb) Sma-Tn554	~180	~211-640	~211-640
ST	30	30[5]/36	5

[1]Antibiotic abbreviations: CLA- clarithromycin; CD – clindamycin; ERY – erythromycin GEN - gentamicin; CIP - ciprofloxacin. [2]Intermediate resistance pattern. [3]Staphylococcal cassette chromosome *mec*. [4]Except B1 and B2; [5]Sequence typing (MLST), only the patterns B1 and B2.

Table 1. Antibiotype and Genotypic characterization of the MRSA clones presented in the Instituto Nacional de Cardiología "Dr. Ignacio Chávez", Mexico (2002-2009).

3.2.2 Hybridization pattern

The hybridization patterns with *mecA* and Tn*554* probes indicated that the MRSA strains accompanying clone A carried the *mecA* gene on a *Sma*I fragment of approximately 180 Kb, while the *mecA* gene of the clones B and C were found on a fragment of approximately 211 Kb (Figure 3-A). One Tn*554* copy was identified usually on the fragment approximately 180 Kb between the isolates of clone A; while the MRSA strains accompanying clones B and C usually carried two identified Tn*554* copies between the *Sma*I fragments of approximately 211 and 640 Kb; only the strains 3CAR, (B1) and 8CAR, (B2), carried the transposon Tn*554* in a fragment of 640 Kb (Figure 3-B).

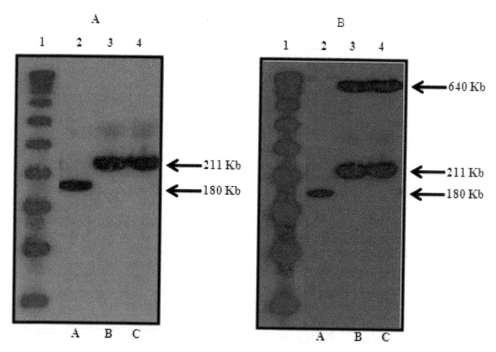

Fig. 3. (A) *SmaI-mecA* and (B) *Sma*I Tn*554* patterns identified among the clones A, B and C of the Instituto Nacional de Cardiología "Dr. Ignacio Chavéz, Mexico. Lane 1, molecular weight markers, lambda ladder; lane 2, 2CAR (pattern A); lane 3, 20CAR (pattern B) and lane 4, 44CAR (pattern C).

Two strains collected in this hospital 3CAR, (pattern B1) and 8CAR, (pattern B2) did not hybridize with the *mecA* DNA probe, interestingly both presented a low susceptibility to oxacillin and these isolates were subtypes of pattern B. The only difference was found in the *Sma*I hybridization fragment, which contains the *mecA* gene: in the two isolates, this fragment had a smaller molecular size (145 instead of 180 Kb) and did not react with the *mecA* probe, indicating a deletion of approximately 35 Kb, which must have included both the *mecA* gene and part of the *mec* element (Figure 4A and 4B). All the isolates accompanying clone A presented SCC*mec* type IV and sequence type 30 (ST30) whereas the

Pulsed Field Gel Electrophoresis in Molecular Typing and Epidemiological Detection of Methicillin Resistant
Staphylococcus aureus (MRSA)

185

MRSA strains of clones B and C had SCC*mec* type II, sequences type 36 and 5 (ST36 and ST5) respectively, except B1 and B2 which did not amplify SCC*mec*, this isolates showed sequence type 30 (ST30) .

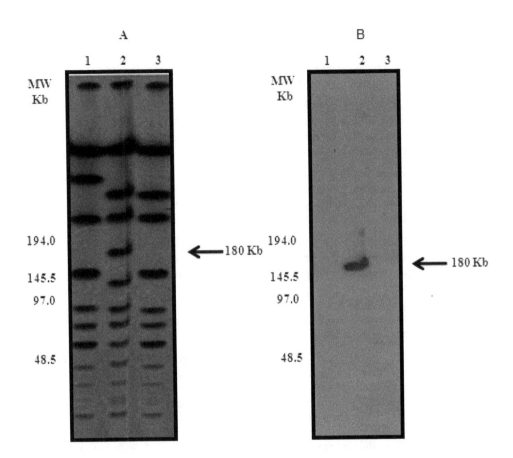

Fig. 4. (A) PFGE patterns of MRSA strains from the Instituto Nacional de Cardiología "Dr. Ignacio Chávez", Mexico. Lane 1, 3CAR, (pattern B1); lane 2, control strain and lane 3, 8CAR (pattern B2).

3.2.3 Homology pattern

One isolate belonging to each type of clones (A, B and C) were compared to strains belonging to previously characterized MRSA clones, i. e., representatives of the pediatric clone and isolates belonging to the New York-Japan clone and also to other international pandemic clones, namely, the Iberian, Brazilian and EMRSA-16 clones (Figure 1). Clone C showed a high degree of similarity to the pediatric (85.5%) and the New York-Japan (89.5%) clones. Clones A and B showed a high degree of similarity to the EMRSA-16 (80%) clone (Figure 2).

4. Discussion

The emergence of strains resistant to methicillin and other antibacterial agents has become a major concern especially in the hospital environment, because of the higher mortality due to systemic methicillin-resistant *S. aureus* infections (Cosgrove et al., 2003; Handberger et al., 2011). Seven major pandemic MRSA (the so-called Brazilian, Hungarian, Iberian, New York-Japan, pediatric, EMRSA 16 and Berlin clone (EMRSA15) have been identified as the cause for the majority of hospital-acquired *S. aureus* infections in the world (Oliveira et al., 2002), indicating that they represent successful clones in terms of their ability to cause infections, persist and spread from one geographic zone to another, including across continents.

The combination of different molecular typing methods used in the present study allowed us to register epidemiologically relevant features of MRSA populations in the Instituto Nacional de Cardiología "Dr. Ignacio Chávez" in Mexico and document the coexistence of MRSA clones of international distribution.

All 90 strains were resistant to at least eleven antibiotics (amoxicillin, cefotaxime, cephalothin, cefazolin, chloramphenicol, imipenem, clindamycin, erythromycin and clarithromycin) in addition to penicillin and oxacillin and 94.4% were resistant to ciprofloxacin as well. The phenotypes of resistance to the antimicrobial agents are shown in the Table 1.

As a response to the emergence and worldwide spread of antibiotic-resistant *S. aureus* there was an urgent need for the creation of international surveillance systems with methodologies that could help hospital infection prevention and control such organisms. MRSA causing nosocomial infections have been reported in other hospitals in Mexico showing a wide geographic spread of MRSA specific clones (Aires de Sousa et al., 2001; Echaniz et al., 2006; Velazquez et al. 2004) similar spread has been observed by other clones in USA and Europe (Da Silva, 2003; Johnson, 2011; Oliveira & de Lencastre 2002).

Interestingly, only three PFGE types were found during the period of the study, designed A, B and C. Previous studies had documented that MRSA clones may spread in and between hospitals, cities and countries and even intercontinental spread may occur (Auken et al., 2002; Nübel et al., 2010). The multiresistant clone C (New York/Japan clone) was present in more than 50% of MRSA that were recovered from a variety of infections sites and hospital wards. Previously, this clone had already been reported in two hospitals in Mexico: Hospital Civil de Guadalajara "Fray Antonio Alcalde" and Hospital de Pediatría del Centro Medico Nacional Siglo XX1-IMSS and it has been circulating in these hospitals since 1999, and 2001 respectively (Echaniz et al., 2006; Velazquez et al. 2004). The results of these studies showed that clone C (New York/Japan clone) had, sequence type 5 and SCC*mec* type II. In this study we found that pattern C was very similar (89.5%) to the multiresistant New York-Japan clone (Figure.2), which correspond to our last year´s results, proving with this the capacity of this clone to persists for long periods of time within the hospitals; as well as its capacity to spread to other hospitals (epidemic clone). whose evidence is the existence of this clone in other hospital of third level in Mexico Instituto Nacional de Cancerología (INCan). It is important to mention that the existence of clone C (New York/Japan clone) had not been present in the INCan before 2006 (Cornejo et al., 2010). All these results are of relevant importance if we consider that the first high-level VRSA (vancomycin-resistant *S. aureus*) (MIC = 1024 μg/mL vancomycin), belonged to the New York lineage (Weigel et al., 2003) and the fact that the descending MRSA strains of this clone are circulating in our

Pulsed Field Gel Electrophoresis in Molecular Typing and Epidemiological Detection of Methicillin Resistant
Staphylococcus aureus (MRSA)

187

population, together with the few means of antibiotic restriction it could represent a potential short term risk for the VRSA appearance in the hospitals of our country. Clone A and B were only found in the isolates in 2002, these clones showed a high degree of similarity to the EMRSA-16 clone, this clone is one of the dominant types of MRSA found in a UK hospital (Moore & Lindsay, 2002) and was widely disseminated in Canada (Simor et al., 2002), Greece (Aires de Sousa et al., 2003) and Mexico (Aires de Sousa et al., 2001). Interestingly, both clones (A and B) are very similar (95%) (Figure 2), nevertheless, clone A showed a reduced resistance profile as clone B, and this is because of the existence of the SCCmec IV in these isolates, this chromosomal cassette was found in relation to isolated MRSA strains in the community (CA-MRSA) (Coombs et al., 2011). Different reports of several infections caused by CA-MRSA in Latin America (Uruguay, Rio de Janeiro, Colombia, Argentina and Mexico) have been published (Alvaréz et al., 2006; Ma et al., 2005; Reyes et al., 2009; Ribeiro et al., 2005; Velazquez et al., 2011). All pattern of PFGE of the clones A, B and C showed subtypes. Probably the PFGE subtypes indicate the continued evolutionary divergence of these clones during its massive geographic expansion.

Relative genetic instability of the mecA element was observed in two strains and this was associated with an apparent deletion of the mec element, these isolates were very similar to profile B (B1 and B2) and presented a low susceptibility to oxacillin. In the literature there are reports of S. aureus strains with low-level methicillin resistance (MIC 2-4 µg/mL) which are not associated to the presence of mecA gene, Tomasz et al. reported one class of borderline methicillin-resistant strains having PBP1 and PBP2 with altered methicillin-binding affinities and overproduction of PBP4 (Tomasz et al., 1989). Another class of low-susceptibility has been reported and was attributed to overproduction of penicillinase (McDougal & Thornsberry, 1986). Hackbarth et al. studied the nucleotide sequence of the PBP2 gene and identified a point mutation near the penicillin-binding motive of transpeptidase (Hackbarth et al., 1995). An MRSA clinical strain with significant methicillin resistance (MIC 64µg/mL) despite absence of mec A was reported (Yoshida et al., 2003).

5. Conclusion

The combination of molecular typing methods (PFGE, mecA, Tn554 probes, SCCmec, and MLST) with epidemiologic and clinical information allows the detection of MRSA clusters and outbreaks and therefore provides a rationale for appropriate infection control intervention. Our study emphasizes the need of national and international collaborations to monitor the spread of current epidemic strains as well as the emergence of new ones in our country. The mechanisms of spread in different areas are poorly understood and further studies are necessary to understand the dynamics involved in the predominance of unique MRSA clones.

6. Acknowledgment

We thank PhD. Lilia Chihu for style review of the paper

7. References

Aires de Sousa, M. Miragaia, M. Santos, I. Avila, S. Adamson, I. Casagrande, S. Brandileone, C. Palacio, R. Dell'Acqua, L. Hortal, M. Camou, T. Rossi, A. Velazquez, ME.

Echaniz, G. Solorzano, F. Heitmann, I. & de Lencastre, H. (2001). Three-year assessment of methicillin-resistant *Staphylococcus aureus* clones in Latin America from 1996 to 1998. *Journal Clinical Microbiology*, Vol.39, No.6, (June 2001), pp. 2197-3205, ISSN 0095-1137

Aires de Sousa, M. Bartzavali, C. Spiliopoulou, I. Santos, I. Crisostomo, I & de Lencastre, H. (2003). Two international methicillin-resistant *Staphylococcus aureus* clones endemic in a university hospital in Patras, Greece. *Journal of Clinical Microbiology*, Vol.41, No.5, (May 2003), pp. 2027-2031, ISSN 0095-1137

Alpuche, C. Avila, C. Espinoza, LE. Gómez, D. & Santos, I. (1989). Perfiles de sensibilidad antimicrobiana de *Staphylococcus aureus* en un hospital pediátrico: prevalencia de resistencia a meticilina. *Boletín Medico del Hospital Infantil de México*, Vol.11, pp. 697-699. ISSN 1665-1146

Alvaréz, C. Barrientes, O. Leal, A. Contreras, G. Barrero, L. Rincón, S. Díaz, L. Vanegas, N. & Arias, C. (2006). Community-associated methicillin-resistant *Staphylococcus aureus*, Colombia. *Emerging Infectious Diseases*, Vol.12, No.12, (December 2006), pp. 2000-2001, ISSN 1080-6040

Appelbaum, P. (2007). Reduced glycopepdide susceptibility in methicillin-resistant *Staphylococcus aureus* (MRSA). *International Journal of Antimicrobial agents*, Vol.30, No.4, (November 2007), pp. 398-408, ISSN 0924-8579

Auken, H. Ganner, M. Murchan, S. Cookson, D. & Johnson, P. (2002). A new UK strain of epidemic methicillin-resistant *Staphylococcus aureus* (EMRSA-17) resistant to multiple antibiotics. *Journal Antimicrobial Chemotheraphy*, Vol.50, No.2, (August 2002), pp.171-175, ISSN 0305-7453

Breurec, S. Zriouil, . Fall, C. Boisier, P. Brisse, S. Djibo, S. Etienne, J. Fonkoua, M Perrier-Gros-Claude, J. Pouillot, R. Ramarokoto, C. Randrianirina, F. Tall, A. Thiberge, J, Working Group on *Staphylococcus aureus* infections, Laurent, F. & Garin, B. (2011) Epidemiology of methicillin-resistant *Staphylococcus aureus* lineages in five major African towns: emergence and spread of atypical clones. *Clinical Microbiology and Infectious Diseases*, Vol.17, No.2, (February 2011), pp. 160-165, ISSN 1198-743X

Calderón, E. Espinoza, LE. & Avila, R. (2002). Epidemiology of drug resistance: The case of *Staphylococcus aureus* and coagulase negative staphylococci infections. *Salud Pública de México*, Vol.44, No.2, (October 2001), pp. 108-12. ISSN 0036-3634

Chávez, B. (2004). Infecciones intrahospitalarias ¿Que ha pasado en 23 años? *Enfermedades Infecciosas y Microbiología*, Vol.24, No.3, (July-September 2004), pp. 89-92.

Chung, M. de Lencastre, H. Matthews, P. Tomasz, A. Adamsson, I. Aires de Sousa, M. Camou, T. Cocuzza, C. Corso, A. Couto, I. Dominguez, A. Gniadkowski, M. Goering, R. Gomes, A. Kikuchi, K. Marchese, A. Mato, R. Melter, O. Oliveira, D. Palacio, R. Sá-Leão, R. Santos, I. Song, J. Tassios, P. & Villari, P. (2000). Molecular typing of methicillin-resistant *Staphylococcus aureus* by pulsed-field gel electrophoresis: comparison of results obtained in a multilaboratory effort using identical protocols and MRSA strains. *Microbial Drug Resistance*, Vol.6, No.3, (October 2000), pp. 189-98, ISSN 1076-6294

[CLSI] Clinical and Laboratory Standards Institute. (2009). Performance Standards for Antimicrobial Susceptibility Testing. Nineteenth Informational Supplement; M100-S19. *Clinical and Laboratory Standards Institute*, Vol.29, No.3, (January 2009), pp. 1-149, ISBN 1-56238-690-5, Wyne, Pennsylvania.

Pulsed Field Gel Electrophoresis in Molecular Typing and Epidemiological Detection of Methicillin Resistant
Staphylococcus aureus (MRSA)

189

Coombs, G. Monecke, S. Pearson, J. Tan, H. Chew, Y. Wilson, L. Ehricht, R. O'Brien, F. & Christiansen, K. (2011). Evolution and diversity of community-associated methicillin-resistant *Staphylococcus aureus* in a geographical region. *BMC Microbiology*, Vol.11, (September 2011), pp. 1-212 ISSN 1471-2180

Cornejo, P. Volkow, P. Sifuentes, J. Echániz, G. Díaz, A. Velázquez, C. Bobadilla M. Gordillo, P. & Velazquez, ME. (2010). Tracing the source of an outbreak of methicillin-resistant *Staphylococcus aureus* in a tertiary-care oncology hospital by epidemiology and molecular methods. *Microbial Drug Resistance*, Vol.16, No.3, (September 2010), pp. 203-208, ISSN 1076-6294

Cosgrove, S. Sakoulas, G. Perencevich, E. Schwaber, M. Karchmer, A. & Carmeli, Y. (2003). Comparison of mortality associated with methicillin-resistant and methicillin-susceptible *Staphylococcus aureus* bacteremia: a meta analysis. *Clinical Infectious Diseases*, Vol.36, No.1, (January 2003), pp. 53-59, ISSN 1058-4838

Creamer, E. Galvin, S. Dolan, A. Sherlock, O. Dimitrov, B. Fitzgerald-Hughes. D. Thomas, T. Walsh, J. Moore, J. Smyth, E. Shore, A. Sullivan. D. Kinnevey, P. O'Lorcain, P. Cunney, R. Coleman, D. & Humphreys, H. (2011). Evaluation of screening risk and nonrisk patients for methicillin-resistant *Staphylococcus aureus* on admission in an acute care hospital. *American Journal of Infection Control*, Vol.30, [Epub ahead of print] ISSN 0196-6553

Da Silva, M. Silva, M. Wisplinghoff, H. Hall, G. Tallent. S. Wallance, S. Edmond, M. Figueiredo, A. & Wenzel. R. (2003). Clonal spread of methicillin-resistant *Staphylococcus aureus* in a large geographic area of the United States. *The Journal of Hospital Infection*, Vol.53, No.2, (February 2003), pp. 103-110, ISSN 0195-6701

de Lencastre, H. Couto, I. Santos, I. Melo, J. Torres, A. & Tomasz, A. (1994). Methicillin-resistant *Staphylococcus aureus* disease in a Portuguese hospital: characterization of clonal types by a combination of DNA typing methods. *European Journal of Clinical Microbiology and Infectious Diseases*, Vol.13, No.1, (January 1994), pp. 64-73, ISSN 0934-9723

Deurenberg, R. Nulens, E. Valvatne, H. Sebastian, S. Driessen, C. Craeghs, J. De Brauwer, E. Heising, B. Kraat, Y. Riebe, J. Stals, F. Trienekens, T. Scheres, J. Friedrich, A. van Tiel, F. Beisser, P. & Stobberingh, E. (2009). Cross-border dissemination of methicillin-resistant *Staphylococcus aureus*, Euregio Meuse-Rhin region. *Emerging infectious diseases*, Vol.15, No.5, (May 2009), pp. 727-734, ISSN 1080-604

Diekema, D. Pfaller, M. Schmitz, F. Smayevsky, J. Bell, J. Jones, R. & Beach, M. (2001). Survey of infections due to *Staphylococcus* species: frequency, occurrence and antimicrobial susceptibility of isolates collected in the United States, Canada, Latin America, Europe and the Western Pacific region for the SENTRY Antimicrobial Surveillance Program, 1997-1999. *Clinical of Infectious Diseases*, Vol.15, No.32, (May 2001), pp. S114-132, ISSN 1058-4838

Echaniz, G. Velazquez, ME. Aires de Sousa, M. Morfín, R. Rodríguez-Noriega, E. Carnalla, N. Esparza, S. & de Lencastre H. (2006). Molecular characterization of a dominant methicillin-resistant *Staphylococcus aureus* (MRSA) clone in a Mexican hospital (1999-2003). *Clinical Microbiology and Infectious Diseases*, Vol.12, No.1, (January 2006), pp. 12-28, ISSN 1198-743X

Enright, M. Day, N. Davies, C. Peacock, S. & Spratt, B. (2000). Multilocus sequence typing for characterization of methicillin-resistant and methicillin-susceptible clones of

Staphylococcus aureus. Journal of Clinical Microbiology, Vol.38, No.3, (March 2000), pp. 1008-1015, ISSN 0095-1137

Enright, M. Robinson, D. Randle, G. Feil, E. Grundmann, H. & Spratt, B. (2002). The evolutionary history of methicillin-resistant *Staphylococcus aureus* (MRSA). *Proceedings of the National Academy of Sciences of the United States of America*, Vol. 99, No.11, (May 2002), pp. 7687-7692, ISSN 0027-8424

Goldstein, F. (2007). The potential clinical impact of low-level antibiotic resistance in *Staphylococcus aureus. Journal Antimicrobial Chemotherapy*, Vol.59, No.1, (January 2007), pp. 1-4, ISSN 0305-7453.

Hackbarth, C. Kocagoz, T. Kocagoz, S. & Chambers. H. (1995). Point mutation in *Staphylococcus aureus* PBP2 gene affects penicillin-binding kinetics and is associated with resistance. *Antimicrobial Agents and Chemotherapy*, Vol.39, No.1, (January 1995), pp. 103-106, ISSN 0066-4804

Hanberger, H. Walther, S. Leone, M. Barie, P. Rello, J. Lipman J. Marshall, J. Anzueto A. Sakr, Y. Pickkers, P. Engoren, M. Vincent, J. & EPIC II Group of investigators. (2011). Increased mortality associated with meticillin-resistant *Staphylococcus aureus* (MRSA) infection in the intensive care united: results from the EPIC II study. *International Journal of Antimicrobial Agents*, Vol.38, No.4, (October 2011), pp. 331-335, ISSN 0924-8579

Johnson, A. (2011). Methicillin-resistant *Staphylococcus aureus*: the European landscape. *The Journal of antimicrobial chemotherapy*, Vol.66, No.4, (May 2011), pp. 43-48, ISSN 0305-7453

Ma, XX. Galiana, A. Pedreira, W. Mowszowicz, M. Christophersen, I. Machiavello, S. Lopez, L. Benaderet, S. Buela, F. Vincentino, W. Albini, M. Bertaux, O. Constenla, I. Bagnulo, H. Llosa, L. Ito, T. & Hiramatsu, K. (2005). Community-acquired methicillin-resistant *Staphylococcus aureus*, Uruguay. *Emerging Infectious Diseases*, Vol.11, No. 6, (June 2005), pp. 973-976, ISSN 1080-604

McDougal, L. & Thornsberry, C. (1986). The role of ß-lactamase in staphylococcal resistance to penicillinase-resistant penicillins and cephalosporins. *Journal of Clinical Microbiology*, Vol.23, No.5, (May 1986), pp. 832-839, ISSN 0095-1137

McDougal, L. Steward, C. Killgore, G. Chaitram, J. McAllister, S. & Tenover, F. (2003). Pulsed-field gel electrophoresis typing of oxacillin-resistant *Staphylococcus aureus* isolates from the United States: establishing a national database. *Journal of Clinical Microbiology*, Vol.41, No.11, (November 2003), pp. 5113-5120, ISSN 0095-1137

Moore, P. & Lindsay. J. (2002). Molecular characterization of the dominant UK methicillin-resistant *Staphylococcus aureus* strain, EMRSA-15 and EMRSA-16. *Journal of Medical Microbiology*, Vol.51, No.6, (June 2002), pp. 516-521, ISSN 0022-2615

Murchan, S. Kaufmann, M. Deplano, A. de Ryck, R. Struelens, M. Zinn, C. Fussing, V. Salmenlinna, S. Vuopio-Varkila, J. El Solh, N. Cuny, C. Witte, W. Tassios, P. Legakis, N. van Leeuwen, W. van belkum, A. Vindel, A. Laconcha, I. Garaizar, J. Haeggman, S. Olsson-Liljequist, B. Ransjo, U. Coombes, G. & Cookson, B. (2003). Harmonization of pulsed-field gel electrophoresis protocols for epidemiological typing of strains of methicillin-resistant *Staphylococcus aureus*: a single approach developed by consensus in 10 European laboratories and its application for tracing the spread of related strains. *Journal of Clinical Microbiology*, Vol.41, No.4, (April 2003), pp. 1574-85, ISSN 0095-1137

Pulsed Field Gel Electrophoresis in Molecular Typing and Epidemiological Detection of Methicillin Resistant
Staphylococcus aureus (MRSA)

191

Nübel, U. Dordel, J. Kurt, K. Strommenger, B. Westh, H. Shukla, S. Zemlicková, H. Leblois, R. Wirth, T. Jombart, T. Balloux, F & Witte, W. (2010). A timescale for evolution, population expansion, and spatial spread of an emerging clone of methicillin-resistant *Staphylococcus aureus*. *PLoS pathogens*, Vol.6, No.4, (April 2010), pp. 1-12, ISSN 1553-7366

Oliveira, D. Tomasz, A. & de Lencastre, H. (2001). The evolution of pandemic clones of methicillin-resistant *Staphylococcus aureus*: identification of two ancestral genetic backgrounds and the associated *mec* elements. *Microbial Drug Resistance*, Vol.7, No.4, (December 2001), pp. 349-361, ISSN 1076-6294

Oliveira, D. & de Lencastre, H. (2002). Multiplex PCR strategy for rapid identification of structural types and variants of the *mec* element in methicillin-resistant *Staphylococcus aureus*. *Antimicrobial Agents and Chemotherapy*, Vol.46, No.7, (July 2002), pp. 2155-2161, ISSN 0066-4804

Oliveira, D. Tomasz, A. & de Lencastre, H. (2002). Secrets of success of a human pathogen: molecular evolution of pandemic clones of methicillin-resistant *Staphylococcus aureus*. *The Lancet infectious diseases*, Vol.2, No.3, (March 2002), pp. 180-189, ISSN 1473-3099

Rebmann, T. & Aureden, K. (2011). Preventing methicillin-resistant *Staphylococcus aureus* transmission in hospitals: an Executive Summary of the Association for Professionals in Infection Control and Epidemiology, Inc, Elimination Guide. *American Journal of Infection Control*, Vol.39, No.7, (September 2011), pp. 595-598, ISSN 0196-6553

Reyes, J. Rincón, S. Díaz, L. Panesso, D. Contreras, G. Zurita, J. Carrillo, C. Rizzi, A. Guzmán, M. Adachi, J. Chowdhury, S. Murray, B. & Arias, C. (2009). Dissemination of methicillin resistant *Staphylococcus aureus* USA 300 sequence type 8 linage in Latin America. *Clinical of Infectious Diseases*, Vol.49, No. 12, (December 2009), pp. 1861-1867, ISSN 1058-4838

Ribeiro, A. Dias, C. Silva-Carvalho, M. Berquó, L. Ferreira, F. Santos, R. Ferreira-Carvalho, B. & Figueiredo, A. (2005). First report of infection with community-acquired methicillin-resistant *Staphylococcus aureus* in South America. *Journal of Clinical Microbiology*, Vol.43, No.4, (April 2005), pp. 1985-1988, ISSN 0095-1137

Simor, A, Ofner-Agostini, M. Bryce, E. McGeer, A. Paton, S. & Mulvey, M. (2002). Laboratory characterization of methicillin-resistant *Staphylococcus aureus* in Canadian hospitals: results of 5 years of national surveillance, 1995-1999. *The Journal of Infectious Disiases*, Vol.186, No.5, (September 2002), pp. 652-660, ISSN 0022-1899

Tenover, F. Arbeit, R. Goering, R. Mickelsen, P. Murray, B. Persing, D. & Swaminathan, B. (1995). Interpreting chromosomal DNA restriction patterns produced by pulsed-field gel electrophoresis: criteria for bacterial strain typing. *Journal of Clinical Microbiology*, Vol.33, No.9, (September 1995), pp. 2233-2239, ISSN 0095-1137

Tenover, F. & Gaynes, R. The epidemiology of *Staphylococcus* infections, In: V. A. Fischetti, R. Novick, J. Ferretti, D. Portnoy, and J. Rood, (Ed.). (2000). *Gram-positive pathogens*. American Society for Microbiology, Washington, DC. pp. 414-421

Tomasz, A. Drugeon, H. de Lencastre, H. Jabes, D. McDougall, D. & Bille, J. (1989). New mechanism for methicillin resistance in *Staphylococcus aureus*: clinical isolates that lack the PBP2a gene and contain normal penicillin-binding proteins with modified penicillin-binding capacity. *Antimicrobial. Agents and Chemotherapy*, Vol.33, No.11, (November 1989), pp. 1869-1874, ISSN 0066-4804

Velazquez, ME. Aires de Sousa, M. Echaniz, G. Solórzano, F. Miranda, G. Silva, J, & de Lencastre, H. (2004). Surveillance of methicillin- resistant *S. aureus* in a pediatric hospital in Mexico City during a 7-year period (1997 to 2003): clonal evolution and impact of infection control. *Journal of Clinical Microbiology*, Vol.42, No.8, (August 2004), pp. 3877-3880, ISSN 0095-1137

Velazquez, ME. Ayala, J. Carnalla, N. Soto, A. Guajardo, C. & Echaniz, G. (2011). First report of community-associated methicillin-resistant *Staphylococcus aureus* (USA300) in Mexico. *Journal of Clinical Microbiology*, Vol.49, No.8, (August 2011), pp. 3099-3100, ISSN 0095-1137

Weigel, L. Clewell, D. Gill, S. Clark, N. McDougal, L. Flannaga, S. Kolonay, J. Shetty, J. Killgore, G. & Tenover, F. (2003). Genetic analysis of a high-level vancomycin-resistant isolate of *Staphylococcus aureus*. *Science*, Vol.302, No.5650, (November 2003), pp. 1569-1571, ISSN 0036-8075

Yoshida, R. Kuwahara-Arai, K. Baba, T. Cui, L. Richardson, J. & Hiramatsu, K. (2003). Physiological and molecular analysis of a *mecA*-negative *Staphylococcus aureus* clinical strain that expresses heterogeneous methicillin-resistance. *The Journal Antimicrobial Chemotherapy*, Vol.51, No.2, (February 2003), pp. 247-255, ISSN 0305-7453

11

Usefulness of Pulsed Field Gel Electrophoresis Assay in the Molecular Epidemiological Study of Extended Spectrum Beta Lactamase Producers

Patrick Eberechi Akpaka[1] and Padman Jayaratne[2]
[1]Department of Para-Clinical Sciences, The University of the West Indies, St. Augustine
[2]Department of Pathology & Molecular Medicine
McMaster University, Hamilton, Ontario
[1]Trinidad & Tobago
[2]Canada

1. Introduction

A major problem in several health institutions, countries and regions is to categorically define or delineate the source or index case of any microbial organism/s during an outbreak of an infection. Understanding bacterial distribution and their relatedness is essential for determining the epidemiology of nosocomial infections and aiding in the design of rational pathogen control methods. The role of bacterial typing is to determine if epidemiologically identical or related isolates are also genetically related **[Singh A et al, 2006]**. Based on phenotypic and genotypic typing methods, multi-drug resistant bacteria organisms such as extended spectrum beta lactamase (ESBL) enzyme producing pathogens e.g. *Escherichia coli* and *Klebsiella pneumoniae* can be traced to have been transferred from one hospital to another, from one country or region to another. Such information and knowledge have greatly assisted clinicians and health care policy makers to determine the best approach of stopping or eliminating such spreads and transfers of the pathogenic organisms involved in the infection.

As noted in the reviews by Singh A et al, the use of molecular methods for typing of nosocomial pathogen has assisted in efforts to obtain a more fundamental assessment of strain interrelationship **[Singh A et al, 2006]**. Establishing clonality of pathogens can aid in the identification of the source (environmental or personnel) of organisms, distinguish infectious from non infectious strains, and distinguish relapse from reinfection. Many of the species that are key hospital-acquired causes of infection are also common commensal organisms, and therefore it is important to be able to determine whether the isolate recovered from the patient is a pathogenic strain that caused the infection or a commensal contaminant that likely is not the source of the infection. Likewise, it is important to know whether a second infection in a patient is due to reinfection by a strain distinct from that causing the initial infection or whether the infection is likely a relapse of the original infection. If the infection is due to relapse, this may be an indication that the initial treatment regimen was not effective, and alternative therapy may be required [Singh A et al, 2006].

Gel electrophoresis and in particular Pulsed-Field-Gel-Electrophoresis (PFGE) is a tool that has made genotyping of bacterial isolates possible. The PFGE is a laboratory technique used for separation of large deoxyribonucleic acid (DNA) molecules if electric current that periodically changes direction is applied to it. The PFGE is the "gold standard "technique used in this discipline of molecular epidemiological studies and it is basically the comparison of large genomic DNA fragments after digestion with a restriction enzyme that cuts infrequently. Since the bacterial chromosome is typically a circular molecule, the digestion by the enzyme yields several linear large DNA molecules. Moving these large DNA molecules posed a problem but Schwartz and Cantor in 1984 introduced a voltage gradient that gave better resolution and the ability to move large molecules [Schwartz DC & Cantor CR, 1984]

Conventional agarose gel electrophoresis can only be used for the separation of DNA fragments that ranges between 20 – 25 base pair (kbp) by using specialized apparatus no matter how long it is run. The distance between DNA bands of a given length is determined by the percent agarose in the gel. The disadvantage of higher concentrations is the long run times (sometimes days). PFGE uses a special type of agarose that has a larger matrix pore sizes even at a higher percentages such as 1%. The most commonly utilized PFGE methods approaches include the contour-clamped homogenous electric field (CHEF) and field inversion gel electrophoresis (FIGE) [Carle, G. F et al, 1986; Finney, M. 1993]. Field inversion gel electrophoresis utilizes a conventional electrophoresis chamber in which the orientation of the electric field is periodically inverted by 180o and has an upper limit of resolution about 200kbp. CHEF uses a more complex electrophoresis chamber with multiple electrodes to achieve highly efficient electric field conditions for separation; typically the electrophoresis apparatus reorients the DNA molecules by switching the electric fields at 120o angles. CHEF can separate even up to 2-3 Mbp.

Interpreting DNA fragment patterns generated by PFGE and relating or associating them into epidemiologically useful information for typing nosocomial pathogens, the clinical microbiologist or researcher must understand how to compare PFGE patterns and how random genetic events can alter these patterns. Ideally, the PFGE isolates representing an outbreak strain will be indistinguishable from each other and distinctly different from those of epidemiologically unrelated strains. If this occurs, the outbreak is relatively easy to identify. A random genetic activity such as mutation in a DNA can occur and when this happens, it will change the restriction fragment profile obtained during the course of the outbreak [Hall LMC, 1994; Quintiliani R., Jr., & P. Courvalin, 1996; Thal LA et al, 1997] These random variations in the fingerprints will depend on the organism and the time period of the outbreak

The aim of this study is to demonstrate the usefulness of PFGE techniques as a tool to be used in identifying outbreaks of bacterial infection and hence can be used as a tool for infection control measures in a hospital. Also to determine its importance in delineating the clonal relatedness or diversity of bacterial strains isolated from several regional hospitals in Trinidad and Tobago. The PFGE has been shown to be useful in the determination of the sources, clonal relatedness and spread of bacterial isolates in hospitals and countries where the isolates have been recovered or encountered.

Usefulness of Pulsed Field Gel Electrophoresis Assay in the Molecular Epidemiological Study of Extended
Spectrum Beta Lactamase Producers

195

2. Materials & methods

More than 230 strains of *Klebsiella pneumoniae* and *Escherichia coli* obtained routinely from three major regional hospitals in Trinidad and Tobago were used for this study. These non consecutive bacterial isolates were identified using standard microbiological methods as had been previously reported [Akpaka PE & Swanston WH; 2008]. The initial screening for ESBL production by these pathogens using MIC values at concentrations and breakpoints recommended by the CLSI for ESBL screening [CLSI 2010] were performed with the automated micro dilution MicroScan WalkAway-96 System (Siemens, USA). Structured standardized questionnaire was used to extract epidemiological information from hospital records of the patients yielding these isolates. Such information included bio data, gender, hospital facilities where the patients were attended to, clinical signs and symptoms, diagnosis, other forms of investigations and treatments, treatment failures and complications.

2.1 Confirmation of ESBL phenotypes

In accordance with the protocols from the manufacturer to phenotypically determine the ESBL production by bacterial isolates, the E-test strips (AB Biodisk, Solna Sweden), a very sensitive and convenient assay to use was employed to confirm ESBL production in the isolates. The control strain for all the phenotypic testing were *E. coli* ATCC 25922 (negative control) and *K. pneumoniae* ATCC 700603 (ESBL positive).

2.2 Multiplex PCR amplification

The detection of gene sequences coding for the TEM, SHV, and CTX-M enzymes were carried out using multiplex PCR as previously described with some modifications [Monstein HJ et al, 2007]. The cycling conditions used in the PCR assays were as previously described [Paterson DL et al, 2003; Boyd DA et al, 2004]. The oligonucleotide primer sets specific for the SHV, TEM and CTX-M genes and the cycling conditions used in the PCR assays were as described previously and are depicted in the Table 1 below.

Gene	Primer	bp Sizes	Reference
*bla*SHV	5'-ATG CGT TAT ATT CGC CTG TG-3' 5'-TGC TTT GTT ATT CGG GCC AA-3'	747-bp	Paterson DL *et al*
*bla*TEM	5'-TCG CCG CAT ACA CTA TTC TCA GAA TGA-3' 5'-ACG CTC ACC GGC TCC AGA TTT AT-3'	445-bp	Boyd DA *et al*
*bla*CTX-M	5'-ATG TGC AGY ACC AGT AAR GTK ATG GC-3' 5'- TGG GTR AAR TAR GTS ACC AGA AYCAGC GG-3'	593-bp	Boyd DA *et al*

Table 1. Showing primers used for amplifications of the genes in ESBL producers

A Multiplex PCR method previously described [Woodford et al, 2006] for detection of *bla*CTX-M alleles was used to identify the CTX-M phylogenetic group of positive isolates. All PCR reactions were carried out using 2µl bacterial cell suspension (density of 70%T in Vitek Colorimeter) as the DNA template. Respective genes were detected by the size separation PCR amplicons by agarose gel electrophoresis.

2.3 DNA Electrophoresis

The molecular genotyping method employed to compare the DNA of the ESBL producing isolates was the PFGE. This was carried out as previously described **[Kaufmann ME, 1998]** with some modifications. Briefly, the bacterial isolate suspensions were embedded in agarose plugs. The cells were lysed and the proteins digested. The plugs were washed to remove cellular debris and they were sectioned. Restriction analysis of chromosomal DNA with *Xba*1 (New England BioLabs, Beverly, MA) was carried out, and separation of the DNA was performed using 1% pulsed-field gel agarose (Bio-Rad Laboratories, La Jolla, CA). The pulsed-field gel electrophoresis was performed using a contour-clamped homogeneous electric field apparatus set (CHEF DRIII, Bio-Rad Hercules, CA, USA) as in Figure 1 below.

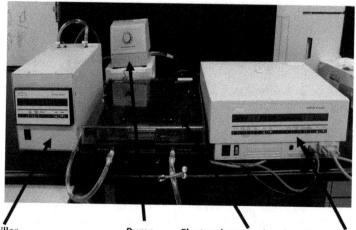

Chiller   Pump   Electrophoresis chamber Programmed Power Supply

The different components of the CHEF system are indicated as electrophoresis chamber, chiller, pump, and the programmed power supply in the above figure. Alternating the electric field between spatially distinct pairs of electrodes causes large and small DNA fragments to re-orient and move at different speeds through the pores in an Agarose gel

Fig. 1. Picture of CHEF DRIII, Bio-Rad Hercules, CA, USA used for the Pulsed field gel Electrophoresis for the microbial agent DNA separation technique.

The gels were stained and images captured on the Gel Doc imaging system using Quantity One software version 4.4.1 (Bio-Rad Laboratories, Hercules CA, USA), Figure 2 below. After viewing the banding patterns, the results were compared and analyzed by manual visualization from a computer monitor following previously established criteria **[Tenover FC et al, 1995]** so as to determine potential outbreak patterns or spread of the infections from one patient to another or hospital facility to another.

The established criteria or guidelines proposed by Tenover *et al.* were used for the interpretation of PFGE **[Tenover FC et al, 1995]**. With these guidelines, a banding pattern difference of up to three fragments could have occurred due to a single genetic event and thus these isolates are classified as highly related, differences of four to six restriction fragments are likely due to two genetic events, and differences of equal to or greater than

Usefulness of Pulsed Field Gel Electrophoresis Assay in the Molecular Epidemiological Study of Extended
Spectrum Beta Lactamase Producers

197

seven restriction fragments are due to three or more genetic events. Isolates that differ by three fragments in PFGE analysis may represent epidemiologically related subtypes of the same strain. Conversely, isolates differing in the positions of more than three restriction fragments may represent a more tenuous epidemiologic relation. Some studies using PFGE and other typing methods indicate that single genetic events, such as those that may alter or create a new restriction endonuclease site or DNA insertions/deletions associated with plasmids, bacteriophages, or insertion sequences, can occur unpredictably even within the time span of a well-defined outbreak (1 to 3 months) **[Arbeit RD et al 1990; Sader HS et al 1993; Tenover FC et al 1995].** With the detection of two genetic variation events by differences in fragment patterns compared to the outbreak strain, the determination of relatedness to an outbreak falls into a gray zone. The results may indicate that these isolates are related (especially if isolates were collected over a long period of time, such as 3 to 6 months), but there is also a possibility that strains are unrelated and not part of the outbreak, hence demonstrating the usefulness of PFGE techniques as a tool in infection control measures in a hospital. PFGE results should always be considered in conjunction with the epidemiologic information and data. The bacterial isolates may also show some degree of clonal relatedness or diversity, thus helping in the determination of the sources, clonal relatedness and spread of the bacterial isolates in hospitals and countries where the isolates have been encountered.

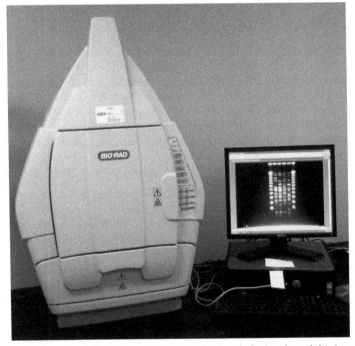

The Gel Doc image system captures picture of the stained gel with the bands and this is transmitted to a computer and monitor for better visualization and analysis

Fig. 2. Picture of Gel Doc (Bio-Rad Hercules, CA, USA) imaging system used to visually analyze images captured after staining the bands formed in the gels.

3. Results

3.1 Bacterial isolates

One hundred and ninety-eight bacterial isolates comprising 120 *K. pneumoniae* and 78 *E. coli* isolates from patients with ages ranging from 2 days old to 85 years had higher MIC values from the E-test assays and thus fulfilled the criteria for further molecular characterization. More than 70% of the isolates were recovered from female patients. The isolates involved in urinary tract infections were 60%. Skin and soft tissue isolates and infections contributed 30% of the isolates and the rest of the isolates were either from respiratory tract system (5%), blood streams (4%) or central nervous systems (1%) respectively. Most of these ESBL isolates were recovered from adult patients admitted into the medical (48%) and surgical (35%) facilities of the hospital. The rest were from patients seen in the paediatrics wards (9%), Obstetrics and gynaecology ward (5%), and intensive care units (3%).

3.2 Multiplex PCR gene detection

The multiplex PCR assay detected the 100% *bla*TEM genes, 25% *bla* SHV and 52% *bla* CTX-M genes among the *E. coli* isolates. Similarly, 94% *bla*TEM, 42% *bla*SHV and 70% *bla*CTX-M genes in the *K. pneumoniae* isolates were detected. All CTX-M genes were identified as alleles belonging to the phylogenetic group I.

3.3 Pulsed-field gel electrophoresis

The PFGE picture of all the isolates used for this study is partly represented in Figure 3 below. As depicted in the figure, the PFGE typing of the ESBL-producing isolates revealed various different and diverse DNA banding profiles among the *E. coli* and *K. pneumoniae* isolates. Bacteriophage lambda ladder PFGE marker (New England Biolabs) is all depicted on the lanes λ. The *E. coli* isolates are demonstrated on lanes 1 – 6, 7 – 12 while lanes 13 – 18 and 19 – 24 shows the *K. pneumonia* isolates. Note that except for lanes 4 and 7, all the lanes containing the isolates have significantly divergent banding patterns. From these results therefore, one would interpret the data as stating that all the *E. coli* or *K. pneumoniae* isolates are distinguishable by the PFGE and divergent from each other (>7 band difference).

There was no major clonal similarity or relatedness of either the *K. pneumoniae* or *E. coli* producing ESBL isolates regardless of which hospital facility the patient was admitted to or specimen the bacterial pathogen was recovered from. In addition, one could notice that from the Figure 2, the bands of the DNA were not separated (i.e. no resolution) for isolates in lanes 4 and 7. This phenomenon is called smearing and occurs when there is a contamination of the nucleases (agarose plug, buffer or reagents), or use of abnormal temperature and concentration of the buffers or wrong conditions which may all affect the enzyme. All these were the case with these isolates because when the tests were repeated with only the two isolates the bands were fully separated.

4. Discussions

The chromosomal DNA is the most fundamental component of identity of the cell and therefore represents a preferred method to assess the relatedness of the strains. And the PFGE method is the gold standard for now in assessing this property of the microbial agents

Usefulness of Pulsed Field Gel Electrophoresis Assay in the Molecular Epidemiological Study of Extended Spectrum Beta Lactamase Producers

199

λ 1 2 3 4 5 6 λ 7 8 9 10 11 12 λ    λ 1314 1516 1718 λ 19 2021 2223 24 λ

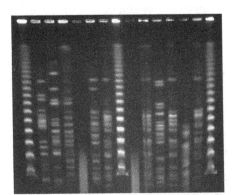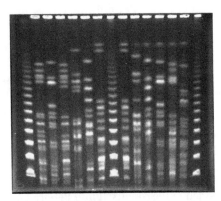

Figure 3 Picture depicting patterns generated by PFGE of *xba*1-digested chromosomal DNA obtained from *bla* TEM, SHV and CTX-M genes produced by *Escherichia coli* and *Klebsiella pneumoniae* isolates. Lane λ, bacteriophage lambda ladder PFGE marker (New England Biolabs), lanes 1 – 6, 7 – 12 *E. coli* isolates and lanes 13 – 18 and 19 – 24, *K. pneumonia* isolates. Smearing phenomenon occurred in lanes 4 and 7 hence the DNA particles were not completely separated or resolved.

Fig. 3. PFGE picture of *Escherichia coli* and *Klebsiella pneumoniae* ESBL producers.

including bacterial cells. All state public health laboratories in the USA as well as Centers for Disease Control and Prevention (CDC) perform molecular epidemiology testing using the PFGE. The PFGE assay can adequately be used to type several organisms including the ones involved in nosocomial infections or pathogens associated with food-borne diseases. PFGE is one of the most reproducible and highly discriminatory typing methods that is available and is a method of choice for many epidemiologic evaluations.

The PFGE typing method used in this study to characterize the ESBL-producing *Klebsiella pneumoniae* and *Escherichia coli* isolates showed various DNA banding profiles. These banding profiles were in no way similar or related to each other indicating their independent origin. This clonal diversity detected among these ESBL-producing isolates suggests that most of the strains have been unable to be maintained or spread in the different wards or facilities of the hospitals from where the bacterial isolates used in this present study were recovered from. This observation may challenge the many conventional thoughts about the nosocomial epidemiology of antibiotic resistance in the hospitals where the isolates were recovered in Trinidad and Tobago. But it is clearly obvious from the PFGE picture that the isolates were in no way closely related as there were different band patterns produced after restriction by the same enzymes under the same physical conditions.

The smearing phenomenon whereby the DNA particles were not completely separated or resolved that occurred and was observed in lanes 4 and 7 in Figure 2 highlights some of the drawbacks to using the PFGE method in studying molecular studies. Once the experimental or laboratory errors are eliminated, results obtained are perfect. Again, an arguement can be made or put forward that the procedures takes several days to be completed. The PFGE process can take less than 48 hours to complete. More time is expended in recovering the bacterial isolates in pure cultures from the clinical specimen because this is the time required for incubation and identification of the bacterial isolate. Thus the turnaround time tends to be long and this can also be a point against the use of PFGE. But despite the longer turnaround time PFGE method in performing molecular epidemiology of bacterial isolates still remains a gold standard for now.

Using the questionnaire to retrospectively review the laboratory and medical records, it was observed that these isolates did not significantly share the same patient demographics and occurrence periods. Despite being isolated mostly from urine of patients admitted in the medical and surgical facilities of the hospitals sharing significant patient demographics and isolate characteristics yet the ESBL enzymes differed. This clearly indicated that most ESBL-producing isolates were not sporadic but that multiple clones were widespread in the hospitals. The occurrence of these ESBL producing pathogens were definitely not from spread from one patient to another or from one ward to another. It must probably therefore be as a result of antibiotic use pressure stemming from the use and overuse of antibiotics such as third generation cephalosporins in these facilities and hospitals as already has been reported in the country **[Pinto Pereira LM et al, 2004; Akpaka PE et al, 2010].** This therefore calls for a need for continuous and active surveillance measures; and effective infection controls practices, most especially antibiotic stewardship which is nonexistent in these hospitals.

This is the first study to report bla*TEM*, bla*SHV* and bla*CTX-M* in the country that reveals that phylogenetic group 1 is the predominant CTX-M types prevalent in the hospitals. This study clearly indicated that CTX-M, mainly CTX-M-1 for ESBL-producing *E. coli* and *K. pneumoniae* was highly prevalent and probably endemic in Trinidad & Tobago. Most ESBL producers were resistant to oxyiminocephalosporins and other non-beta-lactam agents at high levels and exhibited a high rate of the MDR phenotype. The spread of ESBL-producing bacteria appeared to be polyclonal, and none of the major epidemic strains were identified.

5. Conclusions

In summary, this study reports the first extensive study regarding the prevalence and molecular characterization of ESBL genes and the epidemiology of ESBL-producing *E. coli* and ESBL-producing *K. pneumoniae* isolates causing infections in Trinidad and Tobago that was specifically and clearly delineated by the use of the PFGE method.

6. Acknowledgement

The authors wish to acknowledge the several laboratory technicians from the microbiology laboratory of the regional hospitals in Trinidad and Tobago who assisted in collecting the bacterial isolates used in this study.

Conflicts of Interests: None to declare

7. References

Akpaka PE, Legall B, Padman J. 2010. Molecular detection and epidemiology of extended-spectrum beta-lactamase genes prevalent in clinical isolates of *Klebsiella pneumoniae* and *E coli* from Trinidad and Tobago. West Indian Med J. 59(6):591-6.

Usefulness of Pulsed Field Gel Electrophoresis Assay in the Molecular Epidemiological Study of Extended
Spectrum Beta Lactamase Producers

201

Akpaka PE, Swanston WH. 2008. Phenotypic Detection and Occurrence of Extended-Spectrum Beta-Lactamases in Clinical Isolates of *Klebsiella pneumoniae* and *Escherichia coli* at a Tertiary Hospital in Trinidad & Tobago. The Brazilian Journal of Infectious Diseases;12(6):516-520.

Arbeit RD, Arthur M, Dunn R, Kim C, Selander RK, Goldstein R. 1990. Resolution of recent evolutionary divergence among *Escherichia coli* from related lineages: the application of pulsed-field electrophoresis to molecular epidemiology. J Infect Dis; 161:230-35

Boyd DA, Tyler S, Christianson S, McGeer A, Muller MP, Willey BM, Bryce E, Gardam M, Nordmann P, Mulvey MR, and the Canadian nosocomial infection surveillance program, Health Canada; 2004. Complete nucleotide sequence of a 92-kilobase plasmid harbouring the CTX-M-15 extended-spectrum beta-lactamase involved in an outbreak in long-term-care facilities in Toronto, Canada. Antimicrob Agents Chemother; 48:3758-64.

Carle, G. F., M. Frank, and M. V. Olson. 1986. Electrophoretic separations of large DNA molecules by periodic inversion of the electric field. Science 232: 65-68.

Clinical and Laboratory Standard Institute (CLSI) 2010. Performance standards for antimicrobial susceptibility testing: Twentieth Informational Supplement. CLSI document M100-S20; Vol 30 No. 1. CLSI, Wayne, Pa

Finney, M. 1993. Pulsed-field gel electrophoresis, p. 2.5.9–2.5.17. *In* F. M. Ausubel, R. Brent, R. E. Kingston, D. D. Moore, J. G. Seidman, J. A.Smith, and K. Struhl (ed.), Current protocols in molecular biology, vol. 1.Greene-Wiley, New York, N.Y.

Hall, L. M. C. 1994. Are point mutations or DNA rearrangements responsible for the restriction fragment length polymorphisms that are used to type bacteria? Microbiology 140:197–204.

Kaufmann, M E. 1998. Pulsed-field gel electrophoresis. Methods Mol. Med; 15:17-31

Monstein HJ, Östholm-Balkhed Å, Nilsson MV, Dornbusch, Nilsson LE. 2007. Multiplex PCR amplification assay for rapid detection of *bla* SHV, *bla* TEM and *bla* CTX-M genes in Enterobacteriaceae. APMIS; 115:1400-8

Paterson DL, Hujer KM, Hujer AM, Yeiser B, Bonomo MD, Rice LB, Bonomo RA, and the International Klebsiella study group. 2003. Extended spectrum b-lactamases in *Klebsiella pneumoniae* bloodstream isolates from seven countries: dominance and widespread prevalence of SHV-and CTX-M-type b-lactamases. Antimicrob Agents Chemother; 47:3553-60.

Pinto Pereira LM, Phillips M, Ramlal H, Teemu K, Prabhakar P. Third generation cephalosporin use in a tertiary hospital in Port of Spain, Trinidad: need for an antibiotic policy. BMC Infectious Diseases 2004; 4:59 doi: 10.1186/ 1471-2334-4

Quintiliani, R., Jr., and P. Courvalin. 1996. Conjugal transfer of the vancomycin resistance determinant *vanB* between enterococci involves the movement of large genetic elements from chromosome to chromosome. FEMS Microbiol. Lett. 119:359-364.

Sader HS, Pignatari AC, Lemme LL, Burattini MN, Taneresi R, Hollis RJ, Jones RN. 1993. Epidemiologic typing of multiple drug resistant *Pseudomonas aeruginosa* isolated from an outbreak in an intensive care unit. Diagn Microbiol Infect Dis; 17: 13-18

Schwartz DC, Cantor CR. 1984. Separation of yeast chromosome-sized DNAs by pulsed field gradient gel electrophoresis. Cell; 37(1):67-75

Singh A, Goering RV, Simjee S, Foley SL, Zervos MJ. 2006. Application of molecular techniques to the study of hospital infection. Clin Microbiol Reviews; 19 (3) 512 -3

Tenover FC, Abreit RD, Goering RV, Michelsen PA, Murray BE, Persing DH, et al. 1995. Interpreting chromosomal DNA restriction patterns produced by pulsed-field gel electrophoresis: criteria for bacterial strain typing. J Clin Microbiol; 33:2233 — 9.

Thal, L. A., J. Silverman, S. Donabedian, and M. J. Zervos. 1997. The effect of Tn916 insertions on contour-clamped homogenous electrophoresis patterns of the Enterococcus faecalis. J. Clin. Microbiol. 35:969–972.

Woodford N, Fagan EJ, Ellington MJ. 2006. Multiplex PCR for rapid detection of genes encoding CTX-M extended-spectrum B-lactamases. J Antimicrobial Chemotherapy; 57(1):154-5

Part 3

Electrophoresis Application in the Analysis of Protein-Nucleic Acid Interaction and Chromosomal Replication

12

Electrophoretic Mobility Shift Assay: Analyzing Protein – Nucleic Acid Interactions

Carolina Alves and Celso Cunha
Center for Malaria and Tropical Diseases
Institute of Hygiene and Tropical Medicine, New University of Lisbon
Portugal

1. Introduction

Interactions between proteins and nucleic acids mediate a wide range of processes within a cell from its cycle to the maintenance of cellular metabolic and physiological balance. These specific interactions are crucial for control of DNA replication and DNA damage repair, regulation of transcription, RNA processing and maturation, nuclear transport, and translation.

The characterization of protein-nucleic acid interactions is essential not only for understanding the wide range of cellular processes they are involved in, but also the mechanisms underlying numerous diseases associated with the breakdown of regulatory systems. These include, but are far from being limited to, cell cycle disorders such as cancer and those caused by pathogenic agents that rely on or interfere with host cell machinery. More recently, it has been hypothesized that many neurological disorders such as Alzheimer's, Huntington's, Parkinson's, and polyglutamine tract expansion diseases are a consequence, at least in part, of aberrant protein-DNA interactions that may alter normal patterns of gene expression (Jiménez, 2010).

The electrophoretic mobility shift assay (EMSA), also known as gel retardation assay, is a regularly used system to detect protein-nucleic acid interactions. It was originally developed with the aim of quantifying interactions between DNA and proteins (Fried & Crothers, 1981; Garner & Revzin, 1981) and since then evolved to be suitable for different purposes including the detection and quantification of RNA-protein interactions. EMSA is most commonly used for qualitative assays including identification of nucleic acid-binding proteins and of the respective consensus DNA or RNA sequences. Under proper conditions, however, EMSA can also be used for quantitative purposes including the determination of binding affinities, kinetics, and stoichiometry.

EMSA is a commonly used method in the characterization of transcription factors, the most intensely studied DNA-binding proteins, and the largest group of proteins in humans, second only to metabolic enzymes. Their purification and identification is crucial in understanding gene regulatory mechanisms. Transcription factors are sequence specific DNA binding proteins that are usually assembled in complexes formed prior to transcription initiation. They bind discreet and specific DNA sequences in the promoter region functioning either as an activator or repressor of expression of the targeted gene through protein-protein interactions (reviewed by Simicevic & Deplancke, 2010). Transcription factors play essential roles during development and differentiation. It is well established that disruption of normal function of tissue-specific transcription factors, as a result of mutations, is often associated with a number of diseases including most forms of cancer, neurological, hematological, and inflammatory diseases. Additionally, transcription factors are often found differentially expressed in different pathologies suggesting an at least indirect involvement on the onset or progression of diseases. One of the most prominent examples of the involvement of transcription factors in development and progression of diseases is perhaps the p53 protein. p53 is a transcription factor involved in the modulation of expression of several genes that regulate essential cellular processes such as cell proliferation, apoptosis, and DNA damage repair (reviewed by Puzio-Kuter, 2011). Mutations in p53 that cause loss of function were reported in about 50% of all cancers. It is believed that this loss of function makes cancer cells more prone to the accumulation of mutations in other genes thus facilitating and accelerating the formation of neoplasias (reviewed by Goh et al., 2011).

In our laboratory, research is mainly directed to the study of host-pathogen interactions during hepatitis delta virus (HDV) replication and infection. HDV is the smallest human pathogen so far identified and infects human hepatocytes already infected with the hepatitis B virus (HBV). Both viruses have the same envelope proteins that are coded by the HBV DNA genome. HDV is, thus, considered a satellite virus of HBV. The HDV genome consists of a single-stranded, circular, RNA molecule of about 1700 nucleotides. This genome contains only one open reading frame from which two forms of the same protein, the so-called delta antigen, are derived by an editing mechanism catalyzed by cellular adenosine deaminase I. Both forms, small and large delta antigen, were shown to play crucial roles during virus replication: the small delta antigen is necessary for virus RNA accumulation and the large delta antigen plays an important role during envelope assembly (reviewed by Rizzetto, 2009). However, neither protein seems to display any known enzymatic activity. Accordingly, HDV is highly dependent on the host cell machinery for virus replication. It has been shown through EMSA that the small delta antigen binds *in vitro* to RNA and DNA without any specificity, which is in agreement with one of the roles attributed to the protein as a chaperone (Alves et al., 2010). Making use of different experimental approaches it was possible to identify a number of cellular proteins that interact with HDV antigens or RNA (reviewed by Greco-Stewart & Pelchat, 2010). However, the precise role played by most host factors during the virus life cycle remains elusive. Furthermore, it is highly consensual among HDV researchers that many other cellular factors that interact with delta antigens or HDV RNA remain to be identified and it is crucial to find those that interact with HDV RNA for a better insight on its replication and as possible targets for new therapies.

In this chapter we will review the principles of EMSA and its advantages and limitations for the quantitative and qualitative analysis of protein-nucleic acid interactions. The key parameters influencing the quality of protein samples, binding to nucleic acids, complex migration in gels, and sensitivity of detection will be discussed. Finally, an overview of the principles, advantages and disadvantages of methods that are an alternative to gel retardation assays will be provided.

2. Advantages and limitations

Since its first publication, in 1981, several improvements and variant techniques of EMSA were reported. Originally described as a method to qualitatively detect protein-DNA interactions, gel retardation assays rapidly became one of the most popular methods to map interaction sequences and domains not only in DNA but in RNA-protein interactions as well. EMSA was also adapted in order to allow the determination of quantitative parameters including complex stoichiometry, binding kinetics and affinity.

Several features made EMSA one of the most popular methods among researchers that study protein-nucleic acid interactions. Probably, the main advantages of EMSA when compared to other methods, as we will further discuss in the next sections, may be considered as follows: (1) EMSA is a basic, easy to perform, and robust method able to accommodate a wide range of conditions; (2) EMSA is a sensitive method, using radioisotopes to label nucleic acids and autoradiography, it is possible to use very low concentrations (0.1nM or less) and small sample volumes (20 μL or less; Hellman & Fried, 2007). Even though, less sensitive, non-radioactive labels are often used as well. These labels can further be detected using fluorescence, chemiluminescence or immunohistochemical approaches. Although less sensitive then radioisotopes, the wide variety of labels that can be used makes EMSA a very versatile method; (3) EMSA can also be used with a wide range of nucleic acid sizes and structures as well as a wide range of proteins, from small oligonucleotides to heavy transcription complexes; (4) Under the right conditions a gel retardation assay can separate the distribution of proteins between several nucleic acids within a single sample (Fried & Daugherty, 1998) or distinguish between complexes with different protein stoichiometry and/or binding site distribution (Fried & Crothers, 1981); (5) Finally, but not less important, it is possible to use both crude protein extracts and purified recombinant proteins enabling the identification of new nucleic acid-interacting proteins or characterization of specific proteins and its targets.

Despite its sensitivity, versatility and usually easy to perform protocols, EMSA is often considered to bear a number of limitations. Dissociation can occur during electrophoresis since samples are not at equilibrium during the run, thus preventing detection. Additionally, complexes that are not stable in solution may be stable in the gel requiring very short runs so that the observed pattern relates to what happens in solution. EMSA does not provide a straightforward measure of the weights or entities of the proteins as mobility in gels is influenced by several other factors. Also, EMSA does not directly provide information on the nucleic acid sequence the proteins are bound to. However, this problem may usually be overcome using footprinting approaches as described further ahead. Kinetic studies using EMSA are limited since the time resolution for a regular EMSA protocol consists of the time required to mix the binding reaction and for the electrophoretic migration to occur before the mix enters the gel. Only processes that have relaxation times larger than the interval required for solution handling are suitable for kinetic studies.

3. How complexes migrate in gels

In this section, we will start with a simple account of the characteristics of the electrophoretic mobility of nucleic acids alone, and afterwards we will discuss how the formation of protein-nucleic acid complexes alters these characteristics.

In a non-denaturing agarose or polyacrylamide gel and conventional buffer conditions the nucleic acids, being negatively charged, will migrate towards the anode when electric current is applied. The gel will then act as a sieve selectively impeding the migration in proportion to the nucleic acid molecular weight, which is generally proportional to its charge. Therefore, and as the weight is approximately related to chain length, the length of nucleic acid is estimated by its migration. There is though another property that affects gel migration that is the topology of the nucleic acid (conformation, circularity) making the molecules seem longer or shorter than they really are. Secondary and tertiary structures can be removed using denaturing agents (for example, formaldehyde, formamide and urea) allowing for the electrophoretic mobility to become a simple function of molecular weight. Obviously, this denaturing step cannot be applied in a gel retardation assay as it would impede the interaction between the protein and nucleic acid.

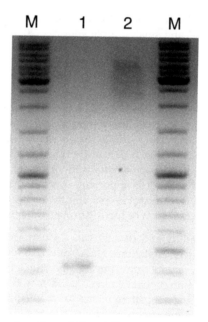

Fig. 1. Example of an electrophoretic mobility shift assay. An unlabeled DNA of 400 base pairs (bp) was incubated in a phosphate buffer (137mM NaCl, 2.7mM KCl, 4.3mM Na$_2$HPO$_4$, 1.5mM KH$_2$PO$_4$, pH 7.4) in the absence (1) or presence (2) of 2μM of small delta antigen. The samples were loaded onto a 1.5% agarose gel and after electrophoresis in TAE buffer (40mM Tris acetate, 1mM EDTA) the DNA was stained with ethidium bromide. (M) represents the molecular weight marker (GeneRuler DNA Ladder mix, Fermentas).

When a protein is added to the mix and interacts with the nucleic acid forming complexes it results in a change in gel migration relative to that of the free nucleic acid. This shift is mainly due to an obvious increase in the molecular weight, the adjustment of charge and eventual changes in the nucleic acid conformation. In figure 1 we give an example of an EMSA study where the small delta antigen was added to DNA. It is clear that the addition of the small delta antigen (Fig.1. well 2) to a 400bp DNA fragment results in the formation of a complex with decreased gel mobility when compared with the unbound DNA (Fig.1. well 1). We can conclude that under our *in vitro* binding conditions, the small delta antigen interacts with the given 400bp DNA fragment causing a clear mobility shift.

It is expected that when protein binds a nucleic acid fragment there will be a decrease in relative mobility and if the protein doesn't induce any appreciable bend on the nucleic acid then the conformational contribution to the decrease is small. Although an increase in the protein molecular weight results in reduction of gel migration it has been reported that the increase of the nucleic acid length can have the opposite effect. This was reported for the Lac repressor bound to DNA fragments of increasing sizes, which resulted in an increase of relative mobility (Fried, 1989). This observation indicates that the ratio of protein and nucleic acid weights is more important in the migration than the absolute weight of the complex. Another interesting study reports that the binding of protein to a nucleic acid can accelerate mobility. This was observed for relatively large linear DNA binding to a protein from the hyperthermophilic *Methanothermus fervidus* that was shown to induce nucleic acid condensation (Sandman et al., 1990). In this case the conformational change of the DNA is a stronger factor than the weight increase, causing acceleration rather than a decrease in relative mobility.

Overall, the conformational features that influence gel migration of protein-nucleic acid complexes are not thoroughly studied and questions are only raised when exceptions emerge such as the ones mentioned above. Nowadays, the EMSA method is almost exclusively used to analyze the interaction between proteins and nucleic acids and to a lesser extent its conformations that can influence gel migration. When exceptions arise and the retardation pattern is not exactly as predicted, it can still point out clearly whether the molecules are interacting or not. In the end, the exact location of the resulting gel bands cannot be predicted but the answer is usually unambiguous.

External factors can also influence the separation of the bound or unbound nucleic acid such as the nature of the gel matrix and temperature during electrophoresis. Generally, the best resolution is obtained with the smallest pore diameter that allows the migration of unbound nucleic acid. However, if large complexes are expected there should be a compromise in pore size so that they can enter the gel matrix. As will be discussed below, polyacrylamide gels offer the best conditions for small complexes and nucleic acid fragments. On the other hand, agarose gels are more suitable for larger aggregates.

The detection of a protein-nucleic acid complex within a gel depends critically on the resolution obtained between unbound nucleic acid and the formed complexes as well as its stability within the gel matrix. In most cases, the gel matrix is expected to stabilize the preformed complex as it impedes the diffusion of dissociating components maintaining the concentration of protein and nucleic acid (and complex) at levels as high or higher than those achieved in the equilibrium binding reaction. This of course is compromised if for instance the salt concentration in the binding reaction differs largely from that in the electrophoresis/gel buffer, resulting in an adjustment in salt concentration that could disrupt the complexes formed. As the gel retardation method is an *in vitro* assay, when extrapolating to the *in vivo* conditions one must be careful as the former may provide favorable binding conditions that are not achieved at physiological concentrations.

4. The method

There are five focal steps in a conventional EMSA protocol that involve different variables susceptible to optimization: (1) preparation of protein sample; (2) synthesis and labeling of nucleic acid; (3) binding reaction; (4) non-denaturing gel electrophoresis and (5) detection of the outcome. In this segment we will discuss each step separately mentioning the key variables in each one and the options available for any given situation. Figure 2 represents schematically the regular steps in a gel retardation assay that will be discussed below. Whenever possible we will also refer to examples in the literature.

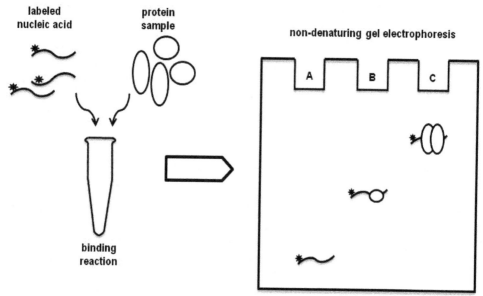

Fig. 2. A schematic representation of a conventional EMSA protocol. The labeled nucleic acid, simplified as lines with a star representing the label, is mixed with the protein sample, represented by the oval shapes, in a binding reaction and then loaded into a non-denaturing gel. After electrophoresis the result is detected according to the label in the nucleic acid. On the schematic gel (A) represents a well on which only the labeled nucleic acid was loaded. The free nucleic acid is expected to have more mobility than the bound molecules. In well (B) is symbolized a labeled nucleic acid binding to one small peptide and in well (C) is binding to two larger proteins. The heavier complex (in C) is expected to display the lowest mobility during electrophoresis and therefore is closer to the beginning of the gel.

4.1 Preparation of the protein sample

Regarding the protein sample, the EMSA can be divided into two categories based on whether the nucleic acid-interacting protein is known or not. Therefore, preparing the protein sample will depend on which category it falls, in order to obtain an optimal performance.

When faced with a putative nucleic acid-binding protein or complex of completely unknown subcellular origin, whole cell extracts must be used. If there is an educated guess on the nature of the protein, it is advisable to isolate nuclear and cytoplasmic proteins from crude extracts improving the results. Particularly, if the binding protein is thought to be nuclear and in low abundance, the isolation of nuclear extracts will prevent the dilution that would occur if whole cell extracts were used, which could render the concentration too low for the protein to be even detected.

Cell extracts are easy and relatively fast to obtain and the methods are commonly derived from the protocol described by Dignam and collaborators almost three decades ago (Dignam et al., 1983). This method isolates both nuclear and/or cytoplasmic proteins suitable for later analysis using EMSA. One disadvantage in preparing cell extracts is its crudeness; they generally degrade faster than purer preparations due to the presence of cellular proteases. To limit protein degradation or alteration the protocol should be performed on ice or at 4°C and protease inhibitors should be added. A control test can easily be performed to assess the viability of the extract by using ubiquitous DNA probes (Kerr, 1995). If these fail than the cell extract might be "dead". Despite its disadvantages cell extracts are needed when the interest lies in identifying new nucleic acid-binding proteins or when a complex of different proteins is needed to interact with the target nucleic acid as sometimes one recombinant protein cannot bind by itself. Tissue samples can also be a source of protein sample for these assays. The same care should be taken as in whole cell extracts to minimize the activity of proteases.

If the nucleic acid-binding protein is known then recombinant proteins can be expressed and purified. Recombinant or heterologous proteins are commonly expressed in bacteria or an eukaryotic cell line of interest. Fusion proteins of the target are generally constructed with a tag to facilitate purification. Common tags, such as glutathione-S-transferase (GST), tandem affinity purification tag (TAP tag), maltose binding protein (MBP) or 6xHistidine, are cloned in frame with the protein. Sometimes it is possible to include a protease cleavage site between the protein of interest and the tag so the latter can be easily removed after purification. Even though a tag can be very helpful, it should be taken into account that it can alter the recombinant protein conformation and even disrupt its binding ability. On the other hand they can be helpful in stabilizing the protein terminus they are close to. A careful study is needed when choosing the tag and usually small peptides are preferred to minimize its impact on the recombinant protein of interest.

There are several systems available for the production of heterologous proteins of which bacterial extracts of Escherichia coli are one of the most widely used. This Gram-negative bacterium remains an attractive host due to its ability to grow rapidly and with high density using inexpensive substrates. Its genetics has been well characterized for quite some time and there is a wide range of cloning vectors as well as mutant host strains that make it such a versatile system. Typically, the heterologous complementary DNA is cloned into a compatible plasmid which is then transfected into the bacteria to achieve a high gene dosage. This doesn't necessarily guarantee the accumulation of high levels of a full-length active form of the recombinant protein but other efforts can be made to improve that. To achieve high-level production in E. coli strong promoters should be used such as the bacteriophage T7 late promoter, and usually the T7 polymerase is also present under IPTG (isopropyl-β-D-1-thiogalactopyranoside)-induction. In the past years several strains have been engineered to improve the recombinant protein yields through efforts to increase mRNA stability as well as improve transcription termination and translational efficiency (reviewed by Baneyx, 1999 and Makino et al., 2011). However, this extensively used system for protein overexpression has an important drawback when studying eukaryotic proteins. The bacterial systems are not able to perform post-translational modifications that would eventually happen *in vivo* in eukaryotic cells.

When working with recombinant nucleic acid-binding proteins it should be taken into account the importance of post-translational modifications on the protein's binding ability. A careful research of previous reports might hint if it is necessary to perform modifications prior to the binding reaction. In some cases post-translational modifications change the sequence-specificity of the binding. For example, genotoxic stress induces modifications on the C-terminus of the tumor suppressor protein p53 that modulate its DNA-binding specificity (Apella & Anderson, 2001). If the modifications are crucial, rather than using bacterial extracts a more biologically relevant host should be considered. Transient gene expression in mammalian cells has become a routine approach to express proteins in cell lines such as human embryonic kidney cells. The benefits are obvious for the production of eukaryotic proteins in mammalian cells as post-translational modifications will likely be native or near-native, solubility and correct folding are more likely to occur as well as expression of proteins in their proper intracellular compartments. These methods, however, tend to be more expensive as cells need a more complex growth media and there is a lower diversity in cloning vectors. To get out of the latter limitation an alternative approach uses baculovirus-infected insect cells. In this method a recombinant virus is produced either by site-specific transposition of an expression cassette into the shuttle vector or through homologous recombination (reviewed by Jarvis, 2009).

When expressing recombinant proteins, sometimes, the heterologous genes interfere severely with the survival of the host cell. For toxic proteins produced in *E. coli* strains there are some techniques available to get around this problem. A highly toxic gene can be defined as a gene that, when introduced into a cell, causes cell death or severe growth and maintenance defects even prior to expression induction. The best solution for expressing a highly toxic gene is to enable the host to tolerate it during the growth phase, so that after induction an efficient expression ensures a rapid and quantitative production of the toxic protein before the cell dies (reviewed by Saida et al., 2006). This can be achieved by different strategies such as manipulation of the gene's transcriptional and translational control elements, for example, by suppressing basal expression of the toxic protein from leaky inducible promoters. Managing the coding sequence to produce reversible inactive forms or controlling the plasmid copy number is also an option as well as selecting less susceptible *E. coli* strains or adding stabilizing sequences.

Cell-free systems are also available to express recombinant proteins including *in vitro* transcription\translation systems such as rabbit reticulocyte systems, wheat germ based systems or *E. coli* cell-free protein expression systems (reviewed by Endo & Sawasaki, 2006). Here, proteins can be expressed directly from cDNA templates obtained through PCR, avoiding subcloning which makes it a faster method by skipping this step, and eventually cheaper. It can also be used to express proteins that seriously interfere with the cell physiology such as the toxic proteins mentioned above. On the other hand these methods usually achieve smaller yields than for instance bacterial extracts approaches.

4.2 Synthesis and labeling of nucleic acids

One of the key advantages of EMSA is its versatility as it can be performed using a wide range of nucleic acid structures and sizes. This method can characterize both double- and single-stranded DNA as well as RNA, triplex and quadruplex nucleic acids or even circular fragments. The probe design and synthesis depends on the application or purpose of the study and is a significant aspect, as it will influence the detection and therefore the sensitivity of the results. There are two main aspects to consider in this step: the length of the nucleic acid and its labeling.

Unlabeled nucleic acids can be used in a gel retardation assay and be detected by post-electrophoretic staining with chromophores or fluorophores that bind nucleic acids or in the "classical way" using ethidium bromide. However the use of labeled nucleic acids is usually preferred as it can facilitate detection and add sensitivity to the method. The most common choice is radioisotope labeling as it offers the best sensitivity without interfering with the structure of the probe. A higher sensitivity makes it ideal for assays that have a limited amount of starting material. The radioisotope, usually ^{32}P, can be incorporated in the nucleic acid during its synthesis, by the use of labeled nucleotides, or afterwards via end labeling using a kinase or a terminal transferase. With a radioactive label the EMSA results can be easily detected by autoradiography. Even if radioisotope labeling confers high sensitivity to the method it implies handling hazardous radioactive material requiring extra safety measures that may not be available. Other labels can be used as alternatives that, even though are less sensitive, are a lot safer to manipulate and more stable such as fluorophores, biotin or digoxigenin (Holden & Tacon, 2011). When these molecules are used detection is achieved by chemiluminescence or immunohistochemistry. Although, in general radioisotope labeling achieves higher sensitivity there are some reports that similar results can be obtained with other labels such as Cyano dye Cy5 (Ruscher et al., 2000).

Although the most common approach is the labeling of the nucleic acid probe there are protocols available that employ protein labeling at the same time. For example, Adachi and co-workers suggest the use of an iodoacetamide derivative labeling of the thiol residue of cysteins (Adachi et al., 2005). Using radioisotope labeled DNA mixed with a nuclear protein extract they perform a conventional EMSA and after detection by autoradiography the complexes are eluted from excised gel bands and treated with 5-iodoacetamidofluorescein for protein labeling. The sample is then loaded onto a denaturing gel and after electrophoresis is transferred to a membrane and detected with anti-fluorescein antibody. This allows the characterization of the proteins in the complex giving information on how many proteins are present and their molecular weight. However it is not able to detect proteins without cystein residues.

Regarding the length of the nucleic acid probe, it depends on what is being studied. If one is looking for specific binding sites, small probes can be used to assess with each segment the protein will interact. The use of short nucleic acids has several advantages as they are easily synthesized and inexpensive to purchase; a small sequence has less non-specific binding sites (it should be particularly advantageous when a protein has low sequence-specificity); the electrophoretic resolution between complexes and free nucleic acid is higher so shorter electrophoresis times can be used. Nevertheless, in a short sequence the binding sites are closer to the molecular ends which can cause aberrant binding and it can be tricky to resolve the free nucleic acid from the complexes formed if these have a very high molecular weight.

On the other hand, the longer nucleic acid targets avoid these problems but will have more non-specific binding sites and the mobility shift is generally smaller requiring longer electrophoresis times as they run more slowly. A compromise needs to be reached depending on what the EMSA study is trying to achieve.

4.3 Binding reaction

The interaction between proteins and nucleic acid is sensitive to salt concentration and pH as it will influence the protein charge and conformation. However, the experimental conditions are very versatile in that different buffers can achieve good results. The most commonly used are Tris based buffers but other options include 4-(2-hydroxyethyl)-1-piperazineethanesulfonic acid (HEPES), 3-(N-morpholino)-propanesulfonic acid (MOPS), and glycine or phosphate buffers. Naturally, it is advisable to provide an environment as close as possible to physiological conditions so the data obtained *in vitro* can be related to what happens *in vivo*.

Additives can be included in the binding reaction either if the interactions require the presence of co-factors or stabilizing agents, or as helpful components to minimize non-specific binding. Glycerol or other small neutral solutes, for example sucrose, can be added to the binding mixture to stabilize labile proteins or enhance the stability of the interaction (Vossen et al., 1997). These solutes are used at final concentrations of 2M or less, as higher concentrations might interfere with the sample's viscosity and complicate handling. Other assays may require the presence of co-factors for a correct interaction such as the presence of cAMP for the *E. coli* CAP protein (Fried & Crothers, 1984) or ATP for human recombinase Rad51 (Chi et al., 2006). Non-ionic detergents are used to maximize protein solubility. In this case, the concentrations used depend on the detergent and system under study. Nuclease and phosphatase inhibitors can be useful as well as protease inhibitors, which as mentioned before, are particularly important when the protein sample comes from cell extracts. These inhibitors are commercially available and the concentration depends on the manufacturer's instructions. Some of the additives mentioned, particularly those involved in stabilizing the formed complexes can be included not only in the binding mixture but also in the gel buffers.

To minimize non-specific loss of protein the addition of a carrier protein (less than 0.1mg/mL) such as bovine serum albumin can be very helpful. The addition of unlabeled competing nucleic acids is suitable when there are secondary binding activities that mask the relevant one. Of course this only works if the protein interacts with the target nucleic acid with greater affinity then its competitor and the secondary binding does not discriminate between the sequences. Since the presence of a competing nucleic acid will always reduce the amount of specific binding, testing different competitors and concentrations is needed to optimize the assay. Another option to circumvent the problem of non-specific binding is the addition of salt at concentrations that will disrupt non-specific ionic bonds but leave the more specific interactions unimpaired.

4.4 Non-denaturing gel electrophoresis

After the binding reaction the free nucleic acid is separated from the formed complexes by non-denaturing gel electrophoresis. EMSA can be performed on polyacrylamide or agarose gels depending mainly on the size of the nucleic acid and desired resolution. The average pore size is estimated to be around 5 to 20nm in diameter for 10 and 4% acrylamide gels respectively (Lane et al., 1992). Typically the higher concentration gels are used for oligonucleotides and small RNAs and the lowest concentration for DNA fragments of around 100bp. A polyacrylamide gradient gel is sometimes preferred over linear gels as the gradient in pore size increases the range of molecular weight fractioned in a single run, which is particularly important when the complex has a much higher weight than the free nucleic acid (Walker, 1994). When complexes of different composition are formed, the gradient gels are also more likely to separate those with close molecular weight.

Agarose gels, on the other hand, have a pore size of around 70 to 700nm (Lane et al., 1992) in diameter and are therefore mostly used in assays with larger nucleic acid fragments or when large protein complexes are expected. Overall, polyacrylamide gels offer a better resolution for nucleic acid-protein complexes with a molecular weight of up to 500,000Da (Fried, 1989 as cited in Hellman & Fried, 2007).

Regarding the electrophoresis buffers, it should be taken into account the fact that the interaction between nucleic acids and proteins involves an ionic component. Therefore, the buffer's ionic strength and pH are important features that play a role in the complex stability. Although this is a very important factor there hasn't been, to our knowledge, any thorough study on the subject. The choice of electrophoresis buffers is varied and generally low ionic strength buffers are preferred and sometimes coincide with the buffer used in the binding reaction. Buffers with a medium salt concentration help stabilize the complexes, generate less heat during electrophoresis and also increase the speed of migration. High salt concentrations not only disrupt the complexes but also interfere with its movement into the gel matrix and lead to significant heating during the electrophoresis. Too low salt concentrations can also disrupt the stability of the preformed complexes as well as separate a double stranded DNA template (Kerr, 1995). The most common buffers are TBE (90mM Tris-Borate, 2mM EDTA, pH 8) and TAE (40mM Tris-Acetate, 1mM EDTA, pH 8). However, there are some complexes that cannot be detected with the classical buffers. For example the complexes formed between phage Mu repressor and its operators have an electrophoresis buffer-dependent stability and require Tris-glycine buffer at pH 9.4. (Alazard et al., 1992 as cited in Lane et al., 1992).

Particularly, in agarose gels it is important to monitor the temperature during electrophoresis to prevent the gel from heating up which could result in dissociation of the nucleic acid-protein complexes. Some cases may require that pre-cooling of the gel or even that the electrophoresis proceeds at lower than room temperatures, which can be achieved with special refrigeration devices.

4.5 Detection

The detection of an EMSA result will naturally depend on the labels used if any has been used. The results uncovered can involve the detection of the mobility shift between free nucleic acid and the complexed form or the detection of the mobility shift of free protein and the complexes.

Looking at the nucleic acid component without any label added the shift in mobility can be detected by staining with molecules that bind nucleic acids. Different products can be used ranging from the classic but hazardous ethidium bromide to other chromophores or fluorophores such as RedSafe DNA Stain (ChemBio) or SYBR® Safe DNA gel stain (Invitrogen). When the nucleic acid has been previously labeled the detection methods depend on the nature of the label. A ^{32}P radioisotope is one of the easiest and most sensitive methods to detect nucleic acids but it's a hazardous material to work with. Other very common labels are biotin, digoxigenin or fluorophores. These labels are innocuous but usually give less sensitive results and the detection procedure can involve extra steps such as transfer to a membrane and incubation with primary and secondary antibodies as well as intermediate washing steps. The results in these cases can be observed by immunohistochemistry or chemiluminescence approaches.

The detection of protein mobility shift involves less direct methods, meaning, extra steps such as a denaturing step and electrotransfer onto a membrane, may be necessary as they are usually immunodetected. If the protein of interest is known, and a specific antibody is available, it can be used in detection. If not, a method such as the one discussed above, proposed by Adachi and colleagues that involves labeling the thiol group of cysteins and using an antibody against the label. Stepwise, the easier way to detect protein in an EMSA is by labeling it with radioisotope, a method designated by reverse EMSA that will be discussed ahead. This procedure has the disadvantage of working with radioactive material but the mobility shift can be visualized by autoradiography.

5. EMSA applications

The gel retardation assay has been used under different conditions in order to achieve specific results. The method is useful in studying not only the interaction between proteins and nucleic acids but also in assessing nucleic acid conformational characteristics. It can be used to characterize bends in the DNA double helix with polyacrylamide gels and comparative measurements (for an example Crothers & Drak, 1992) or to detect complexes formed with super coiled DNA being sometimes designated as topoisomer gel retardation (for examples see Palecek, 1997; Nordheim & Meese, 1988). In this section we mention how a gel retardation assay can help characterize protein-nucleic acid interactions.

5.1 Binding constants

Although EMSA is most commonly used as a qualitative assay it can, under certain conditions, provide quantitative data for relatively stable complexes. One of its earliest applications was in the measurement of kinetic and thermodynamic parameters. The association rates are determined by mixing the complex components at known concentrations and loading them in a running gel at precise intervals (for an example Spinner et al., 2002). For dissociation rates, a time course experiment is done by addition of competing nucleic acid to the preformed complexes (Fried & Crothers, 1981). The binding constant can be determined by the amount of complex formed as a function of protein concentration at equilibrium or as a ratio of the association and dissociation constants (for an example Demarse et al., 2009). An alternative method to measure kinetic and thermodynamic constants is the nitrocellulose filter binding assay that will be mentioned below.

As an example we show in figure 3 the titration of a DNA with the small delta protein to assess binding constants. The binding reaction was done by incubating the samples in a phosphate buffer during the same period of time (10 minutes) and then loading them onto an agarose gel for electrophoresis. It is clear that when the protein is present at only 0.25µM it does not interfere with the DNA mobility (Fig.3. well 2) as the band covered the same distance as the first sample, in which the protein was not present (Fig.3. well 1). But when 1.5µM of the small delta antigen are present in the binding reaction there is almost no free DNA present and the majority of the molecules are bound in a complex (Fig3. well 5). In the intermediate concentrations it can be clearly observed the decreasing presence of free DNA and increasing DNA-protein complexes as the protein concentration raises. We can consider that the dissociation constant can be estimated by quantifying the disappearance of the free DNA band (Demarse et al., 2009). From figure 3 we can say that the apparent dissociation constant is between 1 and 1.5µM.

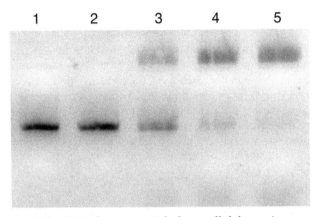

Fig. 3. Titration of a 500bp DNA fragment with the small delta antigen to estimate binding constants. An unlabeled 500bp DNA complementary to part of the HDV RNA was incubated, in a phosphate buffer (137mM NaCl, 2.7mM KCl, 4.3mM Na_2HPO_4, 1.5mM KH_2PO_4, pH 7.4), with increasing concentrations of small delta antigen of 0; 0.25; 0.5; 1; and 1.5µM and samples were loaded onto wells 1, 2, 3,4 and 5, respectively. Electrophoresis was in a 1.5% agarose gel in TAE buffer and the DNA was stained with ethidium bromide.

5.2 Cooperativity

Proteins can bind nucleic acids in a cooperative manner, that is, the complexes formed involve the binding of more than one protein to a specific nucleic acid segment. These multiprotein complexes may be a consequence of direct protein-protein interaction needed for nucleic acid binding, or a protein-induced deformation of the nucleic acid is a prerequisite to facilitate the binding of a second protein, or it may result from the bringing together of molecules bound at distinct sites in the nucleic acid sequence. The cooperativity can be inferred in a gel retardation assay from the underrepresentation of intermediate complexes between the unbound and saturated states. Multiprotein complexes can be comprised of a single protein species forming a homomultimer or of different proteins. The latter can be easily characterized by EMSA by the stability of the complexes formed with one protein in the presence or absence of the other(s).

5.3 Stoichiometry

Determining the important parameter that is stoichiometry is not as easy a task as it seems. The apparent weight changes estimated from the complexes' gel mobility are not applicable in determining the stoichiometry due to complications of charges and conformational effects on gel migration. A different approach is needed. The presence of truncated or extended protein derived from the wild-type but with the same binding and multimerization capacity will originate new bands that can reflect the monomers bound to the nucleic acid (Hope & Struhl, 1987). A similar method that will be discussed in the next segment is the supershift EMSA that uses an antibody specific for the binding protein recognizing an epitope that is accessible while the protein is bound to the nucleic acid. The addition of the antibody to the preformed complex can provide an estimate of the number of proteins bound by the extent of increments in retardation (Michael N & Roizman B, 1991 as cited in Lane & Prentki, 1992).

A more complex approach has been proposed in 1988 to determine a complex's stoichiometry (Granger-Schnarr et al., 1988). After the separation of the free and the complexed nucleic acid on a non-denaturing gel, the proteins are transferred to a membrane after sodium dodecyl sulfate (SDS) denaturation. This then allows the detection of proteins directly or indirectly using a specific antibody. The protein bands as well as the nucleic acids autoradiograph are then quantified by densiometry and the relative stoichiometry can be determined. The need for a specific antibody limits this method to complexes formed by well known proteins with available antibodies.

6. EMSA variants

Over the years variations or coupling of the EMSA protocol with other methods has been proposed to enhance its results or obtain more information from one experiment. Some examples of these EMSA-based approaches will be presented.

6.1 Reverse EMSA (rEMSA)

A reverse EMSA consists in labeling the protein sample rather than the nucleic acid (Filion et al., 2006). This method shows the difference in mobility between the free protein and nucleic acid-bound protein. It is an approach that can facilitate the determination of the protein binding affinity using different nucleic acids. Because the label used is ^{35}S instead of ^{32}P it is less sensitive than the conventional EMSA due to the isotope's energy.

6.2 Supershift EMSA

The supershift EMSA uses the same protocol as a regular EMSA except in that an antibody against the binding protein is added. As a result there is a more marked mobility shift during electrophoresis because the antibody will increase the overall complex molecular weight, hence the term supershift. This method can help identify if the proteins present in the complex have a specific epitope and is also used to validate previously identified proteins. It can also improve resolution when the difference between free nucleic acid and the complex is very small.

6.3 Multiplexed competitor EMSA (MC-EMSA)

The multiplexed EMSA was developed in 2008 by Smith and Humphries to characterize nuclear protein and DNA interactions, namely with transcription factors. In this method the nuclear extract is incubated with a pool of unlabeled DNA consensus competitors prior to adding the labeled DNA probe. An initial EMSA run will determine which cocktail competes with the probe binding to nuclear proteins which will then run individually in another EMSA to determine the precise competitor (Smith & Humphries, 2008). It is a competition-based method to identify uncertain DNA binding proteins requiring only a prior knowledge of transcription factor consensus sequences.

6.4 Two-dimensional EMSA (2D-EMSA)

The two-dimensional EMSA is a process that combines EMSA with proteomic or sequencing techniques to identify the proteins or the nucleic acid sequences that are present in the formed complexes. Two slightly different protocols have been developed to identify the interacting proteins and another method aims at the target nucleic acid sequence.

An initial approach was proposed by Woo and colleagues as they tried to identify and characterize transcription factors (Woo et al., 2002). A crude nuclear extract is partially purified by gel filtration and the resulting fractions are then bound to the nucleic acid probe and analyzed by EMSA. Meanwhile, in parallel, the pI and molecular weight of the putative interacting protein(s) is estimated as the fractions are analyzed by isoelectric focusing or SDS-Polyacrylamide Gel Electrophoresis (SDS-PAGE) in order to characterize possible candidates. Next, spots with the predetermined pI and molecular weight of the candidates are excised from a two-dimensional array of nuclear proteins and the proteins are eluted, renatured and tested for their binding ability through EMSA and the spots are afterwards analyzed by mass spectrometry for protein identification. This method is limited to proteins that can re-form into functional nucleic acid-binding conformations after the denaturing SDS-PAGE step, although EMSA can still show results even if renaturation efficiency is low. Because the final EMSA step that confirms the binding is performed with protein eluted from single spots it is only possible to identify proteins that interact with the nucleic acids as monomers or homomultimers. Proteins that only interact when complexed with other proteins will give a negative result on the validation EMSA.

A similar 2D-EMSA technique has since then been developed that incorporates EMSA into a two-dimensional proteomics approach by replacing the isoelectric focusing with EMSA as the first dimension of the 2D method (Stead et al., 2006). The protein sample, in the presence or absence of the nucleic acid, is separated by native PAGE as in a conventional EMSA. The protein bands from both conditions are then separated in a second dimension by denaturing SDS-PAGE. The proteins showing the nucleic acid dependent shift in mobility can be extracted from the gel for mass spectrometry identification. This approach does not require any previous knowledge of the chemical or physical properties of the binding protein and does not require protein renaturation after gel excision. It is also not limited to identify proteins that bind by themselves or as homomultimers and allows the characterization of complexes composed of different proteins.

These 2D approaches were developed by the two groups to study transcription factors, therefore, double stranded DNA is used as a nucleic acid probe but they can also adapted to other nucleic acid probes making them quite versatile methods to identify nucleic acid-interacting proteins.

Chernov and collaborators have developed a similar protocol with two dimensions but instead of aiming to identify the interacting protein(s) it characterizes and maps the specific protein target sites in regions of the human genome (Chernov et al., 2006). This approach is also based on first separating the complexes from the free nucleic acid in a non-denaturing gel and afterwards separating it under denaturing conditions (Vetchinova et al., 2006). The group used a pool of radioisotope-labeled short DNA sequences covering the genome region of interest and mixed it with a nuclear extract from a specific cell line. The formed complexes were separated in a non-denaturing one-dimensional standard EMSA. The complexes were localized by autoradiography and the gel strip containing them was excised and treated with a denaturing agent, SDS, to disrupt the preformed complexes. The strip is then loaded onto the second-dimension denaturing gel and another electrophoresis is performed. The gel is autoradiographed to determine the location of the freed DNAs, which are afterwards cut from the gel to be analyzed. By pairing this method with high-throughput sequencing the authors were able to identify a multitude of specific protein binding sites within a given genomic region.

6.5 EMSA-three-dimensional-electrophoresis (EMSA-3DE)

A three dimensional approach has very recently emerged to purify nucleic acid binding proteins from complexes separated by EMSA (Jiang et al., 2011). This method focuses on recovering the protein in high yield for subsequent analysis and has been developed to study low abundant transcription factors. In this EMSA-based purification procedure the complexes formed are extracted after a native PAGE retardation assay and applied to two-dimensional electrophoresis, isoelectric focusing and SDS-PAGE. The EMSA conditions are systematically optimized to reduce non-specific binding and increase protein yield. After the three electrophoreses the sample can then be electrotransfered onto a nitrocellulose or polyvinylidene difluoride membrane for southwestern and western blotting analysis to further characterize the complexes. Spots of interest can be cut from the gel or the membrane for protein identification by mass spectrometry.

7. Alternatives to EMSA

There are several alternatives to EMSA used in the analysis of nucleic acid-protein interactions with its own advantages and disadvantages when compared to EMSA.

7.1 Footprinting

Footprinting is essentially a protection assay used to characterize the binding site recognized by a given protein. It relies on the fact that a protein bound to the nucleic acid will protect it and interfere with the modification of the sequence it is bound to. The modification can be chemical or enzymatic and it is usually the endonuclease cleavage of radioisotope-labeled nucleic acid previously mixed with the protein(s) of interest. After cleavage the resulting ladder is analyzed on denaturing polyacrylamide gel and visualized by autoradiography. The gaps in the ladder are indicative of sites protected by the protein or proteins in the mixture (reviewed by Hampshire et al., 2007). This method was originally developed to characterize sequence selectivity but it is also helpful in estimating the binding strength through a footprinting reaction over a range of protein concentrations. For slow binding reactions footprinting can also be applied to assess the reaction kinetics estimating the association and dissociation rates. Although it is a widely used method, there are other approaches that provide higher throughput as the ones described ahead.

A variant on DNA footprinting is the *in vivo* approach, a technique that enables the detection of DNA-protein interactions as they occur in the cell. *In vivo* footprinting also relies on the fact that the bound protein protects the nucleic acid, at its binding site, from cleavage by endonucleases or modification by a chemical agent. The difference is that the cleavage of DNA is carried out within the nucleus following the *in vivo* binding of the proteins to chromatin. Footprints and endonuclease hypersensitive sites that are due to deformations of DNA in chromatin can be detected by this *in vivo* method. This method has been coupled with deep sequencing to identify DNaseI hypersensitive sites in the genome of different cell lines. It enabled the precise identification of a large number of specific cis-regulatory protein binding events with a single experiment (Boyle et al., 2011). Accordingly, the data obtained by this procedure may be more significant and representative of true events when compared with data obtained by the previously described *in vitro* footprinting.

7.2 Nitrocellulose filter binding

Nitrocellulose filter binding assays were developed in the 70s as a rapid enough method to allow kinetic as well as equilibrium studies of DNA-protein interactions (Riggs et al., 1968 and Riggs et al., 1970 as cited in Helwa & Hoheisel, 2010). The manipulation required is rapid enough to allow such measurements. The assay is based on the premise that proteins can bind to nitrocellulose without losing the ability to bind DNA. After the binding reaction the mixture is separated by electrophoresis and then blotted onto a nitrocellulose membrane. Only protein bound DNA remains on the membrane as the free double-stranded DNA will not be retained on nitrocellulose. The amount of DNA on the membrane can be quantified by measuring the label on the nucleic acid. However, this method has its limitations such as the fact that the proteins involved are not identified or the proportion in which they bind DNA. It also provides no information on the DNA sequence the protein interacts with unless well defined nucleic acid fragments are used and is limited to double stranded DNA as single stranded DNA can bind to nitrocellulose under certain conditions resulting in undesirable background.

7.3 Microfluidic mobility shift assay (MMSA)

The capillary microfluidic mobility shift assay (MMSA) is a method that uses fluorescence-based multi-well capillary electrophoresis to characterize protein-nucleic acid interactions. For example, it has been used effectively in characterizing RNA-protein binding in a study of the interaction between human immunodeficiency virus 1 transactivator of transcription and the transactivation-responsive RNA (Fourtounis et al., 2011). This technique requires only nanoliter amounts of sample that are introduced into microscopic channels and separated by pressure-driven flow and application of a potential difference. The free molecules or complexes are visualized by LED-induced fluorescence, discarding the need for hazardous radiolabeling. With the ability to perform 384-well screening this method has an increased capacity over regular EMSA to be compatible with high-throughput screenings.

7.4 Yeast hybrid systems

The yeast one-hybrid is an approach used to identify proteins that bind a given nucleic acid sequence as opposed to the methods that are suited to identify the nucleic acid sequences preferably recognized by a known protein. The protocol is based on a hybrid prey protein fused to a transcription activation domain that allows the expression of a reporter gene when the prey protein interacts with the DNA bait (reviewed by Deplancke et al., 2004). This method allows for a proteome-scale analysis depending on the prey protein library but only detects monomers that bind the target nucleic acid. Although it is an *in vivo* approach it is performed in yeast (*Saccharomyces cerevisiae*), which may not be the endogenous context, and is limited to DNA-protein interactions.

RNA-protein interactions can be studied with a yeast three-hybrid system that involves the expression in yeast cells of not one but three chimerical molecules, which assemble in order to activate two reporter genes (Kraemer et al., 2000). It represents a modification of the yeast two-hybrid system, widely used to identify protein-protein interactions, that was designed to allow high sensitivity *in vivo* detection of RNA-protein interactions. The yeast three-hybrid system includes: a fusion protein consisting of a DNA binding protein and a RNA-binding protein; a hybrid protein consisting of a transcription activating domain and a peptide thought to interact with a particular RNA; a RNA intermediate that promotes the interaction of the two hybrid proteins, this RNA includes the RNA that interacts with the system's RNA-binding protein and the RNA molecule to be investigated. The successful interaction of these 3 components allows the reconstitution of a transcription factor and subsequent activation of reporter genes (Hook et al., 2005 and Wurster & Maher, 2010)

7.5 ChiP assays

Chromatin immunoprecipitation (ChiP) is a commonly used method to study DNA-binding proteins *in vivo* and a standard method for the identification of transcription binding sites and histone modification locations (reviewed by Massie & Mills, 2008). In this method a cross-linking agent (e.g. formaldehyde) is added to cells to covalently bind proteins and chromatin that are in direct contact. Afterwards, the cells are lysed and chromosomal DNA is isolated and fragmented. Specific antibodies are used to immunoprecipitate the targeted proteins with the cross-linked DNA. The bound nucleic acid is released by reverting the cross-linking and then analyzed. Classically, the DNA was characterized by polymerase chain reaction (PCR) which required some previous knowledge of the candidate DNA regions. Nowadays, the DNA bound to protein is more commonly characterized through more powerful tools either coupled with microarrays that represent the genome (ChIP-chip) or state-of-the-art high-throughput sequencing (ChIP-seq). The improvements in DNA sequencing technology allow tens of millions of sequence reads, therefore ChIP-seq has a major advantage of increased sensitivity and resolution to add to the fact that it is not limited to predetermined probe sets as ChIP-chip. The major strength of the ChIP-based approaches is that they capture complexes *in vivo* and the binding reactions can be studied under different cellular conditions and at different time points. However it also has important limitations. The method requires high-quality antibodies that are available only for a limited number of proteins. To circumvent this, epitope-tagged proteins could be used although it usually implies the introduction of modified genes into the endogenous locus in order to obtain expression at physiological levels. This method does not distinguish between proteins that bind directly to the genomic DNA and those that only interact with other proteins that do bind.

7.6 SELEX

The Systematic Evolution of Ligands by Exponential Enrichment (SELEX) is a well established method that enables the selection of enriched sequences from a random library that bind recombinant proteins. This procedure starts with the synthesis of the oligonucleotide library and then incubating the generated sequences with the putative interacting protein(s). The sequences that bind are eluted, amplified by PCR and subjected to more rounds of selection with increasing stringency conditions. This allows the identification of the tightest-binding sequences. It is a widely used approach to obtain transcription factors binding motifs as it requires low amounts of purified proteins (Matys et al., 2006). This approach becomes very complicated to use when large numbers of nucleic acid-binding proteins are analyzed as it then requires multiple rounds of selection. Another limitation is the fact that it is aimed at the identification of the best binding DNA targets *in vitro* and does not allow the characterization of the exact *in vivo* selectivity.

7.7 Protein microarray

A protein microarray is a method that allows high-throughput analysis in which labeled nucleic acids are queried against proteins immobilized on a chip (reviewed by Hu et al., 2011). In a functional protein microarray, thousands of purified recombinant proteins can be immobilized in a glass slide in discrete locations forming a high-density protein matrix, providing a flexible platform to characterize different protein activities. It is a very versatile method as it can perform a semi-quantitative analysis of protein binding to a wide range of molecules (nucleic acids, other proteins, antibodies, lipids, glycans...). In theory, it is feasible to print arrays of all the annotated proteins of a given organism originating a whole proteome microarray. However, it implies the expression and purification of each individual protein and several conditions need to be optimized to render the proteins apt for this method. Since the protein is immobilized it is crucial to guarantee that its structural integrity remains intact especially the binding domains that are to be studied.

7.8 Nucleic acid microarrays

Nucleic acid microarrays can also be used for a direct analysis of protein-nucleic acid interactions. In this case it is the nucleic acid that is immobilized and not the protein. Nucleic acid chips are a powerful and versatile tool in biological research. They consist of high-density arrays of oligonucleotides or complementary DNA that can cover a whole genome (reviewed by Stoughton, 2005). For protein-interaction studies, the protein(s) of interest is expressed usually with an epitope tag, and purified. The tag serves two purposes; it helps to isolate the protein through affinity purification, and allows detection by an epitope-specific reporter antibody. After incubation of the protein with the nucleic acid chip the signal intensities at the several array spots can be measured.

7.9 Ribonucleoprotein Immunoprecipitation – Microarray (RIP-chip)

RNA immunoprecipitation and chip hybridization (RIP) is a protocol very similar to ChIP-chip except that it targets RNA-protein interactions rather than DNA-protein (Keene et al.,

2006). RIP-chip is an approach that consists on a microarray profiling of RNAs obtained from immunoprecipitated RNA-protein complexes. Genome-wide arrays are used to identify messenger RNAs (mRNAs) that are present in endogenous messenger ribonucleoprotein complexes making it a great tool to identify the physiological substrates of mRNAs. The endogenous complexes are immunoprecipitated from cell lysates which limits this study to kinetically stable interactions. Even though it can identify RNA-protein complexes with heteromultimers, at least one of the proteins has to be previously known to be the basis of immunoprecipitation and "fish out" the whole complex.

7.10 Crosslinking and Immunoprecicipation (CLIP) and Photoactivable-Ribonucleoside-Enhanced Crosslinking and Immunoprecipitation (PAR-CLIP)

The RIP-chip method that has just been described is limited to studies of very stable RNA-protein complexes; to remediate this problem another method is available to study RNA-binding proteins. The crosslinking and immunoprecipitation (CLIP) approach uses *in vivo* UV crosslinking prior to the complexes immunoprecipitation to identify less stable interactions (Ule et al., 2003). After immunoprecipitation RNA molecules are separated and cDNA sequencing is carried on. However, this method is not perfect as the commonly used UV 254nm RNA-protein crosslinking has low efficiency and it is difficult to distinguish between crosslinked RNAs from background non-crosslinked fragments that can be detected in the sample due to the presence of abundant cellular RNAs.

A more recent approach tries to further improve the CLIP method using photoreactive ribonucleoside analogs such as 4-thiouridine or 6-thioguanosine (Hafner et al., 2010). In this photoactivatable-ribonucleoside-enhanced crosslinking and immunoprecipitation (PAR-CLIP) protocol the photoreactive nucleosides are incorporated into nascent transcripts within living cells. The irradiation is performed with UV light of 365nm, which induces an efficient crosslink of the labeled cellular RNA to its interacting proteins. The labeled RNAs are isolated after co-immunoprecipitation, and converted into cDNA for deep sequencing. The precise crosslinking position can be identified by mutations in the sequenced cDNA making it possible to distinguish the crosslinked fragments from background.

7.11 High-Throughput Sequencing – Fluorescent Ligand Interaction Profiling (HisT-FLIP)

Very recently a new method was developed to characterize DNA-protein interactions using second-generation sequencing instruments (Nutiu et al., 2011). This method allows high throughput and quantitative measurement of DNA-protein binding affinity. This High-Throughput Sequencing – Fluorescent Ligand Interaction Profiling (HiTS-FLIP) uses the optics of a high-throughput sequencer to visualize *in vitro* binding of a protein to the sequenced DNA in a flow cell. The new method was initially used on a *Saccharomyces cerevisiae* transcription factor. The fluorescently tagged protein was added at different concentrations to a flow cell containing around 88 million DNA clusters, the equivalent of over 160 yeast genomes. The traditional EMSA was used as an independent validation of the dissociation constants obtained and found a high correlation with values obtained with the new method and those from EMSA as reported in literature. This high-throughput method has an obvious advantage in the fact that it can provide hundreds of millions of measurements but is limited to DNA-protein interactions and requires expensive equipment. Another advantage is that the sequencing instrument can measure multiple fluorescent wavelengths allowing hetero and homodimeric forms to be measured in the same run, using distinct tags on individual proteins.

8. Conclusion

Since the first report, 30 years ago, EMSA became one of the most popular methods for detection and characterization of protein-nucleic acid interactions. Hundreds of protocols have been published accommodating modifications in virtually every parameter influencing the experimental outcome. Improvements were made in all EMSA steps including the methods for preparation of protein samples and purification, synthesis and labeling of nucleic acids, and detection. This allowed enlarging and diversifying the applications of EMSA and resulting in a number of variants of the method.

However, despite the large amount of available literature and protocols trial and error will ultimately be the way to optimize the EMSA conditions for the nucleic acid-protein complex to be analyzed. The guidelines discussed above help to provide an initial protocol adjusted to each study but slight changes may be needed to improve binding and detection of the complexes.

In recent years, the use of highthrouhput approaches to detect biologically relevant interactions, including those between proteins and nucleic acids, was reported. Development of these approaches was made possible, at least in part, by the availability of more sensitive and specific equipment and tools. Although EMSA cannot achieve a high throughput level it remains a valuable tool to confirm the detected interactions.

9. Acknowledgements

We are grateful to Dr. Cristina Branco for constructive comments. Work in the authors' laboratory is supported by Fundação para a Ciência e Tecnologia (PTDC/SAU-MII/098314/2008). CA is a recipient of a FCT PhD grant.

10. References

Adachi, Y.; Chen, W.; Shang, W. & Kamata, T. (2005). Development of a direct and sensitive detection method for DNA-binding proteins based on electrophoretic mobility shift assay and iodoacetamide derivative labelling. *Analytical Biochemistry*, Vol.342, pp. 348-351

Alves, C.; Cheng, H.; Roder, H. & Taylor, J. (2010). Intrinsic disorder and oligomerization of the hepatitis delta virus antigen. *Virology*, Vol. 370, pp 12-21

Apella, E. & Anderson, C. (2001). Post-translational modifications and activation of p53 by genotoxic stresses. *European Journal of Biochesmistry*, Vol.268, No.10, pp. 2764-2772

Baneyx, F. (1999). Recombinant protein expression in Escherichia coli. *Current Opinion in Biotechnology*, Vol.10, pp. 411-421

Boyle, A.; Song, L.; Lee, B.; London, D.; Keefe, D.; Birney, E.; Iyer, V.; Crawford, G. & Furey, T. (2011). High-resolution genome-wide in vivo footprinting of diverse transcription factors in human cells. *Genome Research*, Vol.21, No.3, pp. 456-464

Chernov, I.; Akopov, S.; Nikolaev, L. & Sverdlov, E. (2006). Identification and mapping of DNA binding proteins target sequences in long genomic regions by two-dimensional EMSA. *BioTechniques,* Vol.41, pp. 90-96

Chi, P.; Van Komen, S.; Sehorn, M.; Sigurdsson, S. & Sung, P. (2006). Roles of ATP binding and ATP hydrolysis in human Rad51 recombinase function. *DNA Repair,* Vol.5, No.3, pp. 381–391

Crothers, D. & Drak, J. (1992). [3] Global features of DNA structure by comparative gel electrophoresis. *Methods in Enzymology,* Vol 212, pp. 46-71

Demarse, N.; Ponnusamy, S.; Spicer, E.; Apohan, E.; Baatz, J.; Ogretmen, B. & Davies, C. (2009). Direct binding of glutharaldehyde 3-phosphate dehydrogenase to telomeric DNA protects telomeres against chemotherapy-induced rapid degradation. *Journal of Molecular Biology,* Vol.394, No.4, pp. 789-803

Deplancke, B.; Dupuy, D.; Vidal, M. & Walhout, A. (2004). A gateway-compatible yeast one-hybrid system. *Genome Research,* Vol.1, 2093-2101

Dignam, J.; Lebovitz, R. & Roeder, R. (1983). Accurate transcription initiation by RNA polymerase II in a soluble extract from isolated mammalian nuclei. *Nucleic Acids Research,* Vol.11, No.5, pp. 1457-1489

Endo, Y. & Sawasaki, T. (2006). Cell-free expression systems for eukaryotic protein production. *Current Opinion in Biotechnology,* Vol.17, pp. 373-380

Filion, G.; Fouvry, L. & Defossez, P.-A. (2006). Using reverse electrophoretic mobility shift assay to measure an compare protein-DNA binding affinities. *Analytical Biochemistry,* Vol.357, pp. 156-158

Fourtounis, J.; Falgueyret, J.-P. & Sayegh, C. (2011). Assessing protein-RNA interactions using microfluidic capillary mobility shift assays. *Analytical Biochemistry,* Vol.411, pp. 161-163

Fried, M. (1989). Measurement of protein-DNA interaction parameters by electrophoresis mobility shift assay. *Electrophoresis,* Vol. 10, pp. 366-376

Fried, M. & Crothers, D. (1981). Equilibria and kinetics of lac repressor-operator interactions by polyacrylamide gel electrophoresis. *Nucleic Acid Research,* Vol.9, No.23, pp. 6505-6525

Fried, M. & Crothers, D. (1984). Equilibrium studies of the cyclic AMP receptor protein-DNA interaction. *Journal of Molecular Biology,* Vol.172, No.3, pp. 241-262

Fried, M. & Daugherty, M. (1998). Electrophoretic analysis of multiple protein-DNA interactions. *Electrophoreis,* Vol.19, pp. 1247-1253

Garner, M. & Revzin, A. (1981). A gel electrophoresis method for quantifying the binding of proteins to specific DNA regions: application to components of the Escherichia coli lactose operon regulatory system. *Nucleic Acids Research,* Vol.9, No.13, pp. 3047-3060

Goh, A.; Coffil, C. & Lane, D. (2011). The role of mutant p53 in human cancer. *The Journal of Pathology,* Vol.223, No.2, pp. 116-126

Granger-Schnarr, M.; Lloubes, R.; de Murcia, G. & Shnarr, M. (1988). Specific protein-DNA complexes: immunodetection of the protein component after gel electrophoreis and Western blotting. *Analytical Biochemistry,* Vol.174, No.1, pp. 235-238

Greco-Stewart, V. & Pelchat, M. (2010). Interaction of host cellular proteins with components of the hepatitis delta virus. *Viruses,* Vol.2, No.1, pp 189-212

Haffner, M.; Landthaler, M.; Burger, L.; Khorshid, M.; Hausser, J.; Berninger, P.; Rothballer, A.; Ascano, M.; Jungkamp, A.-C.; Munschauer, M.; Ulrich, A.; Wardle, G.; Dewell, S.; Zavola, M. & Tuschi, T. (2010). PAR-CliP – A method to identify transcriptome-wide binding sites of RNA binding proteins. *Journal of Visualized Experiments,* Vol.41, pp.2034

Hampshire, A.; Rusling, D.; Broughton-Head, V. & Fox, K. (2007). Footprinting: a method for determining the sequence selectivity, affinity and kinetics of DNA-binding ligands. *Methods,* Vol.42, pp. 128-140

Hellman, L. & Fried, M. (2007). Electrophoretic mobility shift assay (EMSA) for detecting protein-nucleic acid interactions. *Nature Protocols,* Vol.2, No.8, pp. 1849-1861

Helwa, R. & Hoheisel, J. (2010). Analysis of DNA-protein interactions: from nitrocellulose filter binding assays to microarray studies. *Analytical & Bioanalytical Chemistry,* Vol.398, pp. 2551-2561

Holden, N. & Tacon, C. (2011). Principles and problems of the electrophoretic mobility shift assay. *Journal of Pharmacological and Toxicological Methods,* Vol.63, pp. 7-14

Hope, I. & Struhl, K. (1987). GCN4, a eukaryotic transcriptional activator protein, binds as a dimer to target DNA. *The EMBO Journal,* Vol.6, No.9, pp. 2781-2784

Hook, B.; Bernstein, D.; Zhang, B. & Wickens, M. (2005). RNA-proteins interactions in the yeast three-hybrid system: affinity, sensitivity, and enhanced library screening. *RNA,* Vol.11, No.2, pp. 227-233

Hu, S.; Xie, Z.; Qian, J.; Blackshaws, S. & Zhu, H. (2011). Functional protein microarray technology. *WIREs Systems Biology and Medicine,* Vol.3, pp. 255-268

Jarvis, D. (2009). Baculovirus-insect cell expression systems. *Methods in Enzymology,* Vol.463, pp. 191-222

Jiang, D.; Jia, Y. & Jarrett, H.W. (2011). Transcription factor proteomics: Identification by a novel gel mobility shift-three-dimensional electrophoresis method coupled with southwestern blot and high-performance liquid chromatography-electrospray-mass spectrometry analysis. *Journal of Chromatography A,* Vol.1218, No.39, pp. 7003-7015

Jiang, S.; Macias, M. & Jarrett, H. (2010). Purification and identification of a transcription factor, USF2, binding to E-box elements in the promoter of human telomerase reverse transcriptase (hTERT). *Proteomics,* Vol.10, No.2, pp. 203-211

Jimenez, J. (2010). Protein-DNA interaction at the origin of neurological diseases: a hypothesis. *Journal of Alzheimer's Disease,* Vol.22, No.2, pp. 375-391

Keene, J.; Komisarow, J. & Friedersdorf, M. (2006). RIP-Chip: the isolation and identification of mRNAs, microRNAs and protein components of ribonucleoprotein complexes from cell extracts. *Nature Protocols,* Vol.1, pp. 302-307

Kerr, L. (1995). [42] Electrophoretic mobility shift assay. *Methods in Enzymology,* Vol.254, pp. 619-632

Kraemer, B.; Zhang, B.; SenGupta, D.; Fields, S. & Wickens, M. (2000). Using the yeast three-hybrid system to detect and analyze RNA-protein interactions. *Methods in Enzymology,* Vol.328, pp. 297-321

Lane, D.; Prentki, P. & Chandler, M. (1992). Use of gel retardation to analyze protein-nucleic acid interactions. *Microbiological Reviews,* Vol.56, No.4, pp. 509-528

Makino, T.; Skretas, G. & Georgiou, G. (2011). Strain engeneering for improved expression of recombinant proteins in bacteria. *Microbial Cell Factories,* Vol.10, No.1, pp. 32-42

Massie, C. & Mills, I. (2008). ChIPping away at gene regulation. *EMBO Reports,* Vol.9, No.4, pp. 337-343

Matys, V.; Kel-Margoulis, O.; Fricke, E.; Liebich, I.; Land, S.; Barre-Dirrie, A.; Reuter, I.; Checkmenev, D.; Krull, M.; Hornischer, K.; Voss, N.; Stegmaier, P.; Lewicki-Potapov, B.; Saxel, H.; Kel, A. & Wingender, E. (2006). TRANSFAC and its module TRANSComple: transcriptional gene regulation in eukaryotes. *Nucleic Acids Research,* 1 34 Database issue D108-110

Nordheim, A. & Meese, K. (1988). Topoisomer gel retardation: detection of anti-Z-DNA antibodies bound to Z-DNA within supercoiled DNA minicircles. *Nucleic Acids Research,* Vol.16, No.1, pp. 21-37

Nutiu, R.; Friedman, R.; Luo, S.; Khrebtukova, I.; Silva, D.; Li, R.; Zhang, L.; Schroth, G. & Burge, C. (2011). Direct measurement of DNA affinity landscapes on a high-throughput sequencing instrument. *Nature Biotechnology,* Vol.29, No.7, pp. 659-664

Palecek, E.; Vlk, D.; Stankova, V.; Brazda, V.; Vojtesek, B.; Hupp, T.; Schaper, A. & Jovin, T. (1997). Tumor suppressor protein p53 binds preferentially to supercoiled DNA. *Oncogene,* Vol.15, No.18, pp. 2201-2209

Puzio-Kuter, A. (2011). The role of p53 in metabolic regulation. *Genes & Cancer,* Vol.2, No.4, pp. 385-391

Rizzetto, M. (2009). Hepatitis D: Thirty years after. *Journal of Hepatology,* Vol.50, pp. 1043-1050

Ruscher, K.; Reuter, M.; Kupper, D.; Trendelenburg, G.; Dirnagl, U. & Meisel, A. (2000). A fluorescence based non-radioactive electrophoretic mobility shift assay. *Journal of Biotechnology,* Vol78, pp. 163-170

Saida, F.; Uzan, M.; Odaert, B. & Bontems, F. (2006). Expression of highly toxic genes in E. coli: special strategies and genetic tools. *Current Protein and Peptide Science,* Vol.7, pp. 47-56

Sandman, K.; Krzycki, J.; Dobrinski, B.; Lurz, R. & Reeve, J. (1990). HMf, a DNA-binding protein isolated from the hyperthermophilic archaeon Methanothermus fervidus, is most closely related to histone. *Proceedings of the National Academy of Sciences of the United States of America,* Vol.87, No.15, pp. 5788-5791

Simicevic, J. & Deplancke, B. (2010). DNA-centered approaches to characterize regulatory protein-DNA interaction complexes. *Molecular BioSystems,* Vol.6, No.3, pp. 462-468

Spinner, D.; Liu, S.; Wang, S-W. & Schmidt, J. (2002). Interaction of the myogenic determination factor myogenin with E12 and a DNA target: mechanism and kinetics. *Journal of Molecular Biology,* Vol.317, No.3, pp. 431-445

Stead, J.; Keen, J. & McDowall, K. (2006). The identification of nucleic acid-interacting proteins using a simple proteomics-based approach that directly incorporates the electrophoretic mobility shift assay. *Molecular & Cellular Proteomics,* Vol.5, No.9, pp. 1697-1702

Stoughton, R. (2005). Applications of DNA microarrays in biology. *Annual Review of Biochemistry,* Vol.74, pp. 53-82

Ule, J.; Jensen, K.; Ruggiu, M.; Mele, A.; Ule, A. & Darnell, R. (2003). CLIP identifies Nova-regulated RNA networks in the brain. *Science,* Vol.302, No.5648, pp. 1212-1215

Vetchinova, A.; Akopov, S.; Chernov, I.; Nikolaev, L. & Sverdlov, E. (2006). Two-dimensional electrophoretic mobility shift assay: identification and mapping of transcription factor CTCF target sequences within an FXYD5-COX7A1 region of human chromosome 19. *Analytical Biochemistry,* Vol.354, pp. 85-93

Vossen, K.; Wolz, R.; Daugherty, M. & Fried, M. (1997). Role of macromolecular hydration in the binding of the Escherichia coli cyclic AMP receptor to DNA. *Biochemistry,* Vol.36, No.39, pp. 11640-11647

Walker, J. (1994). Gradient SDS polyacrylamide gel electrophoresis of proteins. *Methods in Molecular Biology,* Vol.32, pp. 35-38

Woo, A.; Dods, J.; Susanto, E.; Ulgiati, D. & Abraham, L. (2002). A proteomics approach for the identification of DNA binding activities observed in the electrophoretic mobility shift assay. *Molecular & Cellular Proteomics,* Vol.1, No.6, pp. 472-478

Wurster, S. & Maher, L. (2010). Selections that optimize RNA display in the yeast three-hybrid system. *RNA,* Vol.16, No.2, pp. 253-258

13

Analysis of Chromosomal Replication Progression by Gel Electrophoresis

Elena C. Guzmán[1] and Enrique Viguera[2]
[1]*Departamento de Bioquímica, Biología Molecular y Genética*
Facultad de Ciencias, Universidad de Extremadura, Badajoz
[2]*Área de Genética, Facultad de Ciencias, Universidad de Málaga, Málaga*
Spain

1. Introduction

An absolute requirement for life is the preservation of genome integrity and the faithful duplication of chromosomes before segregation. A proliferating cell must duplicate its entire complement of DNA with exquisite precision facing a barrage of impediments of different nature. Single-strand breaks (SSBs) in the template DNA, either pre-existing or arising from abnormal DNA structures (folded DNA, cruciform structures or cross-links, etc.), collisions with DNA-bound proteins such as transcription complexes or DNA structural barriers restrain replication progression. For instance, during *Escherichia coli* DNA replication, the two forks initiated at the single origin of replication, *oriC*, move along the chromosome with high probability of pausing, stalling or even collapse (Maisnier-Patin et al., 2001). Replication arrest is a source of genetic instability in all types of living cells (Michel, 2000; Carr, 2002; Kolodner et al., 2002). As a consequence, cells have developed several effective strategies to tackle with replication fork arrest and/or repairing the double strand breaks (DSBs) generated at the stalled replication forks (Bierne et al., 1994; Kuzminov, 1995). Considerable evidence has been accumulated in the past decade demonstrating the involvement of recombination proteins in either direct or bypass repair of the lesions or structures blocking replication fork progression (reviewed in Courcelle et al., 2004; Kreuzer, 2005; Hanawalt, 2007; Michel et al., 2007 and references herein).

Conventional agarose DNA electrophoresis is one of the most frequently used techniques in molecular biology for the isolation or identification of DNA fragments. However, Pulse Field Gel Electrophoresis (PFGE) and Two-Dimensional (2D) Agarose Gel Electrophoresis techniques have been used to study biological processes such as the progression of the replication fork along a DNA fragment. In this work we introduce how these techniques has been used in bacteria to (i) verify and quantify the presence of stalled replication forks (ii) recognize DNA structure at the stalled replication fork, and (iii) understand how the replication fork could be restarted.

2. Pulse Field Gel Electrophoresis (PFGE)

Separation of DNA fragments by standard agarose gel electrophoresis is based on the capacity of the molecules to pass through the pore generated inside the matrix gel. Using this feature as the only separation mechanism, the large DNA molecules cannot be discriminated from each other. The practical range of resolution is up to approximately 50kb; making impossible the direct genomic analysis of large DNA molecules as those generated by the presence of complex DNA structures, or the DSBs involved in stalled replication forks in the *E. coli* chromosome.

In 1983, PFGE was developed as a method to circumvent this limitation, allowing fractionation of very large DNA molecules up to a million base pairs in size (Schwartz & Cantor, 1984; Herschleb et al., 2007). PFGE allows the separation of these large DNA molecules through abrupt electrical perturbations to the paths crawling molecules take trough the gel. In PFGE, the *direction of the electrical field* is periodically changed (usually 120°), requiring electrophoresing molecules to reorientate (Fig. 1). The *time* required to complete the orientation process scales with the size of the DNA, so that increasing the size of the DNA molecules, it takes more time between changes in the direction of the electric field (Fig. 1). These intervals vary depending on the size of the fragments that have to be resolved, a few seconds for small fragments to hours for fragments larger than 5Mb. The principle of PFGE is that large DNA fragments require more time to reverse the direction in an electric field than small DNA fragments. Alternating current direction during gel electrophoresis can resolve DNA fragments of 100 to 1,000 kb.

The *equipment* required to perform PFGE is also different from that used in traditional electrophoresis. The tank contains a set of *electrodes* (6-8), instead of a couple of them, being thicker and disposed to allow the different orientations of the electric field. Maintaining a constant the *temperature* (usually 14°C) during the process is important to avoid temperature variations through the gel, which could affect the resolution of DNA fragments. Accordingly, the system should include a cooling device. Agarose gel preparation does not differ from that reported for conventional electrophoresis.

Due to the fragility of the very large DNA fragments to be separated, preparing the *sample* is the most critical step for PFGE. To avoid breakage of genomic DNA during manipulation, the DNA is not extracted, but the cells are embebed in agarose plugs and then fixed into the wells.

2.1 Verifying replication fork reversal by detection of DNA breakage at the stalled replication forks

The progression of the replication fork can be halted by several causes, including deficiencies in replication enzymes and obstacles such as DNA-bound proteins, transcription complexes, nicks, gaps, DNA damage or topological constrictions. Replication arrest is a source of DNA breakage and rearrangement in all organisms (reviewed in Aguilera & Gómez-González, 2008); consequently stalled replication forks create the need for replication reactivation, and different ways of restarting replication have been proposed (Michel et al., 2004; Michel et al., 2007). In bacteria, the consequences of replication blockage have been studied mainly in *E. coli*. In several *E. coli* replication mutants, the stalled forks generated upon inactivation of the mutant enzyme are reversed and result in the formation of a Holliday junction (HJ) adjacent to a DNA double strand end, a reaction called 'replication fork reversal' (RFR) (Fig. 2A) (Michel et al., 2004; Seigneur et al., 1998; Seigneur et al., 2000). In a *rec* proficient background this intermediary could be processed without

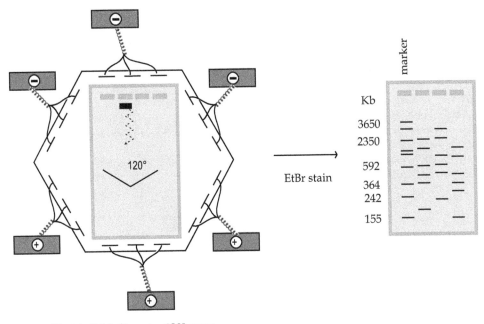

Electric field alternates 120° every
90 seconds for 18 to 24 hours at 14°C

Fig. 1. Schematic diagram of PFGE instrumentation. Contoured clamped homogeneous electric field (CHEF) systems use a hexagonal gel box that alters the angle of the fields relative to the agarose gel. After running the gel by PFGE, DNA fragments are visualized by staining with ethidium bromide.

generating DSBs by using the recombination proteins RecBCD, RecA, and by the HJ-specific resolvase RuvABC (Fig. 2B) (Seigneur et al., 1998). This is a key aspect of the RFR model as it allows restarting of the blocked forks without generating chromosome instability. Nevertheless, in the absence of RecBCD activity (Fig. 2C), resolution of the RFR-produced HJ is done by RuvABC resolvase and leads to fork breakage. These particular DSBs are dependent on RuvABC activity in a *recB* deficient background.

If RFR does not take place at the stalled fork, at least two situations may arise. On the one hand, there would be an increase of DSBs independent of RuvABC activity and generated by another unknown endonuclease (Fig. 2E) as in the case of the thymine starvation (Guarino et al., 2007b). On the other hand, there would be no increase in the amount of DSBs probably because the stalled forks are not susceptible to the endonuclease action, and the restarting of the forks would take place without the generation of fork breakage. This situation has been described in *gyrB* mutants (Grompone et al., 2003), and when replication termination sequences *ter* were placed at ectopic positions on the bacterial chromosome (Bidnenko et al., 2002). Using the system described above, the fate of the stalled replication forks caused by any condition can be studied.

To verify the RFR process by PFGE, a *recB* deficient background should be used (i) to inhibit the degradation or the recombinational repair of the DNA tail created by the regression of the fork (Miranda & Kuzminov, 2003), allowing RuvABC resolvase to transform this tail in a DSB; (ii) to inhibit the repair of the DSBs generated by RuvABC resolvase (Fig. 2C). According to the RFR model, the occurrence of this process at the stalled forks generated under restrictive conditions can be verified by testing whether there is an increase of DSBs in a *recB* deficient background, and determining whether these DSBs are dependent on RuvABC resolvase activity by measuring the amount of DSBs in a *recB* and *recB ruvABC* deficient background (Fig. 2C) (Seigneur et al., 1998). The occurrence of RFR at the stalled forks has been detected by this system in several replication mutants, such as in the helicase mutants *rep* and *dnaBts* (Michel et al., 1997; Seigneur et al., 2000), in *holD*[G10] (Flores et al., 2001, 2002), in *dnaEts* at 42°C and in the *dnaNts* mutant at 37°C (Grompone et al., 2002) and finally in *nrdA101ts* (Guarino et al., 2007a, 2007b).

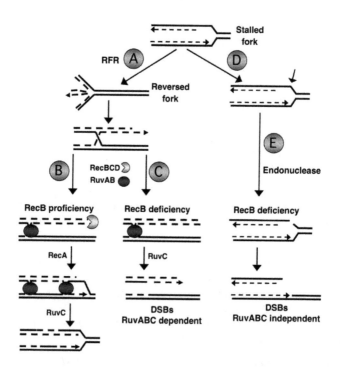

Fig. 2. The fate of the stalled forks. In the first step (A), the replication fork is arrested, causing fork reversal. The reversed fork forms a HJ (two alternative representations of this structure are shown – open X and parallel stacked X). In Rec+ cells (B), RecBCD initiates RecA-dependent homologous recombination, and the resulting double HJ is resolved by RuvABC. In the absence of RecBCD (C), resolution of the HJ by RuvABC leads to DSBs at the stalled replication fork. Alternatively, the replication fork is arrested without being regressed (D) and it is susceptible to be cut by an endonuclease, generating DSBs at the stalled replication fork (E). Continuous line (parental chromosome); dashed lines (newly synthesized strands); disk (RuvAB); incised disk (RecBCD). Adapted from Guarino et al., 2007b.

The amount of linear DNA resulting from DSBs can be estimated by using PFGE combined with cell lysis in agarose plugs (Michel et al., 1997). Briefly, cultures of *recB* and *recB ruvABC* strains growing in M9 minimal medium are labelled by addition of 5 µCi/ml [*methyl*-^{3}H] thymidine (100 Ci/mmol). When cultures reached 0.2 OD$_{450nm}$, 1ml of cells were collected, washed in cold minimal medium and resuspended in 100µl of TEE buffer. Cells were incubated at 37°C for ten minutes, mixed with 100 µl of low melting agarose 2% in TEE at 55°C and poured into the mould. Once agarose had solidified, cell lysis was performed in the plugs. This ensures only linear chromosomes to enter the gels, while circular molecules remain in the wells (Michel et al., 1997; Seigneur et al., 1998). Plugs were incubated with lysozyme at 5 mg/ml and sarcosyl 0.05% in TEE for 2 h with gently shaking. Then, plugs were retrieved and incubated with lysis solution (1 mg/ml Proteinase K , 1% SDS in TEE) at 56°C overnight. PFGE were run for 48 h at 4°C as described (Seigneur et al., 1998); initial run 500 sec, final run 500 sec, 3 volts/cm and 106° reorientation angle. DNA was visualized by ethidium bromide staining. Lanes were cut into slices and the proportion of migrating DNA was determined by calculating the amount of tritium present in each slice with respect to the total amount of tritium present in the corresponding lane plus the well (Fig. 3A). All the PFGE linear DNA data were analyzed by the least-squares statistical approach, considering measures as highly significantly different if p< 0.01.

A typical profile of gel migration for the different strains analyzed is shown in figure 3B. In this case, results indicate that the amount of DSBs in the strain *nrdA101 recB* was greater than in the strain *nrdA+ recB*, suggesting an increase in the number of the stalled forks induced by the presence of a defective ribonucleotide reductase (RNR) at the permissive temperature. To establish the possible origin of the DSBs induced by the *nrdA101 recB* background we investigated whether the formation of DSBs resulted from the action of the RuvABC resolvase (Fig. 2). The DSBs levels estimated in *nrdA101 recB ruvABC* and *nrdA+ recB ruvABC* strains were markedly lower than in the respective Ruv+ counterpart strains. As RuvABC is a specific resolvase for HJ, according the RFR model (Fig. 2), it generates DSBs at arrested replication forks in *recB* deficient background (Seigneur et al., 1998); these results indicated the occurrence of replication fork reversal in *nrdA101* mutant. As RFR is one of the mechanisms to restart the stalled replication forks, we could infer that the *nrdA101* strain growing at 30°C increases the number of stalled replication forks that would proceed with the help of RFR process in a Rec+ proficient context (Guarino et al., 2007a).

2.2 Replication fork collapse at natural arrest sites

In *E. coli*, replication termination occurs by the encounter of two opposite migrating forks at the terminus region or at specific arrest sites named *Ter*, when one of the forks reaches the terminus before the other. Tus protein binds the *Ter* sites forming a complex that acts as a polar replication fork barrier by preventing the action of the DnaB helicase (Neylon et al., 2005).

PFGE was used to determine the analysis of replication forks blocked at terminator sequences *Ter* inserted at ectopic positions on the bacterial chromosome (Bidnenko et al., 2002). This strain requires the RecBCD pathway of homologous recombination for viability, although replication forks blocked at *Ter* are not broken nor reversed (Fig. 2A, D). The analysis of the structure of the chromosomes by PFGE showed linear fragments of about 2 Mb, corresponding to the distance between the origin and the ectopic *Ter* sites. A model of a collapse of replication forks at terminator sequences was proposed in which the blocked replication forks at Ter/Tus are stable, but they are re-replicated in a new replication round, generating 2 Mb linear fragments. These results suggest that natural and accidental replication arrest sites are processed differently in the cell (Bidnenko et al., 2002).

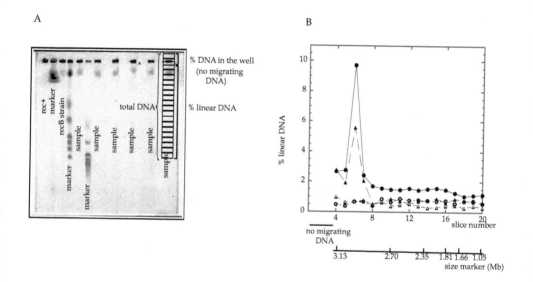

Fig. 3. (A) Visualization of a representative agarose gel stained with ethidium bromide (B) Typical profile of a PFGE experiment. *nrdA+ recB* (▲), *nrdA101, recB* (●), *nrdA+ recB ruvABC* (Δ), and *nrdA101 recB ruvABC* (O). Agarose plugs were prepared as described in the text. Gels were cut in 3 mm slices and the amount of [*methyl*-³H]thymidine present in each slice was measured, and the ratio of the total amount of [*methyl*-³H]thymidine in the lane was calculated for each slice. The gel origin is not shown; only the migrating DNA is shown. The position of size marker is shown (*Hansenula wingei*, Bio-Rad). The amount of linear DNA was calculated from slice 4 to 12. Total proportions of migrating DNA in this experiment from slice 4 to 12 were 16.65% *nrdA+ recB*, 25.21% *nrdA101, recB*, 5.59% *nrdA+ recB ruvABC*, and 5.73% *nrdA101 recB ruvABC*. Adapted from Guarino et al., 2007a.

2.3 Detection of branched DNA structures by PFGE

In PFGE, linear DNA migrates according to its size; however, circular chromosomes or branched DNA structures do not enter in the gel and remain trapped in the wells (Bidnenko et al., 2002). Moreover, replication or recombination intermediates, which are Y or X structures, prevent migration of linear DNA fragments in PFGE. These molecules remain also trapped in the wells after PFGE so that the measurement of the amount of such DNA fragments allows quantification of the formation of abnormal structures in a DNA region.

2.3.1 Reinitiation events under thymine starvation

Thymine starvation results in cellular death in thymine requiring strains. This is a phenomena know as *thymineless death* (TLD), first described in the 50's (Cohen & Barner 1954). Some proposals postulated the formation of branched DNA as the source of the toxic effect of thymine starvation (Nakayama, 2005). Nakayama and co-workers demonstrated the presence of complex DNA structures by digesting DNA from cultures under thymine starvation with the restriction enzyme *XbaI* and separating it by PFGE (Nakayama et al., 1994). They called these structures "non-migrating DNA" (nmDNA), defined as the DNA that is unable to enter the gel and gets stuck in the well. The nmDNA was characterized as having single-stranded tails or gaps and branching with single-stranded arms. TLD has been related to DNA replication (Maaloe & Hanawalt, 1961; Hanawalt & Maaloe, 1961); nevertheless, ongoing replication does not appear to be required for TLD as same lethality is observed under thymine starvation in the presence of hidroxyurea (Morganroth & Hanawalt, 2006), an inhibitor of the DNA synthesis. By contrast, TLD is suppressed by the addition of rifampicin or cloramphenicol, both inhibitors of the new initiation events of the *E. coli* chromosomal DNA (Hanawalt, 1963).

To study whether the formation of nmDNA correlates with TLD under the above described replication conditions, we analyzed the generation of nmDNA after thymine starvation in the presence or absence of rifampicin, cloramphenicol and hidroxyurea (Mata & Guzmán, 2011). Mid-exponentially growing culture of *thyA* mutant MG1693 was starved for thymine in the presence or absence of rifampicin, cloramphenicol, or hydroxyurea. Two hours after the treatment, cells were collected, washed, embedded in agarose plugs, gently lysed and plugs treated with *XbaI* (50 U/100 µl) for two hours before being used for PFGE (Matushek et al., 1996; Gautom, 1997). The visualization of DNA bands was achieved by ethidium bromide staining (Fig. 4A). The amount of nmDNA was quantified by densitometry of the PFGE by using the *Imagen J* program. The nmDNA values were expressed as the ratio (%) between the arbitrary densitometric units of the gel well and those of the gel line plus the well (Table 1, Figure 4B). By using this experimental approach, we showed that nmDNA was generated under thymine starvation, and it was absent in the presence of rifampicin or chloramphenicol, as previously reported (Nakayama et al., 1994). This might suggest that TLD correlates with the generation of nmDNA. However, we found no nmDNA under thymine starvation in the presence of hydroxyurea, indicating that the generation of nmDNA is not a requirement for TLD (Mata & Guzmán, 2011).

Treatment	% nmDNA[1]	Treatment effect[2]	Lethality
None -Exponential culture	8.6 ± 3.6	1	no
-Thymine	21.4 ± 4.7	2.48	yes
-Thy+ 75 mM hydroxyurea	11.2 ± 3.8	1.30	yes
-Thy+ 150 µg/ml rifampicin	9.0 ± 3.9	1.04	no
-Thy+ 200 µg/ml chloramphenicol	9.5 ± 4.4	1.10	no

Table 1. Percentage of nmDNA in MG1693 cells after 2 h of thymine starvation in the presence or absence of hydroxyurea, rifampicin or chloramphenicol. [1] The percentage of nmDNA is expressed as the mean ± standard deviation. [2] The percentage of nmDNA relative to the exponential culture.

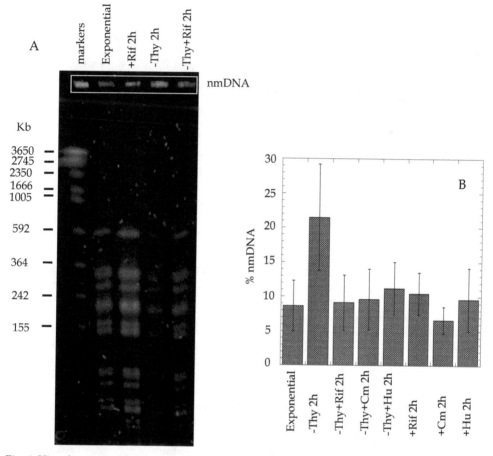

Fig. 4. Visualization and quantification of nmDNA by PFGE after treatment with *XbaI*. (A) Visualization of a representative agarose gel stained with ethidium bromide. (B) Quantification of the amount of nmDNA after different treatments of thymine starvation in the absence or the presence of the drugs, by using the data obtained by densitometry analysis. The standard deviation from 2-3 independent experiments is shown.

2.3.2 Collision between replication and transcription machines

PFGE has been also used for the characterization of collisions of the replication and transcription complexes. Genetic instability following head-on collisions of replication and transcription has been described in bacteria (Vilette et al., 1995) and yeast (Torres et al., 2004; Prado & Aguilera, 2005). Replication fork barriers are present on the rDNA of all eucaryotic cells described until now, for example in yeast (Brewer & Fangman, 1988; Kobayashi et al., 1992; López-Estraño et al., 1998; Sánchez-Gorostiaga et al., 2004), *Xenopus* (Wiesendanger et al., 1994) and plants (Hernández et al., 1993; López-Estraño et al., 1998) and they bound specific proteins. In yeast, these barriers are polar since they block replication forks facing transcription from the pre-rRNA 35S (Brewer & Fangman, 1988; Linkens & Huberman,

1988). It has been proposed that the function of those barriers could be to prevent head-on collisions between replication forks and the highly expressed rRNA genes (Takeuchi et al., 2003).

In order to avoid head-on collisions, ribosomal operons (*rrn*) are transcribed in the direction of replication in bacteria. By using genetic approaches together with PFGE analysis, the laboratory of B. Michel (Boubakri et al., 2010) identified that the three *E. coli* DNA helicases DinG, Rep and UvrD are recruited to the replication fork to allow replication across oppositely oriented highly transcribed ribosomal operons. Strains containing an inversion of an *rrn* operon were used in such a way that a region of increased head-on collisions between replication and transcription were created. Increased level of DNA trapping at wells in PFGE experiments was correlated with a high level of *rrn* transcription, suggesting the formation of abnormal structures in certain genetic backgrounds (Boubakri et al., 2010).

3. Two-dimensional agarose gel electrophoresis

The movement of a DNA molecule through an agarose gel is determined either by factors intrinsic to the electrophoretic conditions (agarose concentration, the strength of the electric field, the presence of intercalating agents, etc.) as well as the size and shape of the molecule.

The most evident example of the influence of the shape of a DNA molecule on the electrophoretic mobility in an agarose gel is observed when circular DNA molecules are analyzed: supercoiled DNA molecules and the corresponding relaxed-nicked DNA forms do not migrate necessarily at the same position than a linear DNA molecule of the same mass.

Taking into account this property, neutral/neutral 2D agarose gel electroforesis technique was developed to study the shape of recombination intermediates (Bell & Byers, 1983). Later on, it was adapted to study the DNA replication intermediates (RIs) (Brewer & Fangman, 1987). Since then, 2D agarose gel electrophoresis was used to map and characterize replication origins (Brewer & Fangman, 1988; Gahn & Schildkraut, 1989; Liu & Botchan, 1990; Schvartzman et al., 1990; Linskens & Huberman, 1990 b; Friedman & Brewer, 1995; Bach et al.; 2003), to analyze the progression of DNA replication along a DNA fragment (Azvolinsky et al., 2006), to characterize replication fork barriers (Brewer & Fangman, 1988; Linskens & Huberman, 1988; Hernandez et al., 1993; Wiesendanger et al., 1994; Samadashwily et al., 1997, López-Estraño et al., 1998, Possoz et al., 2006; Mirkin et al., 2006, Boubakri et al., 2010), replication termination (Zhu et al., 1992; Santamaría et al., 2000a,b), origin replication interference (Viguera et al., 1996), RIs knotting (Viguera et al., 1996; Sogo et al., 1999), fork reversal (Viguera et al.; 2000; Fierro-Fernandez et al., 2007a) or the topology of partially replicated plasmids (Martín-Parras et al., 1998; Lucas et al., 2001). See (Schvartzman et al., 2010) for an excellent review in plasmid DNA replication analyzed by 2D-gel.

2D agarose gel electrophoresis consists of two successive electrophoreses in which the second dimension occurs perpendicular to the first. Two different migration conditions are used so that the first dimension conditions (low voltage, low agarose concentration) minimize the effect of molecular shape on electrophoretic mobility, whereas this effect is maximized during the second dimension (high voltage and high agarose concentration, in the presence of an intercalating agent) (Friedman & Brewer, 1995). As a consequence, a branched DNA molecule like a recombination or a replication intermediate is separated from a linear molecule of the same mass during the second dimension.

As DNA replication is a continuous process, a sample of DNA isolated from an exponentially growing culture should contain all the replication intermediates (RIs), ranging from the linear non-replicative forms (named 1.0X) to molecules almost completely replicated (2.0X) (Fig. 5A). See (Krasilnikova & Mirkin, 2004), for a detailed protocol of isolation of RIs in *E. coli* and *S. cerevisiae*.

The different migration patterns of a RI digested with a specific restriction enzyme are revealed after southern blotting hybridization with a specific probe and it indicates the mode it has been replicated (Fig. 5). Electrophoresis conditions must be adapted to the fragment size in order to obtain a good separation of the different patterns (Friedman & Brewer, 1995). Different situations can be discerned by using 2D gels. (i) A single fork that moves from one end to the other end of the fragment generates a simple-Y pattern indicating that the DNA fragment is replicated passively and does not contain neither a replication origin nor a replication terminus (Fig. 5A). (ii) Two forks that move convergently generate a double-Y pattern, indicating that replication termination occurs within the analyzed fragment (Fig. 5C); and (iii) two forks that have initiated at some specific point in the analyzed fragment and progress divergently, generate a bubble pattern, indicating that DNA replication has been initiated inside this fragment (Fig. 5B).

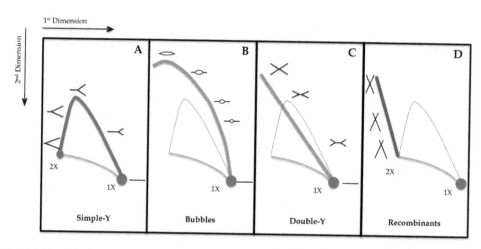

Fig. 5. 2-D gel hybridization patterns generated by replication and recombination intermediates after two-dimensional agarose gel electrophoresis. Replication and recombination intermediates of the restriction fragment are shown above the different 2D-gel pattern. In panels B, C and D, the simple-Y arc is presented as a reference. See text for details.

Moreover, the relative proportion of a particular RI in the population increases as a consequence of the stalling of the replication forks at a specific site. This accumulation is detected as a discrete signal on top of the corresponding arc produced by the RIs. In order to map the region where replication is paused, different restriction fragments must be analyzed to confirm that the signal corresponding to the accumulated molecules move along the arc.

Recently, this technique was used to get insight into the nature of the elements that causes the trapping of the DNA in PFGE experiments in *E. coli* mutant strains containing an inverted *rrn* operon (Boubakri et al., 2010). No RIs were detected in the non-inverted strains or the inversion mutants that express all helicases. However, a simple-Y arc that corresponds to the accumulated Y-shape restriction intermediates was detected in all *dinG*, *rep* and *uvrD* helicase mutants in which the Inv-fragment was trapped in PFGE wells. Moreover, an intense elongated spot was observed over the simple-Y arc. These results indicate that a specific accumulation of RIs occur at the 3′ end of the *rrn* operon (Figure 6).

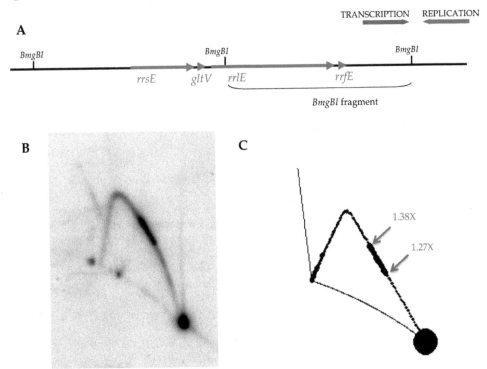

Fig. 6. Replication forks are arrested in inverted *rrn*. 2D-Gels were used to examine DNA replication in restriction fragments containing a large 3′ region of *rrnA* in *InvA* mutants and of *rrnE* in *InvBE* mutants (Adapted from Boubakri et al., 2010). (A) Schematic representation of the restriction fragment used for 2D gels (only *InvBE* is presented in this figure). The position of rrn and restriction sites is shown. (B) DNA from *InvBE dinG rep* mutant was digested with *BmgBI* , analyzed by 2D gels and probed for the sequence just downstream of *rrnE*. A simple-Y arc is clearly detected. On top of this arc, an enlarged signal corresponding to arrested forks is detected. (C) A simulation of replication arrest in this fragment of about 500 pb around the *rrn* transcription terminator sequence was obtained by using the 2D-Gel computer program (Viguera et al., 1998).

4. Concluding remarks

Replication arrest is a source of genetic instability in all types of living cells. As a consequence, cells have developed several effective strategies to tackle with replication fork arrest and/or repairing the double DSBs generated at the stalled replication forks. We have reviewed how PFGE and 2D gels can be used to elucidate some features related to the progression of the replication forks. By using these two non-conventional electrophoresis it can be verified the presence of stalled replication forks, understanding how they have been generated and how they could be restarted.

5. Acknowledgements

We are very grateful to Bénédicte Michel for bacterial strains and continuous support and advice. We especially thank Estrella Guarino, Israel Salguero, Carmen Mata, Encarna Ferrera and Hasna Boubakri for their works and technical help. This work was supported by grants BFU2007-63942 to EG and BFU2007-64153, and P09-CVI-5428 to EV from the Ministerio de Ciencia e Innovación and Junta de Andalucía. EV is grateful to Dr. JB Schvartzman for training in 2D electrophoresis and helpful discussions along the years. EV is grateful to Dr. JB Schvartzman for training in 2D electrophoresis and helpful discussions.

6. References

Aguilera, A. & Gómez-González, B. (2008). Genome instability: a mechanistic view of its causes and consequences. *Nat Rev Genet* 9: 204-217.

Azvolinsky, A., Dunaway, S., Torres, J., Bessler, J.B., & Zakian, V. (2006). *Genes & Development* 20: 3104–3116.

Bach, T., Krekling, M.A., & Skarstad, K. (2003). Excess SeqA prolongs sequestration of *oriC* and delays nucleoid segregation and cell division. *EMBO J* 22: 315-323.

Bell, L. & Byers, B. (1983). Separation of branched from linear DNA by two-dimensional gel electrophoresis. *Analytical Biochemistry* 130: 527-535.

Bierne, H. & Michel, B. (1994). When replication forks stop. *Mol. Microbiol.* 13: 17-23.

Brewer, B.J. & Fangman, W.L. (1987). The localization of replication origins on ARS plasmids in S. cerevisiae. *Cell* 51: 463-471.

Brewer, B.J. & Fangman, W.L. (1988). A replication fork barrier at the 3' end of yeast ribosomal RNA genes. *Cell* 55: 637-643.

Bidnenko, V., Ehrlich, S.D., & Michel, B. (2002). Replication fork collapse at replication terminator sequences. *EMBO J.* 21:3898-3907.

Boubakri, H., de Septenville, A.L., Viguera, E., & Michel, B. (2010) The helicases DinG, Rep and UvrD cooperate to promote replication across transcription units *in vivo*. *EMBO J* 29: 145–157.

Cohen, S.S. & Barner, H.D. (1954). Studies on unbalanced growth in *Escherichia coli*. *Proc Natl Acad Sci USA* 40: 885-893.

Courcelle, J., Belle, J.J., & Courcelle, C.T. (2004). When replication travels on damaged templates: bumps and blocks in the road. *Research in Microbiol.* 155: 231-237.

Fierro-Fernandez, M., Hernández, P., Krimer, D.B., & Schvartzman, J.B. 2007a. Replication fork reversal occurs spontaneously after digestion but is constrained in supercoiled domains. *Journal of Biological Chemistry* 282: 18190–18196.

Flores, M.J., Bierne, H., Ehrlich, S.D., & Michel. B. (2001). Impairment of lagging strand synthesis triggers the formation of a RuvABC substrate at replication forks. *EMBO J.* 20: 619-629

Flores, M.J., Ehrlich, S.D., & Michel, B. (2002). Primosome assembly requirement for replication restart in the *Escherichia coli* holDG10 replication mutant. *Mol. Microbiol.* 44: 783-792.

Friedman, K.L. & Brewer, B.J. (1995). Analysis of replication intermediates by two-dimensional agarose gel electrophoresis. In *DNA Replication*, Campbell, J.L., (Ed.), 613-627, Academic Press Inc. (San Diego, CA).

Gahn, T.A. & Schildkraut, C.L. (1989). The Epstein-Barr virus origin of plasmid replication, *oriP*, contains both the initiation and termination sites of DNA replication. *Cell* 58: 527-535.

Gautom, R.K. (1997). Rapid pulsed-field gel electrophoresis protocol for typing of *Escherichia coli* O157:H7 and other gram-negative organisms in 1 day, *J. Clin. Microbiol.* 35: 2977-2980.

Grompone, G., Seigneur, M., Ehrlich, S.D., & Michel, B. (2002). Replication fork reversal in DNA polymerase III mutants of *Escherichia coli*: a role for the β clamp. *Mol. Microbiol.* 44:1331-9.

Grompone, G., Ehrlich, S.D., & Michel, B. (2003). Replication restart in *gyrB Escherichia coli* mutants. *Mol. Microbiol.* 48:845-54.

Guarino, E., Jiménez-Sánchez, A., & Guzmán, E.C. (2007a). Defective ribonucleoside diphosphate reductase impairs replication fork progression in *Escherichia coli*. *J Bacteriol* 189: 3496-501.

Guarino, E., Salguero, I., Jiménez-Sánchez, A., & Guzmán, E.C. (2007b). Double-strand break generation under deoxyribonucleotide starvation in *Escherichia coli*. *J Bacteriol* 189: 5782-6.

Hanawalt, P.C., Maaloe, O., Cummings, D.J., & Schaechter, M. (1961). The normal DNA replication cycle. II. *J. Mol. Biol.* 3: 156-165.

Hanawalt, P.C. (1963). Involvement of synthesis of RNA in thymineless death. *Nature* 198: 286.

Hanawalt, P.C. (2007) Paradigms for the three Rs: DNA Replication, Recombination and repair. *Mol. Cell* 28: 702-707.

Hernández, P., Martín-Parras, L., Martínez-Robles, M.L., & Schvartzman, J.B. (1993). Conserved features in the mode of replication of eukaryotic ribosomal RNA genes. *EMBO J* 12: 1475-1485.

Herschleb, J., Ananiev, G., & Schwartz D.C. (2007). Pulsed-field gel electrophoresis. *Nature Protocols* 2: 677-684.

Horiuchi, T., Nishitani, H., & Kobayashi, T. (1995). A new type of *E. coli* recombinational hotspot which requires for activity both DNA replication termination events and the Chi sequence. *Adv. Biophys.* 31:133-47.

Kobayashi, T., Hidaka, M., Nishizawa, M., & Horiuchi, T. (1992). Identification of a site required for DNA replication fork blocking activity in the rRNA gene cluster in *Saccharomyces cerevisiae*. *Mol Gen Genet.* 233: 355-62.

Krasilnikova, M. & Mirkin, S. (2004). Analysis of Triplet Repeat Replication by Two-Dimensional Gel Electrophoresis. *Methods in Molecular Biology.* 277: 19-28

Kreuzer, K.N. (2005). Interplay between DNA Replication and Recombination n prokaryotes. *Ann. Rev. Microbiol.* 59: 43-67.

Kuzminov, A. (1995). Instability of inhibited replication forks in *E. coli. Bioessays* 17:733-41.

Linskens, M.H.K. & Huberman, J.A. 1988. Organization of replication of ribosomal DNA in *Saccharomyces cerevisiae. Molecular and Cellular Biology* 8: 4927–4935.

Linskens, M.H.K. & Huberman, J.A. (1990b). The two faces of eukaryotic DNA replication origins. *Cell* 62: 845–847.

Liu, Y. & Botchan, M. (1990). Replication of bovine papillomavirus type-1 DNA initiates within an E2-responsive enhancer element. *Journal of Virology* 64: 5903–5911.

López-Estraño, C., Schvartzman, J.B., Krimer, D.B., & Hernández, P. (1998). Colocalization of polar replication fork barriers and rRNA transcription terminators in mouse rDNA. *J Mol Biol* 277: 249.

Lucas, I., Germe, T., Chevrier-Miller, M., & Hyrien, O. (2001). Topoisomerase II can unlink replicating DNA by precatenane removal. *EMBO J* 20: 6509-6519.

Maaloe, O. & Hanawalt, P.C. (1961). Thymine deficiency and the normal DNA replication cycle. I, *J. Mol. Biol.* 3: 144-155.

Maisnier-Patin, S., Nordstrom, K., & Dasgupta, S. (2001). Replication arrests during a single round of replication of the *Escherichia coli* chromosome in the absence of DnaC activity. *Mol. Microbiol.* 42: 1371-1382.

Martin-Parras, L., Lucas, I., Martínez-Robles, M.L., Hernandez, P., Krimer, D.B., Hyrien, O., & Schvartzman, J.B. (1998). Topological complexity of different populations of pBR322 as visualized by two-dimensional agarose gel electrophoresis. *Nucleic Acids Res* 26: 3424-3432.

Mata Martín, C. & Guzmán, E.C. (2011). DNA replication initiations as a key element in thymineless death. *DNA Repair* 10: 94-101.

Matushek, M.G., Bonten, M.J. & Hayden, M.K. (1996). Rapid preparation of bacterial DNA for pulsed-field gel electrophoresis. *J Clin Microbiol.* 34: 2598-2600.

Michel, B., Ehrlich, S.D. & Uzest, M. (1997). DNA double-strand breaks caused by replication arrest. *EMBO J.* 16: 430-8.

Michel, B., Grompone, G., Flores, M.J. & Bidnenko, V. (2004). Multiple pathways process stalled replication forks. *Proc. Natl. Acad. Sci. U S A* 101: 12783-8.

Michel, B., Boubakri, H., Baharoglu, Z., Lemasson, M. & Lestini, R. (2007). Recombination proteins and rescue of arrested replication forks. *DNA Repair (Amst).* 6: 967-80.

Miranda, A. & Kuzminov, A. (2003). Chromosomal lesion suppression and removal in *Escherichia coli* via linear DNA degradation. *Genetics* 163:1255-71.

Mirkin, E., Castro, D., Nudler, E. & Mirkin S. (2006). Transcription regulatory elements are punctuation marks for DNA replication. *Proc. Natl. Acad. Sci. USA* 103: 7276-7281

Morganroth, P.A. & Hanawalt, P.C. (2006). Role of DNA replication and repair in thymineless death in *Escherichia coli. J. Bacteriol.* 188: 5286-5288.

Nakayama, K., Kusano, K., Irino, N. & Nakayama, H. (1994). Thymine starvation-induced structural changes in Escherichia coli DNA. Detection by pulsed field gel electrophoresis and evidence for involvement of homologous recombination. *J. Mol. Biol.* 243: 611-620.

Nakayama, H. (2005). *Escherichia coli* RecQ helicase: a player in thymineless death. *Mutat. Res.* 577: 228-236.

Neylon, C., Kralicek, A.V., Hill, T.M. & Dixon, N.E. (2005). Replication termination in *Escherichia coli*: structure and antihelicase activity of the Tus-Ter complex. *Microbiol Mol Biol Rev.* 69:501-26.

Prado, F. & Aguilera, A. (2005). Impairment of replication fork progression mediates RNA polII transcription-associated recombination. *EMBO J* 24: 1267–1276

Possoz, C., Filipe, S., Grainge, I. & Sherratt D. (2006) Tracking of controlled Escherichia coli replication fork stalling and restart at repressor-bound DNA *in vivo*. *EMBO J.* 25: 2596-2604

Samadashwily, G., Raca, G. & Mirkin, S. (1997). Trinucleotide repeats afect DNA replication *in vivo*. *Nature Genetics*. 17: 298-304

Sánchez-Gorostiaga, A., López-Estraño, C., Krimer DB, Schvartzman J.B. & Hernández P. (2004). Transcription terminator factor reb1p causes two replication fork barriers at its cognate sites in fission yeast rDNA *in vivo*. *Molecular and Cellular Biology* 24: 398–406

Santamaría, D., Hernández, P., Martínez-Robles, M.L., Krimer, D.B., & Schvartzman, J.B. (2000a). Premature termination of DNA replication in plasmids carrying two inversely oriented ColE1 origins. *J Mol Biol* 300: 75-82.

Santamaría, D., Viguera, E., Martínez-Robles, M.L., Hyrien, O., Hernandez, P., Krimer, D.B., & Schvartzman, J.B. (2000b). Bi-directional replication and random termination. *Nucleic Acids Res* 28: 2099-2107.

Schwartz, D.C. & Cantor, C.R. (1984). Separation of yeast chromosome-sized DNAs by pulsed field gradient gel electrophoresis. *Cell*. 37: 67-75.

Schvartzman, J.B., Adolph, S., Martín-Parras, L., & Schildkraut, C.L. (1990). Evidence that replication initiates at only some of the potential origins in each oligomeric form of Bovine Papillomavirus Type 1 DNA. *Mol Cell Biol* 10: 3078-3086.

Schvartzman J.B., Martínez-Robles, M.L., Hernandez, P. & Krimer, D.B. (2010). Plasmid DNA replication and topology as visualized by two-dimensional agarose gel electrophoresis. *Plasmid* 63: 1–10.

Seigneur, M., Bidnenko, V., Ehrlich, S.D., & Michel, B. (1998). RuvAB acts at arrested replication forks. *Cell* 95:419-30.

Seigneur, M., Ehrlich, S.D. & Michel, B. (2000). RuvABC-dependent double-strand breaks in *dnaBts* mutants require RecA. *Mol. Microbiol.* 38:565-74.

Sogo, J.M., Stasiak, A., Martínez-Robles, M.L., Krimer, D.B., Hernández, P. & Schvartzman, J.B. (1999). Formation of knots in partially replicated DNA molecules. J Mol Biol 286, 637-643.

Takeuchi, Y., Horiuchi, T. & Kobayashi, T. (2003) Transcription-dependent recombination and the role of fork collision in yeast rDNA. *Genes Dev* 17: 1497–1506.

Torres, J.Z., Schnakenberg, S.L. & Zakian, V.A. (2004) *Saccharomyces cerevisiae* Rrm3p DNA helicase promotes genome integrity by preventing replication fork stalling: viability of rrm3 cells requires the intra-S-phase checkpoint and fork restart activities. *Mol Cell Biol* 24: 3198–3212

Viguera, E., Hernández, P., Krimer, D.B., Boistov, A.S., Lurz, R., Alonso, J.C., & Schvartzman, J.B. (1996). The ColE1 unidirectional origin acts as a polar replication fork pausing site. *J Biol Chem* 271: 22414-22421.

Viguera, E., Rodríguez, A., Hernandez, P., Krimer, D.B., Trelles, O., & Schvartzman J.B. (1998) A computer model for the analysis of DNA replication intermediates by two-dimensional agarose gel electrophoresis. *Gene* 217: 41–49.

Viguera, E., Hernandez, P., Krimer, D.B., & Schvartzman J.B. (2000). Visualisation of plasmid replciation intermediates containing reversed forks. *Nuc. Acids Res.* 28: 498-503.

Vilette, D., Ehrlich, S.D., & Michel, B. (1995). Transcription-induced deletions in *Escherichia coli* plasmids. *Mol Microbiol* 17: 493–504.

Voineagu, I., Narayanan, V., Lobachev, K., & Mirkin, S. (2008). Replication stalling at unstable inverted repeats: Interplay between DNA hairpins and fork stabilizing proteins. *Proc. Natl. Acad. Sci. USA* 105: 9936-9941.

Wiesendanger, B., Lucchini, R., Koller, T., & Sogo, J.M. (1994). Replication fork barriers in the *xenopus* rDNA. *Nucleic Acids Res* 22: 5038-5046.

Zhu, J., Newlon, C.S., & Huberman, J.A. (1992). Localization of a DNA replication origin and termination zone on chromosome III of *Saccharomyces cerevisiae*. *Mol Cell Biol* 12: 4733-4741.

Permissions

The contributors of this book come from diverse backgrounds, making this book a truly international effort. This book will bring forth new frontiers with its revolutionizing research information and detailed analysis of the nascent developments around the world.

We would like to thank Sameh Magdeldin, MVSc, PhD, for lending his expertise to make the book truly unique. He has played a crucial role in the development of this book. Without his invaluable contribution this book wouldn't have been possible. He has made vital efforts to compile up to date information on the varied aspects of this subject to make this book a valuable addition to the collection of many professionals and students.

This book was conceptualized with the vision of imparting up-to-date information and advanced data in this field. To ensure the same, a matchless editorial board was set up. Every individual on the board went through rigorous rounds of assessment to prove their worth. After which they invested a large part of their time researching and compiling the most relevant data for our readers. Conferences and sessions were held from time to time between the editorial board and the contributing authors to present the data in the most comprehensible form. The editorial team has worked tirelessly to provide valuable and valid information to help people across the globe.

Every chapter published in this book has been scrutinized by our experts. Their significance has been extensively debated. The topics covered herein carry significant findings which will fuel the growth of the discipline. They may even be implemented as practical applications or may be referred to as a beginning point for another development. Chapters in this book were first published by InTech; hereby published with permission under the Creative Commons Attribution License or equivalent.

The editorial board has been involved in producing this book since its inception. They have spent rigorous hours researching and exploring the diverse topics which have resulted in the successful publishing of this book. They have passed on their knowledge of decades through this book. To expedite this challenging task, the publisher supported the team at every step. A small team of assistant editors was also appointed to further simplify the editing procedure and attain best results for the readers.

Our editorial team has been hand-picked from every corner of the world. Their multi-ethnicity adds dynamic inputs to the discussions which result in innovative outcomes. These outcomes are then further discussed with the researchers and contributors who give their valuable feedback and opinion regarding the same. The feedback is then collaborated with the researches and they are edited in a comprehensive manner to aid the understanding of the subject.

Apart from the editorial board, the designing team has also invested a significant amount of their time in understanding the subject and creating the most relevant covers. They scrutinized every image to scout for the most suitable representation of the subject and create an appropriate cover for the book.

The publishing team has been involved in this book since its early stages. They were actively engaged in every process, be it collecting the data, connecting with the contributors or procuring relevant information. The team has been an ardent support to the editorial, designing and production team. Their endless efforts to recruit the best for this project, has resulted in the accomplishment of this book. They are a veteran in the field of academics and their pool of knowledge is as vast as their experience in printing. Their expertise and guidance has proved useful at every step. Their uncompromising quality standards have made this book an exceptional effort. Their encouragement from time to time has been an inspiration for everyone.

The publisher and the editorial board hope that this book will prove to be a valuable piece of knowledge for researchers, students, practitioners and scholars across the globe.

List of Contributors

Loretto Contreras-Porcia and Camilo López-Cristoffanini
Universidad Andrés Bello, Faculty of Ecology and Natural Resources, Department of Ecology and Biodiversity, Santiago, Chile

María Esther Rodríguez, Laureana Rebordinos, Eugenia Muñoz-Bernal, Francisco Javier Fernández-Acero and Jesús Manuel Cantoral
Microbiology Laboratory, Faculty of Marine and Environmental Sciences, University of Cadiz, Puerto Real, Spain

Zeynep Cetecioglu and Orhan Ince
Istanbul Technical University, Environmental Engineering Department, Maslak, Istanbul, Turkey

Bahar Ince
Bogazici University, Institute of Environmental Science, Rumelihisarustu-Bebek, Istanbul, Turkey

Gizella Jahnke, János Májer and János Remete
University of Pannonia Centre of Agricultural Sciences, Research Institute for Viticulture and Enology, Badacsony, Hungary

Maria de Lourdes T. M. Polizeli, Simone C. Peixoto-Nogueira and Tony M. da Silva
Biology Department, Faculty of Philosophy Sciences and Letters of Ribeirão Preto, São Paulo University, Brazil

Alexandre Maller
Biochemistry and Immunology Department, School of Medicine of Ribeirão Preto, São Paulo, São Paulo University, Brazil

Hamilton Cabral
Science Pharmaceutical Department, School of Pharmaceutical Science of Ribeirão Preto, São Paulo University, Brazil

Tsai-Hsin Chiu, Yi-Cheng Su, Hui-Chiu Lin and Chung-Kang Hsu
Department of Food Science, National PengHu University of Science and Technology, Taiwan
Seafood Research and Education Center, Oregon State University, USA
Penghu Marine Biology Research Center, Fisheries Research Institute, COA, EY, Taiwan

Laura Bonofiglio, Noella Gardella and Marta Mollerach
Department of Microbiology, Immunology and Biotechnology, University of Buenos Aires, Argentina

Denise Feder
Laboratório de Biologia de Insetos, GBG Universidade Federal Fluminense-UFF, Rio de Janeiro, RJ, Brazil

Danielle Misael, Cristina S. Silva, Alice H. Ricardo-Silva, Jacenir R. Santos-Mallet and Teresa Cristina M. Gonçalves
Laboratório de Transmissores de Leishmanioses, Setor de Entomologia Médica e Forense IOC-FIOCRUZ-Rio de Janeiro, RJ, Brazil

André L. S. Santos
Laboratório de Estudos Integrados em Bioquímica Microbiana, Instituto de Microbiologia Paulo de Góes (IMPG), Bloco E-subsolo, Universidade Federal do Rio de Janeiro (UFRJ), Rio de Janeiro, RJ, Brazil

Suzete A. O. Gomes
Laboratório de Biologia de Insetos, GBG Universidade Federal Fluminense-UFF, Rio de Janeiro, RJ, Brazil
Laboratório de Transmissores de Leishmanioses, Setor de Entomologia Médica e Forense IOC-FIOCRUZ-Rio de Janeiro, RJ, Brazil

Carlos Garrido, María Carbú, Victoria E. González-Rodríguez and Eva Liñeiro
Microbiology Laboratory, Faculty of Marine and Environmental Sciences, University of Cádiz, Puerto Real, Spain

Velazquez-Meza Maria Elena
Instituto Nacional de Salud Pública, Cuernavaca Morelos, México D. F.

Vázquez-Larios Rosario, Hernández Dueñas Ana Maria and Rivera Martínez Eduardo
Instituto Nacional de Cardiología "Dr. Ignacio Chávez", México D. F.

Patrick Eberechi Akpaka
Department of Para-Clinical Sciences, the University of the West Indies, St. Augustine, Trinidad & Tobago

Padman Jayaratne
Department of Pathology & Molecular Medicine, McMaster University, Hamilton, Ontario, Canada

Carolina Alves and Celso Cunha
Center for Malaria and Tropical Diseases, Institute of Hygiene and Tropical Medicine, New University of Lisbon, Portugal

Elena C. Guzmán
Departamento de Bioquímica, Biología Molecular y Genética, Facultad de Ciencias, Universidad de Extremadura, Badajoz, Spain

Enrique Viguera
Área de Genética, Facultad de Ciencias, Universidad de Málaga, Málaga, Spain